Some of the most beautiful mathematical objects found in the last forty years are the sporadic simple groups, but gaining familiarity with these groups presents problems for two reasons. Firstly, they were discovered in many different ways, so to understand their constructions in depth one needs to study lots of different techniques. Secondly, since each of them is, in a sense, recording some exceptional symmetry in space of certain dimensions, they are by their nature highly complicated objects with a rich underlying combinatorial structure.

Motivated by initial results on the Mathieu groups which showed that these groups can be generated by highly symmetrical sets of elements, the author develops the notion of symmetric generation from scratch and exploits this technique by applying it to many of the sporadic simple groups, including the Janko groups and the Higman–Sims group.

ENCYCLOPEDIA OF MATHEMATICS AND ITS APPLICATIONS

FOUNDING EDITOR G.-C. ROTA

Editorial Board

P. Flajolet, M. Ismail, E. Lutwak

ENCYCLOPEDIA OF MATHEMATICS AND ITS APPLICATIONS

Symmetric Generation of Groups

With Applications to Many of the Sporadic Finite Simple Groups

ROBERT T. CURTIS

University of Birmingham, UK

CAMBRIDGE
UNIVERSITY PRESS

CAMBRIDGE UNIVERSITY PRESS
Cambridge, New York, Melbourne, Madrid, Cape Town, Singapore, São Paulo

Cambridge University Press
The Edinburgh Building, Cambridge CB2 8RU, UK

Published in the United States of America by Cambridge University Press, New York

www.cambridge.org
Information on this title: www.cambridge.org/9780521857215

First published 2007

Printed in the United Kingdom at the University Press, Cambridge

A catalogue record for this publication is available from the British Library

ISBN 978-0-521-85721-5 hardback

This book is dedicated to my mother, Doreen Hannah (née Heard),
1912–2006, and to my wife, Ahlam.

Contents

Preface

The book is aimed at postgraduate students and researchers into finite groups, although most of the material covered will be comprehensible to fourth year undergraduates who have taken two modules of group theory. It is based on the author's technique of symmetric generation, which seems able to present many difficult group-theoretic constructions in a more elementary manner. It is thus the aim of the book to make these beautiful, but combinatorially complicated, objects accessible to a wider audience.

The stimulus for the investigations which led to the contents of the book was a question from a colleague of mine, Tony Gardiner, who asked me if the Mathieu group M_{24} could contain two copies of the linear group $L_3(2)$ which intersect in a subgroup isomorphic to the symmetric group S_4. He needed such a configuration in order to construct a graph with certain desirable properties. I assured him that the answer was almost certainly yes, but that I would work out the details. I decided to use copies of $L_3(2)$ which are maximal in M_{24} and found that the required intersection occurred in the nicest possible way, in that one could find subgroups $H \cong K \cong L_3(2)$, with $H \cap K \cong S_4$, and an involution t such that $C_{M_{24}}(H \cap K) = \langle t \rangle$ and $H' = K$. This means that t has seven images under conjugation by H, and the maximality of H together with the simplicity of M_{24} mean that these seven involutions must generate M_{24}. The symmetry of the whole set-up enables one to write down seven corresponding involutory permutations on 24 letters directly from a consideration of the action of $L_3(2)$ on 24 points.

Applying the same ideas with $L_3(2)$ replaced by the alternating group A_5, or more revealingly the projective special linear group $PSL_2(7)$ replaced by $PSL_2(5)$, I found that in an analogous manner the smaller Mathieu group M_{12} is generated by five elements of order 3 which can be permuted under conjugation within the large group by a subgroup isomorphic to A_5.

From here the generalization to other groups became clear and many of the sporadic simple groups revealed themselves in a pleasing manner. This book concentrates on groups of moderate size, and it is satisfying to see how the symmetry of the generating sets enables one to verify by hand claims that would appear to be beyond one's scope. With groups such as the smallest Janko group J_1, the Higman–Sims group HS and the second Janko group HJ, I have included the full manual verification, so that the reader can appreciate what can be achieved. However, in writing the book I have come more and more to make use of the double coset enumerator which was produced by John Bray and myself specifically for groups defined by what we now call a *symmetric presentation*. The program implementing this algorithm is written in MAGMA, which has the advantage that it is very easy to read what the code is asking the machine to do. Thus, even when a hand calculation is possible, and indeed has been completed, I have often preferred to spare the reader the gory details and simply include the MAGMA output. Of course, some of the groups which are dealt with in this manner are out of range for all but the doughtiest reckoner!

As is made clear in the text, every finite simple group possesses definitions of the type used in this book. However, I have not seen fit to include those groups which are plainly out of range of mechanical enumeration, or where a description of the construction introduces additional complicated ideas. Nonetheless, 19 of the 26 sporadic groups are mentioned explicitly and it is hoped that the definitions given are quite easily understood. The book is in three Parts.

Part I: Motivation

Part I, which assumes a rather stronger background than Part II and which could, and perhaps should, be omitted at a first reading, explains where the ideas behind symmetric generation of groups came from. In particular, it explains how generators for the famous Mathieu groups M_{12} and M_{24} can be obtained from easily described permutations of the faces of the dodecahedron and Klein map, respectively. This not only ties the approach in with classical mathematics, but demonstrates a hitherto unrecognized link with early algebraic geometry. Although Part I is, in a sense, independent of what follows, the way in which combinatorial, algebraic and geometric constructions complement one another gives an accurate flavour of the rest of the book.

Part I is essentially background and does not contain exercises.

Part II: Involutory symmetric generators

Part II begins by developing the basic ideas of symmetric generation of finite groups in the most straightforward case: when the generators have order 2. The preliminary topics of *free products of cyclic groups* and *double cosets* are defined before the notions of *symmetric generating sets*, *control subgroup*, *progenitor*, *Cayley diagrams* and *coset stabilizing subgroups* are introduced and fully explained through elementary but important examples. It is shown that every finite simple group can be obtained in the manner described, as a quotient of a progenitor. Through these elementary examples the reader becomes adept at handling groups defined in terms of highly symmetric sets of elements of order 2.

At this stage we demonstrate how the algebraic structure can be used to do the combinatorial work for us. The Fano plane emerges as a by-product of the method, and the famous isomorphisms $A_5 \cong PSL_2(5)$ and $PSL_2(7) \cong PSL_3(2)$ are proved. Further, $PSL_2(11)$ emerges in its exceptional Galois action on 11 points, and the 11-point biplane is revealed. An easy example produces the symmetric group S_6 acting non-permutation identically on two sets of six letters, and the outer automorphism of S_6 reveals itself more readily than in other constructions known to the author; the isomorphism $\mathrm{Aut}A_6 \cong P\Gamma L_2(9)$ follows. The method is also used to exhibit the exceptional triple covers of A_6 and A_7.

There follows a systematic computerized investigation of groups generated by small, highly symmetrical sets of involutory generators, and it is seen that classical and sporadic groups emerge alongside one another. The results of this investigation are presented in convenient tabular form, as in Curtis, Hammas and Bray [36].

Having familiarized the reader with the methods of symmetric generation, we now move on to more dramatic applications. Several sporadic simple groups are defined, and in many cases constructed by hand, in terms of generating sets of elements of order 2.

Part II concludes by describing how the methods of symmetric generation afford a concise and amenable way of representing an element of a group as a permutation followed by a short word in the symmetric generators. Thus an element of the smallest Janko group J_1 can be written as a permutation of eleven letters, in fact an element of $L_2(11)$, followed by a word of length at most four in the eleven involutory symmetric generators. A manual algorithm for multiplying elements represented in this manner, and for reducing them to canonical form, has been computerized in Curtis and Hasan [37].

Part III: Symmetric generators of higher order

In Part III we extend our investigations to symmetric generators of order greater than 2. It soon becomes apparent that this leads us into a consideration of monomial representations of our so-called control subgroup over finite fields. The resulting progenitors are slightly more subtle objects than those in Part II, and they reward our efforts by producing a fresh crop of sporadic simple groups.

Nor is it necessary to restrict our attention to finite fields of prime order. A monomial representation over, say, the field of order 4 may be used to define a progenitor in which each 'symmetric generator' is a Klein fourgroup. It turns out that this is a natural way to obtain the Conway group Co_1 and other sporadic groups.

The classification of finite simple groups is one of the most extraordinary intellectual achievements in the twentieth century. It states that there are just 26 finite simple groups which do not fit into one of the known infinite families. These groups, which range in size from the smallest Mathieu group of order 7920 to the Monster group of order around 10^{53}, were discovered in a number of unrelated ways and no systematic way of constructing them has as yet been discovered. Symmetric generation provides a uniform concise definition which can be used to construct surprisingly large groups in a revealing manner. Many of the smaller sporadic groups are constructed by hand in Parts II and III of this book, and computerized methods for constructing several of the larger sporadics are described. It is our aim in the next few years to complete the task of providing an analogous definition and construction of each of the sporadic finite simple groups.

Acknowledgements

I should first of all like to thank John Conway for introducing me to those beautiful objects the Mathieu groups, and for the many hours we spent together studying other finite groups as we commenced work on the ATLAS [25]. In many ways, all that I came to understand at that time has fed into the present work. I am also indebted to my colleague, Tony Gardiner, for asking me the question mentioned in the Preface which sparked the central ideas in this book.

Since that time, several of my research students have worked on topics arising out of these ideas and have thus contributed to the contents of this book. I shall say a few words about each of them in the chronological order in which they submitted their dissertations.

Ahmed Hammas (1991) from Medina in Saudi Arabia carried out the first systematic search for images of progenitors with small control subgroups. Perhaps his most startling and satisfying discovery was the isomorphism

$$\frac{2^{\star 5} : A_5}{[(0\ 1\ 2\ 3\ 4)t_0]^{\overline{7}}} \cong J_1,$$

the smallest Janko group. **Abdul Jabbar** (1992) from Lahore in Pakistan also joined the project early on, having worked with Donald Livingstone on (2,3,7)-groups. He concentrated on symmetric presentations of subgroups of the Conway group and, in particular, those groups in the Suzuki chain. **Michelle Ashworth**, in her Masters thesis (1997), explored the manner in which the hexads of the Steiner system $S(5,6,12)$ can be seen on the faces of a dodecahedron, and how the octads of the Steiner system $S(5,8,24)$ appear on the faces of the Klein map. **John Bray** (1998), who now works at Queen Mary, University of London, made a massive contribution to the project, both as my research student and later as an EPSRC research fellow. His thesis contains many results which I have not included in this book and far more details than would be appropriate in a text of this nature. His formidable computational skills came to fruition in his programming and improvement of the *double coset enumerator* which had evolved out of my early hand calculations. **Stephen Stanley** (1998), who now works for a software company in Cambridge, UK, was mainly concerned with monomial representations of finite groups and their connection with symmetric generation. One of his most interesting achievements was a faithful 56-dimensional representation of the covering group $2^{2\cdot}L_3(4)$ over $\mathbb{Z}_8$, the integers modulo 8. **Mohamed Sayed** (1998) from Alexandria in Egypt produced an early version of the double coset enumerator, which worked well in some circumstances but was probably too complicated to cope with larger groups. **Sean Bolt** (2002) works for the Open University in Coventry; he made a comprehensive study of a symmetric presentation of the largest Janko group J_4, which eventually led us to the definition described in Part II of this book. **John Bradley** (2005), who is presently teaching in the University of Rwanda, verified by hand the symmetric presentations we had for the McLaughlin group McL and the Janko group J_3. This brings us up to the present day with **Sophie Whyte** (2006), who has just graduated having followed on from the work of Stephen Stanley. In particular she has identified all faithful irreducible monomial representations of the covering groups of the alternating groups, and has used them as control subgroups in interesting progenitors.

I should also like to thank my colleague **Chris Parker** for his collaboration with me on a very successful project to extend symmetric generation to the larger sporadic groups, and I should like to thank the EPSRC for supporting that project. Chris's commitment and infectious enthusiasm for mathematics made it a pleasure to work with him. The School of Mathematics is to be thanked for providing additional funding which enabled us to take on two research assistants: John Bray and **Corinna Wiedorn**. Corinna came to us from Imperial College, where she had been working with Sasha Ivanov. She possessed geometric skills which proved invaluable as we dealt with larger and larger objects. Among her many achievements was a verification by hand of the symmetric presentation for J_1 mentioned above as being found by Ahmed Hammas. The skills of the four people involved meshed perfectly and led to a very productive period. Sadly Corinna died in 2005, and her exceptional talent is lost to mathematics.

Part I

Motivation

Introduction to Part I

In Part I we use the two smallest non-abelian finite simple groups, namely the alternating group A_5 and the general linear group $L_3(2)$ to define larger permutation groups of degrees 12 and 24, respectively. Specifically, we shall obtain highly symmetric sets of generators for each of the new groups and use these generating sets to deduce the groups' main properties. The first group will turn out to be the Mathieu group M_{12} of order $12 \times 11 \times 10 \times 9 \times 8 = 95\,040$ [70] and the second the Mathieu group M_{24} of order $24 \times 23 \times 22 \times 21 \times 20 \times 16 \times 3 = 244\,823\,040$ [71]; they will be shown to be quintuply transitive on 12 and 24 letters, respectively. These constructions were first described in refs. [31] and [32].

1

The Mathieu group M_{12}

1.1 The combinatorial approach

As is well known, the alternating group A_5 contains $4! = 24$ 5-cycles; these are all conjugate to one another in the symmetric group S_5, but, since 24 does not divide 60, they fall into two conjugacy classes of A_5 with 12 elements in each. Let $A \cong A_5$ act naturally on the set $Y = \{1, 2, 3, 4, 5\}$, and let $a = (1\ 2\ 3\ 4\ 5) \in A$ be one of these 5-cycles. Then the two classes may be taken to be

$$\Lambda = \{a^g \mid g \in A\} \quad \text{and} \quad \bar{\Lambda} = \{(a^2)^g \mid g \in A\}.$$

We shall define permutations of the set Λ, and eventually extend them to permutations of the set $\Lambda \cup \bar{\Lambda}$. In Table 1.1 we write the elements of Λ so that each begins with the number 1, and for convenience we label them using the projective line $P_1(11) = \{\infty, 0, 1, \ldots, X\} = \{\infty\} \cup \mathbb{Z}_{11}$, where X stands for '10'. The other conjugacy class $\bar{\Lambda}$ is then labelled with the set $\{\bar{\infty}, \bar{0}, \bar{1}, \ldots, \bar{X}\}$, with the convention that if $\lambda \in \Lambda$ is labelled n, then $\lambda^2 \in \bar{\Lambda}$ is labelled $\bar{n}$. Clearly, for $g \in A$, conjugation of elements of Λ by g yields a permutation of the 12 elements of Λ, and thus we obtain a transitive embedding of $A \cong A_5$ in the symmetric group S_{12}. Indeed, since A_5 is a simple group, it must be an embedding in the alternating group A_{12}.

We now define a new permutation of Λ, which we shall denote by s_1. It will be clear that permutations $s_2, \ldots, s_5$ can be defined similarly, by starting each of the 5-cycles in the definition of s_i with the symbol i. For $(1\ w\ x\ y\ z) \in \Lambda$ we define the following:

$$s_1 : (1\ w\ x\ y\ z) \mapsto (1\ w\ x\ y\ z)^{(x\ y\ z)} = (1\ w\ y\ z\ x).$$

We note that s_1 *is* a function from Λ to Λ; after all, the image of a given 5-cycle is certainly another 5-cycle and, since the permutation $(x\ y\ z)$ is even, it is in Λ rather than $\bar{\Lambda}$. Moreover, we see that s_1^3 acts as the identity

3

Table 1.1. Labelling of the 24 5-cycles with elements of the 12-point projective line

	Λ				$\bar{\Lambda}$		
∞	(1 2 3 4 5)	0	(1 5 4 3 2)	$\bar{\infty}$	(1 3 5 2 4)	$\bar{0}$	(1 4 2 5 3)
1	(1 3 2 5 4)	2	(1 4 5 2 3)	$\bar{1}$	(1 2 4 3 5)	$\bar{2}$	(1 5 3 4 2)
9	(1 5 2 4 3)	7	(1 3 4 2 5)	$\bar{9}$	(1 2 3 5 4)	$\bar{7}$	(1 4 5 3 2)
4	(1 3 5 4 2)	8	(1 2 4 5 3)	$\bar{4}$	(1 5 2 3 4)	$\bar{8}$	(1 4 3 2 5)
3	(1 5 3 2 4)	6	(1 4 2 3 5)	$\bar{3}$	(1 3 4 5 2)	$\bar{6}$	(1 2 5 4 3)
5	(1 4 3 5 2)	X	(1 2 5 3 4)	$\bar{5}$	(1 3 2 4 5)	$\bar{X}$	(1 5 4 2 3)

on Λ, and so s_1 possesses an inverse (namely s_1^2) and is a permutation. But s_1 does not fix any 5-cycle, and so it has cycle shape 3^4 on Λ. It turns out that, if $\hat{a}$ denotes the image of a as a permutation of Λ, then $\hat{a}$ and s_1 generate a subgroup of A_{12} of order 95 040. In fact, we have the following:

$$\langle \hat{a}, s_1 \rangle = \langle s_1, s_2, s_3, s_4, s_5 \rangle \cong M_{12},$$

the Mathieu group [70], which was discovered in 1861. Explicitly, we see that

$$\hat{a} = (1\ 9\ 4\ 3\ 5)(2\ 7\ 8\ 6\ X) \text{ and } s_1 = (\infty\ 8\ X)(0\ 3\ 9)(1\ 4\ 7)(2\ 6\ 5).$$

It turns out that M_{12} is remarkable in that it can act non-permutation identically on two sets of 12 letters, and so acts intransitively on 24 letters with two orbits of length 12; it possesses an outer automorphism which can act on this set of 24 letters interchanging the two orbits. Not surprisingly, for us the two sets of size 12 will be Λ and $\bar{\Lambda}$. The element $\hat{a}$, by a slight abuse of notation, can be interpreted as an element of the alternating group A_{24} acting by conjugation on each of the two sets, with cycle shape 3^4 on each of them. Our new element s_1, however, requires a slight adjustment, and we define

$$s_1 : (1\ w\ x\ y\ z) \mapsto (1\ w\ x\ y\ z)^{(x\ y\ z)} = (1\ w\ y\ z\ x) \text{ if } (1\ w\ x\ y\ z) \in \Lambda$$
$$\mapsto (1\ w\ x\ y\ z)^{(z\ y\ x)} = (1\ w\ z\ x\ y) \text{ if } (1\ w\ x\ y\ z) \in \bar{\Lambda}.$$

This yields

$$s_1 = (\infty\ 8\ X)(0\ 3\ 9)(1\ 4\ 7)(2\ 6\ 5)(\bar{\infty}\ \bar{3}\ \bar{5})(\bar{0}\ \bar{8}\ \bar{7})(\bar{1}\ \bar{6}\ \bar{9})(\bar{2}\ \bar{4}\ \bar{X}),$$

and in a similar way we obtain all five generators given in Table 1.2.

If we now define $S \cong S_5$ to be the set of all permutations of Y, then the odd permutations of S interchange the two sets Λ and $\bar{\Lambda}$ by conjugation. From the definition of the five elements $\{s_i \mid i = 1, \ldots, 5\}$, it is not surprising that conjugation by even elements of $\hat{S}$ simply permutes their subscripts in the natural way; however, odd elements of $\hat{S}$ permute *and invert*. These statements could be verified directly by conjugating the permutations given in Table 1.2, by generators for $\hat{A}$ and $\hat{S}$; however, we prefer to prove them formally. Thus we have Lemma 1.1.

Table 1.2. Action of the five symmetric generators of M_{12} on $\Lambda \cup \bar{\Lambda}$

$$s_1 = (\infty\ 8\ \text{X})(0\ 3\ 9)(1\ 4\ 7)(2\ 6\ 5)(\tilde{\infty}\ \bar{3}\ \bar{5})(\bar{0}\ \bar{8}\ \bar{7})(\bar{1}\ \bar{6}\ \bar{9})(\bar{2}\ \bar{4}\ \bar{\text{X}})$$
$$s_2 = (\infty\ 6\ 2)(0\ 5\ 4)(9\ 3\ 8)(7\ \text{X}\ 1)(\tilde{\infty}\ \bar{5}\ \bar{1})(\bar{0}\ \bar{6}\ \bar{8})(\bar{9}\ \bar{\text{X}}\ \bar{4})(\bar{7}\ \bar{3}\ \bar{2})$$
$$s_3 = (\infty\ \text{X}\ 7)(0\ 1\ 3)(4\ 5\ 6)(8\ 2\ 9)(\tilde{\infty}\ \bar{1}\ \bar{9})(\bar{0}\ \bar{\text{X}}\ \bar{6})(\bar{4}\ \bar{2}\ \bar{3})(\bar{8}\ \bar{5}\ \bar{7})$$
$$s_4 = (\infty\ 2\ 8)(0\ 9\ 5)(3\ 1\ \text{X})(6\ 7\ 4)(\tilde{\infty}\ \bar{9}\ \bar{4})(\bar{0}\ \bar{2}\ \bar{\text{X}})(\bar{3}\ \bar{7}\ \bar{5})(\bar{6}\ \bar{1}\ \bar{8})$$
$$s_5 = (\infty\ 7\ 6)(0\ 4\ 1)(5\ 9\ 2)(\text{X}\ 8\ 3)(\tilde{\infty}\ \bar{4}\ \bar{3})(\bar{0}\ \bar{7}\ \bar{2})(\bar{5}\ \bar{8}\ \bar{1})(\bar{\text{X}}\ \bar{9}\ \bar{6})$$

LEMMA 1.1 For s_i a permutation of $\Lambda \cup \bar{\Lambda}$ defined as above and $\pi \in S$, we have the following:

$$s_i^{\hat{\pi}} = s_{i\pi} \text{ if } \pi \in A; \quad s_i^{\hat{\pi}} = s_{i\pi}^{-1} \text{ if } \pi \in S \setminus A.$$

Proof Let $\lambda = (a_0\ a_1\ a_2\ a_3\ a_4) \in \Lambda$ and let $\pi \in A$. Then we have

$$
\begin{aligned}
\lambda^{\hat{\pi}^{-1} s_j \hat{\pi}} &= (a_0^{\pi^{-1}}\ a_1^{\pi^{-1}}\ a_2^{\pi^{-1}}\ a_3^{\pi^{-1}}\ a_4^{\pi^{-1}})^{s_j \hat{\pi}} \\
&= (a_i^{\pi^{-1}}\ a_{i+1}^{\pi^{-1}}\ a_{i+2}^{\pi^{-1}}\ a_{i+3}^{\pi^{-1}}\ a_{i+4}^{\pi^{-1}})^{s_j \hat{\pi}} && \text{(where } j = a_i^{\pi^{-1}}\text{)} \\
&= (a_i^{\pi^{-1}}\ a_{i+1}^{\pi^{-1}}\ a_{i+3}^{\pi^{-1}}\ a_{i+4}^{\pi^{-1}}\ a_{i+2}^{\pi^{-1}})^{\hat{\pi}} && \text{(where } j^\pi = a_i\text{)} \\
&= (a_i\ a_{i+1}\ a_{i+3}\ a_{i+4}\ a_{i+2}) && \text{(where } j^\pi = a_i\text{)} \\
&= (a_i\ a_{i+1}\ a_{i+2}\ a_{i+3}\ a_{i+4})^{(a_{i+2}\ a_{i+3}\ a_{i+4})} && \text{(where } j^\pi = a_i\text{)} \\
&= \lambda^{s_{a_i}} = \lambda^{s_{j\pi}}.
\end{aligned}
$$

A similar calculation holds for $\lambda \in \bar{\Lambda}$, and so we have $s_j^{\hat{\pi}} = s_{j\pi}$.

Further suppose that $\lambda = (a_0\ a_1\ a_2\ a_3\ a_4) \in \Lambda$ and let $\sigma \in S \setminus A$. Then we have

$$
\begin{aligned}
\lambda^{\hat{\sigma}^{-1} s_j \hat{\sigma}} &= (a_0^{\sigma^{-1}}\ a_1^{\sigma^{-1}}\ a_2^{\sigma^{-1}}\ a_3^{\sigma^{-1}}\ a_4^{\sigma^{-1}})^{s_j \hat{\sigma}} \\
&= (a_i^{\sigma^{-1}}\ a_{i+1}^{\sigma^{-1}}\ a_{i+2}^{\sigma^{-1}}\ a_{i+3}^{\sigma^{-1}}\ a_{i+4}^{\sigma^{-1}})^{s_j \hat{\sigma}} && \text{(where } j = a_i^{\sigma^{-1}}\text{)} \\
&= (a_i^{\sigma^{-1}}\ a_{i+1}^{\sigma^{-1}}\ a_{i+4}^{\sigma^{-1}}\ a_{i+2}^{\sigma^{-1}}\ a_{i+3}^{\sigma^{-1}})^{\hat{\sigma}} && \text{(where } j^\sigma = a_i\text{)} \\
&= (a_i\ a_{i+1}\ a_{i+4}\ a_{i+2}\ a_{i+3}) && \text{(where } j^\sigma = a_i\text{)} \\
&= (a_i\ a_{i+1}\ a_{i+2}\ a_{i+3}\ a_{i+4})^{(a_{i+4}\ a_{i+3}\ a_{i+2})} && \text{(where } j^\sigma = a_i\text{)} \\
&= \lambda^{s_{a_i}^2} = \lambda^{(s_{j\sigma})^{-1}},
\end{aligned}
$$

where the third line follows because $\lambda^{\sigma^{-1}} \in \bar{\Lambda}$. As above, a similar calculation follows for $\lambda \in \bar{\Lambda}$, and so we have $s_j^{\hat{\sigma}} = (s_{j\sigma})^{-1}$. $\qquad\square$

The reader would be right to wonder why we chose to conjugate our 5-cycles $\lambda = (1\ w\ x\ y\ z)$ by the 3-cycle $(x\ y\ z)$ rather than by one of the other possibilities. In fact, we could have chosen any one of $(x\ y\ z), (y\ z\ w), (z\ w\ x)$ or $(w\ x\ y)$ and conjugated every element of Λ by it and every element of $\bar{\Lambda}$ by its inverse. In this way, we obtain four copies of the group M_{12} acting on $\Lambda \cup \bar{\Lambda}$, each of which contains the original group $\hat{A}$. A calculation involving normalizers, which is given explicitly in Section 1.3, shows that these are the only ways in which a copy of the alternating group A_5 acting transitively on 12 points can be extended to a copy of M_{12}.

In order better to understand the relationship between these four copies of M_{12}, it is useful to consider the normalizers of our groups $\hat{S}$ and $\hat{A}$ in the symmetric group Σ acting on the $12 + 12 = 24$ letters which $\hat{S}$ permutes. Now, the normalizer of $\hat{S}$ in Σ, factored by the centralizer of $\hat{S}$ in Σ, must be isomorphic to a subgroup of the automorphism group of S_5, which is just S_5 (since all automorphisms of S_5 are inner and its centre is trivial). Thus,

$$|N_\Sigma(\hat{S})| \le |C_\Sigma(\hat{S})| \times 120;$$

so we wish to find all permutations of Σ which commute with $\hat{S}$. Before proceeding we recall the following elementary result.

LEMMA **1.2** A permutation which commutes with a transitive group must be regular (i.e. has all its disjoint cycles of the same length), and a permutation which commutes with a doubly transitive group of degree greater than 2 must be trivial.

Proof Let $\pi \ne 1$ commute with a transitive group H. If π has cycles of differing length, then some non-trivial power of π possesses fixed points and, of course, commutes with H. But conjugation by H would then imply that every point must be fixed by this power of π, which is thus the identity. So we conclude that π could not have had cycles of differing lengths.

Suppose now that π commutes with the doubly transitive H and that $\pi : a_1 \mapsto a_2$ with $a_1 \ne a_2$. Choose $a_3 \notin \{a_1, a_2\}$. Then there exists a $\rho \in H$ with $a_1^\rho = a_1$ and $a_2^\rho = a_3$, and so $\pi = \pi^\rho : a_1 \mapsto a_3$. Thus we have a contradiction unless the degree is less than 3.   □

Now, $\hat{S}$ acts transitively on $\Lambda \cup \bar{\Lambda}$, and so any permutation which commutes with it must be regular. Moreover, $\hat{S}$ has blocks of imprimitivity of size 4, namely the sets $\{\lambda, \lambda^2, \lambda^3, \lambda^4\}$, and it acts doubly transitively on these six blocks (as the projective general linear group $\mathrm{PGL}_2(5)$). Thus a permutation centralizing $\hat{S}$ must fix each block, and there can be at most four such permutations. We now define

$$\tau : \lambda \mapsto \lambda^2 \ \text{ for } \lambda \in \Lambda \cup \bar{\Lambda}.$$

Clearly τ has order 4 and fixes each block. Moreover, we have

$$\lambda^{\hat{\sigma}\tau} = (\lambda^\sigma)^\tau = (\lambda^\sigma)^2 = (\lambda^2)^\sigma = (\lambda^\tau)^\sigma = \lambda^{(\tau\hat{\sigma})},$$

and so τ commutes with $\hat{S}$. We conclude that $C_\Sigma(\hat{S}) = \langle \tau \rangle$. We can now readily observe the following.

LEMMA **1.3** Conjugation by the element τ cycles the four copies of M_{12} which extend $\hat{S}$ within Σ.

Proof For $\lambda = (1\ w\ x\ y\ z) \in \Lambda$, we have

$$(1\ w\ x\ y\ z)^{\tau^{-1}s_1\tau} = [(1\ w\ x\ y\ z)^3]^{s_1\tau} = (1\ y\ w\ z\ x)^{s_1\tau}$$
$$= (1\ y\ x\ w\ z)^\tau = (1\ x\ z\ y\ w) = (1\ w\ x\ y\ z)^{(w\ x\ z)},$$

and similarly for $\lambda \in \bar{\Lambda}$. (Note that $\lambda^{\tau^{-1}} \in \bar{\Lambda}$.) Of course, this argument can be repeated for all four possible definitions of the generators s_i. □

For convenience, we give τ as a permutation of the 24 points of $\Lambda \cup \bar{\Lambda}$ as labelled in Table 1.1; thus

$$\tau = (\infty \; \bar{\infty} \; 0 \; \bar{0})(1 \; \bar{1} \; 2 \; \bar{2})(9 \; \bar{9} \; 7 \; \bar{7})$$
$$(4 \; \bar{4} \; 8 \; \bar{8})(3 \; \bar{3} \; 6 \; \bar{6})(5 \; \bar{5} \; \text{X} \; \bar{\text{X}}).$$

1.2 The regular dodecahedron

If we consider the group of rotational symmetries of the regular dodecahedron acting on its 12 faces, then the Orbit-Stabilizer Theorem soon tells us that the group has $12 \times 5 = 60$ elements. As we shall see later in this section, the 20 vertices of the dodecahedron fall (in two different ways) into five sets of four, each of which forms the vertices of a regular tetrahedron. These five tetrahedra are permuted by the group of rotational symmetries and all even permutations of them are realized; so the group is isomorphic to the alternating group A_5. Thus the transitive (but imprimitive) 12-point action of A_5 can be seen as rotational symmetries of the 12 faces. Before describing how our generators of order 3 appear acting on the faces, we show how a dodecahedron may be constructed from our group A.

For the sake of visual impact, we choose to replace the members of the set $Y = \{1, 2, 3, 4, 5\}$ by colours; thus, for example, we replace

1 by black,
2 by yellow,
3 by red,
4 by blue,
5 by green.

Now, for each of the 5-cycles $\lambda \in \Lambda$ we take a regular pentagon with its vertices coloured clockwise in the order in which the colours appear in λ. We now have a child's puzzle: can you piece these pentagons together, three at each vertex, so that the colours all match up? If we start with $\infty = (1\,2\,3\,4\,5)$, in the notation of Table 1.1, we have to ask which pentagon should be placed on its '23' edge. This must be $0 = (3\,2\,1\,5\,4)$, $1 = (3\,2\,5\,4\,1)$ or $5 = (3\,2\,4\,1\,5)$, but the first of these is clearly impossible as it would require a pentagon with two black vertices. Thus there are just two possibilities, and once that choice has been made the rest of the solution is forced. We thus obtain two dodecahedra with their 20 vertices labelled with five colours. Had we started with the 5-cycles of $\bar{\Lambda}$ rather than Λ, we should have obtained two more. In order to obtain generators for the usual version of M_{12} with the above labelling of the faces, we choose to place $1 = (3\,2\,5\,4\,1)$ on edge '23', and we obtain the solution shown in Figure 1.1.

Note that inverse elements λ and λ^{-1} correspond to opposite faces and that any two vertices having the same colour are the same distance apart.

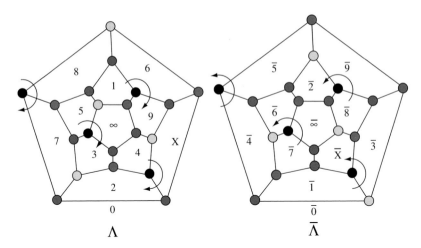

Figure 1.1. Two dodecahedra, each with its 20 vertices labelled using five colours.

In fact, if you move from any vertex along an edge, take the right fork at the first junction and the left fork at the second, then you will arrive at a vertex of the same colour. Thus the four vertices labelled with the same colour form the vertices of a regular tetrahedron, and we have partitioned the 20 vertices of the dodecahedron into five disjoint tetrahedra. There are in fact two such partitions and, had we chosen the option $5 = (3\ 2\ 4\ 1\ 5)$ instead, we should have obtained the other one. The two possible colourings furnished by $\bar{\Lambda}$ also correspond one each to these two partitions. Of course, A, the group of rotational symmetries of the dodecahedron, permutes these five tetrahedra in its natural action.

We are now in a position to read off the action of our 'black' generator s_1 on the dodecahedron shown in Figure 1.1. Recall that s_1 acts as $(\infty\ 8\ X)(0\ 3\ 9)(1\ 4\ 7)(2\ 6\ 5)$ on the faces. So we see that the rule is as follows.

> *Note that the three edges from a vertex lead to three faces; for each black vertex rotate these three faces clockwise.*

In order to see the outer automorphism of M_{12} and to appreciate the four possible sets of five symmetric generators of order 3, it is necessary to consider both the dodecahedra shown in Figure 1.1. Our canonical generator s_1 acts on both dodecahedra in the manner described in italics above, except that it 'twists' the latter in an anticlockwise sense. Note that the element τ, which cycles the four extensions, conjugates s_1 into

$$s_1^\tau = (\infty\ 4\ 9)(0\ 6\ X)(1\ 8\ 5)(2\ 3\ 7)$$
$$(\bar{\infty}\ \bar{8}\ \bar{X})(\bar{0}\ \bar{3}\ \bar{9})(\bar{1}\ \bar{4}\ \bar{7})(\bar{2}\ \bar{6}\ \bar{5}),$$

which twists *not* the three faces joined by an edge to a vertex, but the three faces incident with a vertex. If we call the first type of twist a *deep* twist and the second type a *shallow* twist, then the four possible extensions of A to a copy of M_{12} are characterized as follows.

> *Choose one of the two partitions of the 20 vertices into five regular tetrahedra; then choose either deep twists or shallow twists.*

Thus one can see from Figure 1.1 that the conjugate generator s_1^7 corresponds to the other partition into disjoint tetrahedra and a shallow twist.

1.3 The algebraic approach

In this section we shall assume certain knowledge of the Mathieu group M_{12} as given in the ATLAS (see ref. [25], p. 33) and use this to show that the permutations produced in the preceding ways do indeed generate the group; in Section 1.4 we shall prove that our permutations generate a group with the familiar properties of M_{12} *without* assuming the existence of such a group.

Firstly note that $M \cong M_{12}$ contains a class of transitive subgroups isomorphic to the projective special linear group $L_2(11)$, and recall that this group contains (two classes) of transitive subgroups isomorphic to the alternating group A_5. Let A be such a subgroup and note that the stabilizer of a point in A is cyclic of order 5 and so subgroups of A isomorphic to A_4 act transitively, and so regularly, on the 12 points. Now, the normalizer in M_{12} of such an $H \cong A_4$ is a maximal subgroup of shape $A_4 \times S_3$, and so there is an element s_1 (of class $3B$ and cycle shape 3^4) commuting with H. Thus, under conjugation by A, s_1 will have five images which we may label $\{s_1, s_2, \ldots, s_5\}$. If $a \in A$ is chosen to have order 5, then we may choose our labels so that $s_i^a = s_{i+1}$, where $i = 1, \ldots, 4$ and $s_5^a = s_1$. The subgroup $\langle s_1, \ldots, s_5 \rangle$ is normalized by A and, since the only proper subgroups properly containing A are isomorphic to $L_2(11)$ (as can be seen from the table of maximal subgroups in the ATLAS [25]) in which A_4 subgroups have trivial centralizer, we must have $\langle s_1, s_2, \ldots, s_5 \rangle \cong M$.

Suppose now we start with a group $A \cong A_5$ acting transitively on 12 letters, and thus embedded in the symmetric group $\Sigma \cong S_{12}$. In order to obtain the configuration which we know exists in M_{12}, we must produce elements of order 3 which commute with subgroups of A isomorphic to A_4. Now, as above a subgroup $H \cong A_4$ of A must act regularly on the 12 points, since the point stabilizer in A is cyclic of order 5. We must seek the centralizer in Σ of H. But, as proved in Lemma 1.2, any permutation commuting with a transitive group must itself be regular, and so $C_\Sigma(H)$ has order at most 12. Moreover if H is realizing, say, the *left* regular representation of A_4, then it certainly commutes with the *right* regular representation. This is simply a consequence of the associativity of multiplication, for if L_x and R_y denote left multiplication by x and right multiplication by y, respectively, then

$$g(L_x R_y) = (gL_x)R_y = (xg)R_y = (xg)y$$
$$= x(gy) = (gy)L_x = (gR_y)L_x = g(R_y L_x),$$

where g is any element of the group. Thus $C_\Sigma(H)$ is another copy of A_4 which contains precisely four cyclic subgroups of order 3. Conjugating these by the group A we obtain four sets of five generators. Certainly at least one of these sets must generate M_{12}, since we know that this configuration exists inside it. In order to see that each set generates a copy of M_{12}, we note that the normalizer in Σ of A has the following shape:

$$(2 \times A_5){}^{\cdot}2,$$

a slightly subtle group which contains no copy of S_5: every element in the outer half squares to the central involution times an element of A. The argument used in Section 1.1 applies, and we see that the normalizer of A in Σ factored by the centralizer must be isomorphic to a subgroup of S_5. But A acts imprimitively on the 12 letters with blocks of size 2, and acts doubly transitively (as $L_2(5)$) on the six blocks. Thus the only non-trivial element of Σ centralizing A is an element of order 2 interchanging each of the pairs which constitute the blocks. This, of course, corresponds to the central reflection of the dodecahedron which interchanges opposite faces. This shows that the normalizer has maximal order $(2 \times 5!)$. We can obtain it by adjoining τ times an odd permutation of $\hat{S}$ to our group $\hat{A}$ of Section 1.1. Thus,

$$\tau(4\ 5) = (\infty\ 7\ 0\ 9)(1\ \mathrm{X}\ 2\ 5)(3\ 4\ 6\ 8)$$

can be readily checked to normalize our $\hat{A}$, which acts as rotational symmetries of the dodecahedron and is generated by

$$\hat{A} = \langle (1\ 9\ 4\ 3\ 5)(2\ 7\ 8\ 6\ \mathrm{X}), (\infty\ 1)(7\ \mathrm{X})(0\ 2)(9\ 5)(3\ 6)(4\ 8) \rangle.$$

The four sets of five generators are conjugate under the action of the above element of order 4 and, since one set at least had to generate a copy of M_{12}, they all do. We note that, if the four copies of M_{12} containing our initial A_5 are placed at the vertices of a square so that conjugation by the element $\tau(4\ 5)$ above rotates the square through 90°, then adjacent copies intersect in subgroups isomorphic to $L_2(11)$, while diagonally opposite copies intersect in just the initial A_5.

1.4 Independent proofs

In this section we define $M = \langle s_1, s_2, \ldots, s_5 \rangle$ to be the subgroup of Σ, the symmetric group on 12 letters, generated by the s_i as defined in the unbarred part of Table 1.2, and deduce the well known properties of M_{12}. Thus the s_i are as displayed in Table 1.3 and

$$\hat{a} = (1\ 9\ 4\ 3\ 5)(2\ 7\ 8\ 6\ \mathrm{X}).$$

Firstly we show the following lemma.

Table 1.3. Action of the five symmetric generators on Λ only

$$s_1 = (\infty\ 8\ \text{X})(0\ 3\ 9)(1\ 4\ 7)(2\ 6\ 5)$$
$$s_2 = (\infty\ 6\ 2)(0\ 5\ 4)(9\ 3\ 8)(7\ \text{X}\ 1)$$
$$s_3 = (\infty\ \text{X}\ 7)(0\ 1\ 3)(4\ 5\ 6)(8\ 2\ 9)$$
$$s_4 = (\infty\ 2\ 8)(0\ 9\ 5)(3\ 1\ \text{X})(6\ 7\ 4)$$
$$s_5 = (\infty\ 7\ 6)(0\ 4\ 1)(5\ 9\ 2)(\text{X}\ 8\ 3)$$

LEMMA 1.4 $M = \langle s_1, \ldots, s_5 \rangle = \langle \hat{a}, s_1 \rangle$.

Proof Since $\hat{a}$ cycles the s_i by conjugation, it is clear that $\langle s_1, \ldots, s_5 \rangle \leq \langle \hat{a}, s_1 \rangle$. However, we see that $\hat{a}s_1 = (\infty\ 8\ 5\ 4\ 9\ 7\ \text{X}\ 6)(0\ 3\ 2\ 1)$, of order 8. Thus,

$$1 = (\hat{a}s_1)^8 = \hat{a}^8 s_1^{\hat{a}^7} s_1^{\hat{a}^6} s_1^{\hat{a}^5} s_1^{\hat{a}^4} s_1^{\hat{a}^3} s_1^{\hat{a}^2} s_1^{\hat{a}} s_1 = \hat{a}^3 s_3 s_2 s_1 s_5 s_4 s_3 s_2 s_1,$$

and so $\hat{a} \in \langle s_1, s_2, \ldots, s_5 \rangle$. $\qquad\square$

Now let $A \cong A_5$ denote the group of rotational symmetries of the dodeca-hedron of Figure 1.1 on which our generators s_i are defined to act. Then we may readily see that the following lemma holds.

LEMMA 1.5 $M = \langle s_1, s_2, \ldots, s_5 \rangle \geq A$.

Proof This follows easily from Lemma 1.4, but it is also useful to observe that the permutation

$$(s_1 s_2^{-1})^2 = (\infty\ \text{X}\ 8)(0\ 5\ 4)(1\ 9\ 6)(2\ 7\ 3)$$

corresponds to a rotation through $120°$ about an axis through the vertex incident with faces 1, 6 and 9 and the vertex incident with faces 2, 3 and 7. Similarly, the elements of form $(s_i s_j^{-1})^2$ give all the elements of order 3 in A. Clearly, since A is a simple group, these generate A. $\qquad\square$

In order to prove that the group M has the desired properties, we intro-duce $\mathcal{S}$, a collection of subsets of the faces of the regular dodecahedron as labelled in Figure 1.1. These sets, each of which contains six faces, will be known as *special hexads* or simply *hexads*.[1] We start with the hexad h whose faces are labelled $\{\infty, 2, 7, 8, 6, \text{X}\}$; it consists of the ring of faces around the face 0 (but not 0) together with the face opposite 0, namely ∞. There are clearly 12 such hexads, permuted transitively by A; we refer to them as *parasols*.

---

[1] Ashworth [5] has investigated all ways in which the hexads of the Steiner system $S\ (5, 6, 12)$ can appear on the faces of the dodecahedron.

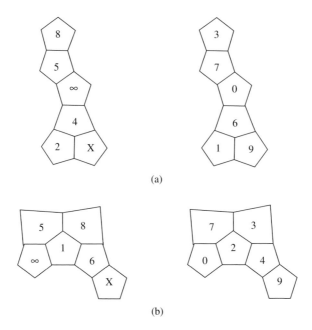

(a)

(b)

Figure 1.2. The 132 special hexads on the dodecahedron. (a) A lugworm and its complement. (b) A molar and its complement.

If we let our generator s_1 and its square act on h, we obtain the hexads shown in Figure 1.2; note that in each case the hexad and its complement are congruent configurations. Since no rotational symmetry fixes either shape, or by simply counting, we see that there are 60 hexads of each type and that the group A permutes each set transitively. We refer to the first type as *lugworms* and to the second as *molars* (note that they are more properly *right lugworms* and *right molars*). We now define S to be the union of the 12 parasols, the 60 lugworms and the 60 molars; thus $|S| = 132$ and the group A has three orbits on the members of S.

LEMMA 1.6 The collection of hexads S is invariant under the action of M and M acts transitively upon it.

Proof Since the set of generators $\{s_1, \ldots, s_5\}$ is preserved by $A \cong A_5$, the group of rotational symmetries of the dodecahedron, it suffices to show that h, h^{s_1} and $h^{s_1^2}$ (representatives for the three orbits of A acting on S) are taken to members of S by each of the s_i. For instance, $h^{s_3} = \{X, 9, \infty, 2, 4, 7\}$ is clearly a molar with 'root' at face 7. The remaining 11 checks can be carried out in a few moments. That the group M fuses the three orbits of A on S into one orbit follows immediately from the facts that h^{s_1} is a lugworm and $h^{s_1^{-1}}$ is a molar.   □

We are now in a position to prove Theorem 1.1.

THEOREM 1.1 The group $M = \langle s_1, \ldots, s_5 \rangle$ has order 95 040.

Proof From the way in which $\mathcal{S}$ was defined, it is clear that M acts transitively on the 132 hexads. We must work out the order of the stabilizer of a hexad in order to obtain $|M|$. Now, the element $\hat{a}$ fixes h and has cycle shape 1.5 on its six faces. Moreover, the element

$$(\hat{a}s_1)^4 = (\infty\ 9)(8\ 7)(5\ X)(4\ 6)(0)(3)(2)(1)$$

acts with cycle shape $1^4.2$ on the lugworm $\{1, 9, \infty, 3, 2, 0\}$ and with cycle shape 2^3 on the complementary lugworm $\{5, 7, 8, 6, X, 4\}$. Thus the stabilizer of a hexad is at least doubly transitive and contains a transposition. So it contains all transpositions and acts as the symmetric group S_6 on the faces of the hexad. Suppose now that $\rho \in M$ fixes every face of some hexad, which, without loss of generality, may be taken to be h, but has non-trivial action on its complement h^c. But the action of the stabilizer of h^c on the six faces of h^c must also be S_6, since it is itself a hexad, and so the conjugates of ρ under this stabilizer must generate a normal subgroup of S_6. But the only non-trivial normal subgroups of S_6 are A_6 and S_6 itself, and so we may assume that M contains all elements fixing every face of a hexad h and acting with cycle shape $1^3.3$ on its complement h^c; such an element is (3 4 5). But the parasol $\{\infty, 5, 7, 2, 4, 6\}$ acted on by (3 4 5) is $\{\infty, 3, 7, 2, 5, 6\}$, which does not belong to $\mathcal{S}$. We conclude that the stabilizer of a hexad acts faithfully on the six faces of that hexad, and so $|M| = 132 \times 6! = 95\,040$. ☐

Note It is clear from the above proof that the stabilizer of a hexad is acting non-permutation identically on a hexad and its complement, realizing both actions of the group S_6 on six letters.

The collection of 132 hexads, labelled $\mathcal{S}$ above, is an important and much-studied combinatorial structure which deserves more discussion.

DEFINITION 1.1 A *Steiner system* $S(l, m, n)$ is a collection of m-element subsets of an n-element set Λ such that every l elements of Λ appear together in precisely one of them.

When a Steiner system $S(l, m, n)$ exists, it is clear that it must contain precisely $\binom{n}{l} / \binom{m}{l}$ special m-element subsets; in particular, this number must be an integer.

LEMMA 1.7 The set $\mathcal{S}$ is a Steiner system $S(5, 6, 12)$.

Proof Suppose that some 5-element subset of Λ is contained in two members of $\mathcal{S}$; by the transitivity of M, we may assume one of these to be h, and the 5-element subset may be assumed to be the ring of faces adjacent to 0. But it is clear that this set of faces is contained in no lugworm, no molar and no other parasol. Thus no two hexads intersect in five faces or, equivalently, no 5-element subset is contained in more than one hexad. So

132×6 distinct 5-element subsets of Λ are contained in hexads of $\mathcal{S}$. But $132 \times 6 = 792 = \binom{12}{5}$, and so every 5-element subset of Λ is contained in precisely one hexad of $\mathcal{S}$. □

We are now in a position to deduce Theorem 1.2.

Theorem 1.2 *The group* M *acts sharply quintuply transitively on the elements of* Λ.

    Proof Let $\{x_1, x_2, \ldots, x_5\}$ and $\{y_1, y_2, \ldots, y_5\}$ be two ordered subsets of five elements in Λ. We shall show that there exists an element $\pi \in M$ such that $x_i^{\pi} = y_i$ for $i = 1, \ldots, 5$. Let the hexads containing these two 5-subsets be k and l, respectively. Then, by the transitivity of M on hexads, there exists an element $\sigma \in M$ such that $k^{\sigma} = l$. But the stabilizer of l acts as S_6 on its six elements, and so there exists an element ρ in this stabilizer such that $x_i^{\sigma\rho} = y_i$ for $i = 1, \ldots, 5$, as required. Finally, note that $12 \times 11 \times 10 \times 9 \times 8 = 95\,040$; so there is no element other than the identity fixing five points, and M acts *sharply* quintuply transitively on the 12 points of Λ. □

2

The Mathieu group M$_{24}$

Our approach to obtaining generators for the large Mathieu group M$_{24}$ will be remarkably analogous to that for M$_{12}$. In fact, we found this method of obtaining generators for M$_{24}$ first and adapted the approach to produce M$_{12}$, but it seems more natural in this book to deal with the smaller group before the larger one.

2.1 The combinatorial approach

In place of the group A considered in Chapter 1, we take $L \cong L_3(2)$, the general linear group in three dimensions over the field of order 2 acting as a permutation group on seven letters; L can be defined to be the set of permutations of the set $\mathcal{P} = \{0, 1, 2, 3, 4, 5, 6\}$ which preserve $\mathcal{L}$, the seven lines of the *Fano plane* given in Figure 2.1. These lines are labelled in the standard way by the set of quadratic residues modulo 7, i.e. $\{1, 2, 4\}$, and its translates.

Now L possesses two conjugacy classes of elements of order 7, which in this permutation representation are just 7-cycles. These classes contain 24 elements each and, if we let $a = (0\ 1\ 2\ 3\ 4\ 5\ 6) \in L$, then the class containing a is given by

$$\Omega = a^L = \left\{ (a_0\ a_1\ a_2\ a_3\ a_4\ a_5\ a_6) \mid \{a_i, a_{i+1}, a_{i+3}\} \in \mathcal{L} \text{ for } i = 0, 1, \ldots, 6 \right\},$$

where the subscripts are to be read modulo 7. Of course, the other conjugacy class consists of the inverses of the elements of Ω.

The elements of Ω are displayed in Table 2.1; they are written so that each 7-cycle starts with 0 and are labelled with the points of the projective plane $P_1(23)$. As in Chapter 1, we now define a permutation t_0 of Ω:

$$t_0 : (0\ u\ v\ w\ x\ y\ z) \mapsto (0\ u\ v\ w\ x\ y\ z)^{(v\ z)(x\ y)} = (0\ u\ z\ w\ y\ x\ v),$$

for $(0\ u\ v\ w\ x\ y\ z) \in \Omega$.

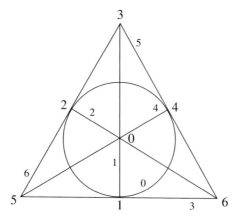

Figure 2.1. The Fano plane.

Table 2.1. Labelling of the 24 7-cycles with the elements of the projective line $P_1(23)$

∞	$(0\ 1\ 2\ 3\ 4\ 5\ 6)$	17	$(0\ 2\ 4\ 6\ 1\ 3\ 5)$	22	$(0\ 4\ 1\ 5\ 2\ 6\ 3)$
0	$(0\ 6\ 4\ 2\ 3\ 1\ 5)$	11	$(0\ 3\ 6\ 1\ 4\ 5\ 2)$	6	$(0\ 4\ 3\ 5\ 6\ 2\ 1)$
18	$(0\ 5\ 3\ 4\ 2\ 6\ 1)$	10	$(0\ 2\ 5\ 6\ 3\ 1\ 4)$	12	$(0\ 3\ 2\ 1\ 5\ 4\ 6)$
3	$(0\ 2\ 1\ 6\ 4\ 5\ 3)$	4	$(0\ 4\ 2\ 5\ 1\ 3\ 6)$	5	$(0\ 1\ 4\ 3\ 2\ 6\ 5)$
20	$(0\ 5\ 6\ 4\ 1\ 3\ 2)$	2	$(0\ 1\ 5\ 3\ 6\ 2\ 4)$	9	$(0\ 6\ 1\ 2\ 5\ 4\ 3)$
8	$(0\ 5\ 2\ 4\ 3\ 1\ 6)$	16	$(0\ 3\ 5\ 1\ 2\ 6\ 4)$	19	$(0\ 2\ 3\ 6\ 5\ 4\ 1)$
14	$(0\ 1\ 6\ 3\ 5\ 4\ 2)$	7	$(0\ 5\ 1\ 4\ 6\ 2\ 3)$	21	$(0\ 6\ 5\ 2\ 1\ 3\ 4)$
15	$(0\ 4\ 6\ 5\ 3\ 1\ 2)$	13	$(0\ 3\ 4\ 1\ 6\ 2\ 5)$	1	$(0\ 6\ 3\ 2\ 4\ 5\ 1)$

Note that for $\lambda = (0\ u\ v\ w\ x\ y\ z) \in \Omega$, the permutation $(v\ z)(x\ y)$ is an element of L, and so t_0 is a function from Ω into Ω. Since $t_0^2 = 1$, we see that t_0 is an *onto* function and thus a permutation of Ω. We may work out the action of t_0 on these 7-cycles, and using the notation of Table 2.1 we find that

$$t_0 = (1\ 9)(2\ 5)(3\ 19)(4\ 15)(6\ 22)(7\ 18)(8\ 20)(10\ 17)(11\ 12)(13\ 16)(14\ \infty)(21\ 0).$$

In this manner, we obtain the seven permutations $\mathcal{T} = \{t_0, t_1, \ldots, t_6\}$ shown in Table 2.2, one for each point of the plane. Of course the group L permutes the set Ω transitively by conjugation and, as before, we let $\hat{L}$ denote the image of L as a transitive subgroup of the symmetric group S_{24}. Conjugation by elements of $\hat{L}$ permutes the elements of $\mathcal{T}$ by permuting their subscripts in the natural way; thus,

$$t_i^{\hat{\pi}} = t_{i^\pi} \quad \text{for} \quad \pi \in L,$$

Table 2.2. The seven involutions seen on the Klein map κ

$t_0 = (1\ 9)(2\ 5)(3\ 19)(4\ 15)(6\ 22)(7\ 18)(8\ 20)(10\ 17)(11\ 12)(13\ 16)(14\ \infty)(21\ 0)$
$t_1 = (1\ 18)(2\ 0)(3\ 13)(4\ 17)(5\ 10)(6\ 19)(7\ 11)(8\ 14)(9\ 16)(12\ 22)(15\ \infty)(20\ 21)$
$t_2 = (1\ 8)(2\ 17)(3\ 6)(4\ 9)(5\ 22)(7\ 19)(10\ 13)(11\ 20)(12\ 21)(14\ 15)(16\ 18)(0\ \infty)$
$t_3 = (1\ 5)(2\ 19)(3\ 7)(4\ 11)(6\ 14)(8\ 10)(9\ 22)(12\ 20)(13\ 21)(15\ 0)(16\ 17)(18\ \infty)$
$t_4 = (1\ 11)(2\ 10)(3\ \infty)(4\ 14)(5\ 8)(6\ 9)(7\ 17)(12\ 15)(13\ 20)(16\ 21)(18\ 0)(19\ 22)$
$t_5 = (1\ 7)(2\ 15)(3\ 18)(4\ 16)(5\ 0)(6\ 10)(8\ 11)(9\ 14)(12\ 19)(13\ 17)(20\ \infty)(21\ 22)$
$t_6 = (1\ 22)(2\ 7)(3\ 20)(4\ 12)(5\ 21)(6\ 13)(8\ \infty)(9\ 18)(10\ 14)(11\ 17)(15\ 19)(16\ 0)$

the proof of this statement being directly analogous to that of Lemma 1.1. It turns out that if $\hat{a}$ denotes the image of a in $\hat{L}$, then $\hat{a}$ and t_0 generate a subgroup of the alternating group A_{24} of order $244\,823\,040$. In fact, we have the following:

$$\langle \hat{a}, t_0 \rangle = \langle t_0, t_1, \ldots, t_6 \rangle \cong M_{24},$$

the large Mathieu group [71], which was discovered in 1873.

As in the M_{12} case, we have a number of choices as to what we should conjugate the 7-cycles by. Indeed, if $\omega = (0\ u\ v\ w\ x\ y\ z) \in \Omega$, then the three lines of the projective plane containing 0 are $0uw, 0vz$ and $0xy$. The Klein fourgroup of L fixing every line through 0 consists of the identity together with the three permutations $\{(v\ z)(x\ y), (x\ y)(u\ w), (u\ w)(v\ z)\}$. By conjugating elements of Ω by each of these in turn, we obtain three sets of generators $\{\mathcal{T}_1, \mathcal{T}_2, \mathcal{T}_3\}$, say, and these generate three copies of M_{24}. Moreover, we may obtain a further three sets of generators which correspond not to the points $\mathcal{P}$, but to $\mathcal{L}$, the lines of the projective plane. Explicitly, we fix a line, say $0 = \{1, 2, 4\}$, and write the elements of Ω so that 1, 2 and 4 appear in the first, second and fourth positions. Then we may define

$$t_0 : (\alpha, \beta, w, \gamma, x, y, z) \mapsto (\alpha, \beta, w, \gamma, x, y, z)^{(w\ x)(y\ z)}$$

for each $\omega = (\alpha, \beta, w, \gamma, x, y, z) \in \Omega$, where $\{\alpha, \beta, \gamma\} = \{1, 2, 4\}$. As was the case for the point-type generators, we can conjugate by any one of $(w\ x)(y\ z), (w\ y)(x\ z)$ or $(w\ z)(x\ y)$, these being the three involutions in the Klein fourgroup fixing every point on the line 0.

In this way, we obtain six copies of M_{24} containing a given copy of $L_3(2)$ acting transitively on 24 points. As before, we prove in Section 2.3 that these are the only ways in which such a copy of $L_3(2)$ may be extended to M_{24}. Now the normalizer of our $\hat{L}$ in S_{24} will permute these six copies of M_{24} by conjugation. In order to identify this normalizer we first note that $\hat{L}$ acts transitively, but imprimitively with blocks of size 3, on the points of Ω. Its action on the eight blocks is that of $L_2(7)$ on the projective line $P_1(7)$ and is thus doubly transitive. So the lemmas proved in Chapter 1 show that any permutation of S_{24} commuting with $\hat{L}$ must be regular (that is to say, must have all disjoint cycles of the same length) and have trivial action on the blocks; so a non-trivial such element would have cycle shape 3^8. But the element

$$\zeta : \omega \mapsto \omega^2 \text{ for } \omega \in \Omega$$

clearly commutes with $\hat{L}$ and has the required cycle shape. It is easily seen that conjugation by ζ cycles the three sets of generators of point-type and the three sets of line-type. Moreover, the outer automorphism of $L_3(2)$ is realized within S_{24} – although it does not, of course, preserve Ω – and we obtain a copy of $L_3(2) : 2$ which interchanges points and lines and so interchanges sets of generators of the two types. The latter group commutes with our element ζ, and so we have

$$N_\Sigma(\hat{L}) \cong \mathrm{PGL}_2(7) \times C_3,$$

and the action on the six copies of M_{24} is cyclic.

2.2   The Klein map

2.2.1   A geometric interpretation of the seven generators

We now wish to construct a geometric object on whose faces our generators for M_{24} act in an analogous manner to the way in which our generators for M_{12} act on the faces of a regular dodecahedron. As before we let the elements of the set $\mathcal{P} = \{0, 1, \ldots, 6\}$ being permuted by the group $L \cong L_2(7)$ be replaced by colours; thus, for instance, we let

0 ∼ black,
1 ∼ blue,
2 ∼ yellow,
3 ∼ pink,
4 ∼ brown,
5 ∼ green,
6 ∼ red.

Now, for each 7-cycle in the conjugacy class Ω, take a heptagon with its edges coloured clockwise in the order defined by the cycle, so we have 24 heptagons, each of which has an edge of each colour. We now attempt to assemble these faces with edges of the same colour abutting and three faces at each vertex, in such a way that the resulting object is preserved by the permutations of the group L with which we started. Of course, we cannot expect to obtain a 3-dimensional regular solid as in the previous case, as there is no platonic solid with heptagonal faces. So we shall have to allow our heptagons to be rather flexible!

Imagine then that we start with the face corresponding to the element $a = (0\ 1\ 2\ 3\ 4\ 5\ 6)$ and ask what colour the third edge at the vertex between

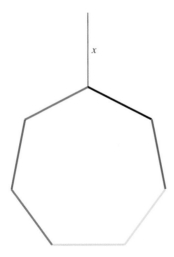

Figure 2.2. The heptagon puzzle.

this heptagon's red (6) and black (0) edges can be. This is equivalent to asking what colour the edge marked x in Figure 2.2 can be. Of course, if the object we are building is to be preserved by the group L, then once we have coloured this edge the colour of the seven corresponding edges is determined by the 7-fold rotational symmetry about our initial face.

Clearly this edge cannot itself be red or black as this would give rise to a heptagon with two edges of the same colour. Nor can it be coloured green (5), as the 7-fold symmetry realized by a would then require the third edge at the black/blue (0/1) vertex of face a to be coloured red (6). This would lead to consecutive faces of a heptagon being coloured red, blue, green (6,1,5), which is a line of the plane and thus not possible in a 7-cycle. It is slightly more difficult to dismiss the possibility of colouring this edge brown (4) or pink (3), but it is left to the reader to verify that both of these would lead to the two edges incident with the other end of this brown or pink edge requiring the same colour. This leaves just two possibilities, namely yellow (2) and blue (1). Both these lead to perfectly consistent coloured figures involving all the faces, and once this first colour has been chosen the whole structure and colouring is uniquely defined. In Figure 2.3 we have coloured this edge blue.

The remarkable figure which we have constructed in this manner is the celebrated *Klein map*, which we shall denote by κ; it has 24 faces, and must have $(24 \times 7)/3 = 56$ vertices and $(24 \times 7)/2 = 84$ edges. The generalized Euler formula tells us that if V is the number of vertices, E the number of edges, and F the number of faces, then

$$V - E + F = 2(1 - g),$$

where g is the lowest genus of a surface on which the map can be drawn. So in this case we see that $(56 - 84 + 24) = -4 = 2(1 - 3)$, and the genus is 3.

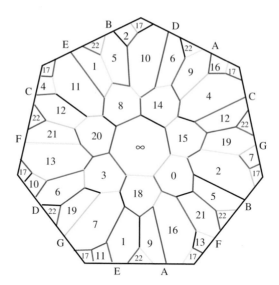

Figure 2.3. The Klein map κ.

Figure 2.3 shows a 14-gon with its edges labelled in pairs by the letters
A, B, C, D, E, F, G. It is drawn to show the 7-fold symmetry of the situation, but each edge of the enfolding heptagon corresponds to two edges
of this defining 14-gon. As in the familiar diagrammatic representation
of the torus as a rectangle with its opposite edges identified, it is understood here that if one leaves this figure across face X then one reappears
through the other edge labelled X. Thus, for example, if we leave the face
labelled 13 across edge F on the extreme left of Figure 2.3, then we rejoin
the figure through the other edge F (to be found bottom right) still on
face 13.

It is a consequence of our construction that the Klein map is obtained
with its 84 edges coloured with seven colours in such a way that every face
has an edge of each colour. Both colourings are preserved by the group L$_2$(7)
acting as rotational symmetries of the map, with the colours corresponding
to blocks of imprimitivity, but the two colourings are visibly different from
one another. In the case shown in Figure 2.3 the three edges at the vertex
first considered are coloured red (6), black (0) and blue (1), which is *not*
a line; but in the other case (see Figure 2.4) the three edges are coloured
red (6), black (0) and yellow (2), which *is* a line of the 7-point projective
plane.

Just as it was easy to read off generators for M$_{12}$ from the vertex-colouring
of the dodecahedron, we may now read off generators for M$_{24}$ from the
edge-colouring of the Klein map. Corresponding to each colour we define a
permutation as follows.

Choose a colour and interchange every pair of faces which are sepa-
rated by an edge of that colour.

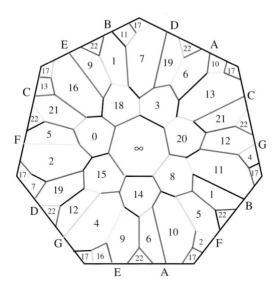

Figure 2.4. Alternative colouring of the Klein map.

Thus, if we choose the colour black in Figure 2.3, then we obtain the following permutation:

$$t_0 = (1\ 9)(2\ 5)(3\ 19)(4\ 15)(6\ 22)(7\ 18)(8\ 20)(10\ 17)(11\ 12)(13\ 16)(14\ \infty)(21\ 0),$$

as in the combinatorial approach. Corresponding to the seven colours, we again obtain the set of generators listed in Table 2.4. Indeed, we can use the colouring to define a different set of permutations as follows.

Choose a colour and interchange every pair of faces which are joined by an edge of that colour.

So, if the colour is black again, we obtain the following permutation:

$$s_0 = (\infty\ 11)(0\ 10)(1\ 3)(2\ 13)(4\ 20)(5\ 16)(6\ 7)(8\ 15)(9\ 19)(12\ 14)(17\ 21)(18\ 22).$$

This element and its images generate a second copy of M_{24}, which intersects the first copy in the group L with which we started. Finally, we note that t_0 and s_0 commute with one another and so the product $t_0 s_0$ is also an involution. If we take the seven images of this as our generators, we obtain a third copy of M_{24}. Now, as has been remarked, our group $L_2(7)$ acts as rotations on the Klein map. But, just as the dodecahedron possesses reflective symmetries, so does κ. Indeed, a reflection in a vertical line through the centre of Figure 2.3 certainly preserves incidence of faces and acts on them as

$$d = (8\ 14)(20\ 15)(3\ 0)(5\ 6)(1\ 9)(4\ 11)(7\ 16)(19\ 21)(2\ 13).$$

This element extends the group L to $PGL_2(7)$. Now, if we conjugate each of t_0, s_0 and $t_0 s_0$ by d and take all images under the group L, we obtain three more sets of generators for M_{24}, making six copies of M_{24} in total. As is shown later, these are the only ways in which a group isomorphic to $L_2(7)$ acting transitively on 24 letters can be extended to a copy of M_{24}. They are cycled by the element of order 6:

$$zd = (1\ 2\ 15\ 9\ 13\ 20)(3\ 6\ 4\ 0\ 5\ 11)(7\ 8\ 21\ 16\ 14\ 19)(10\ 18\ 12)(17\ 22\ \infty),$$

where z is an element of order 3 commuting with L.

But what about our other construction of the Klein map which led to a colouring with a line of colours at each vertex? In fact, the permutation defined by the rule which gave us t_0 above now has the following form:

$$r_0 = (\infty\ 3)(0\ 15)(1\ 11)(2\ 7)(4\ 17)(5\ 22)(6\ 13)(8\ 10)(9\ 14)(12\ 19)(16\ 18)(20\ 21),$$

which preserves the blocks of imprimitivity of size 3 defined by L. This element r_0, together with L, generates a subgroup of S_{24} of shape

$$\langle r_0, L \rangle \cong 3^7 : (2 \times 2^3 : L_2(7)).$$

As the reader will deduce, the elementary abelian subgroup of shape 3^7 preserves the eight blocks and is normalized by an element of order 2 which inverts each of its elements. The action on the blocks is given by the homomorphic image of shape $2^3 : L_3(2)$.

2.2.2   More about the Klein map

When we coloured the vertices of the dodecahedron so that the faces corresponded to a conjugacy class of 5-cycles, we found that diametrically opposite faces corresponded to inverses of one another. In this case there are triangles of faces at maximal distance from one another in the map; they correspond to a 7-cycle, its square and its fourth power. Thus, for instance, we have

$$\{\infty, 17, 22\} \leftrightarrow \{(0\ 1\ 2\ 3\ 4\ 5\ 6), (0\ 2\ 4\ 6\ 1\ 3\ 5), (0\ 4\ 1\ 5\ 2\ 6\ 3)\}.$$

Now, any 3-cycle on three non-collinear points of a plane of order 2 extends uniquely to an element of the group $L_3(2)$ acting with cycle shape 1.3^2 on both points and lines, and so fixes a unique point (and a unique line). Thus, the rotation of order 3 about a vertex of the Klein map fixes a unique point of the plane; so we can label the vertices with the seven points. Now there are 56 vertices, and so eight of them correspond to each point of the plane. Such a set of eight vertices of the Klein map is preserved by a subgroup of $L_2(7)$ isomorphic to S_4, acting on them as the rotational symmetries of a cube. Thus, just as the 20 vertices of the dodecahedron fall into five disjoint regular tetrahedra, the 56 vertices of the Klein map may be thought of as falling into seven disjoint cubes.

It is of course impossible to make a 3-dimensional model of κ so as to reveal all its symmetries. However, Shulte and Wills [78] have produced a beautiful computer-assisted model of the dual map, which is denoted $\{3, 7\}_8$

in the notation of Coxeter. This model, which has 56 triangular faces and 24 vertices, has the 12 rotational symmetries of a regular tetrahedron. Indeed, it is based on a tetrahedron with a hole in each of its faces and a hollow centre, an object which is clearly topologically equivalent to a doughnut with three holes.

2.2.3 The connection with the Klein quartic

The Klein map, exhibited in Figure 2.3 on a surface of genus 3, was discovered in 1878 by Felix Klein in connection with his celebrated quartic curve [64]. The equation of the curve is given by

$$F(x, y, z) = xy^3 + yz^3 + zx^3 = 0;$$

thus we are considering the set of all non-trivial triples of complex numbers (x, y, z) which satisfy the above equation, so points of the curve are 1-dimensional subspaces. The obvious symmetry of the equation shows us that cycling the coordinates x, y and z preserves the curve, which is to say the linear tranformation of the space $\mathbb{C}^3$ defined by the matrix given by

$$\beta = \begin{pmatrix} 0 & 1 & 0 \\ 0 & 0 & 1 \\ 1 & 0 & 0 \end{pmatrix}$$

maps points of the curve to points of the curve. If we let $\eta = e^{2\pi i/7}$, a complex seventh root of unity, then we see that the matrix given by

$$\alpha = \begin{pmatrix} \eta & 0 & 0 \\ 0 & \eta^2 & 0 \\ 0 & 0 & \eta^4 \end{pmatrix}$$

also preserves the curve. Now it is easily seen that $\alpha^\beta = \alpha^4$, and so $\langle \alpha, \beta \rangle$ is a Frobenius group of order 21. But it turns out that the curve $F = 0$ possesses far more symmetries than these rather obvious monomial transformations. In fact, following Baker [8], we let $a = \eta + \eta^{-1}$, $b = \eta^2 + \eta^{-2}$ and $c = \eta^4 + \eta^{-4}$, and find that these complex numbers are the roots of the cubic $x^3 + x^2 - 2x - 1 = 0$. It is readily checked that they satisfy the following relations:

$$a^2 = b + 2, \qquad b^2 = c + 2, \qquad c^2 = a + 2,$$
$$bc + c + 1 = 0, \quad ca + a + 1 = 0, \quad ab + b + 1 = 0,$$
$$a^{-1} = a + b, \qquad b^{-1} = b + c, \qquad c^{-1} = c + a,$$

and using these we may verify that the curve $F = 0$ is preserved by the following matrix:

$$\gamma = d^{-1} \begin{pmatrix} 1 & a^{-1} & b \\ a^{-1} & b & 1 \\ b & 1 & a^{-1} \end{pmatrix},$$

where $d^{-1} = -(1 + a^{-1} + b)$.

The famous Hurwitz bound asserts that a group preserving a surface of genus g cannot have order exceeding $84(g-1)$. In our case, we have $g = 3$, and so the maximal order of the group of linear transformations preserving the curve is 168. But we may readily verify that the elements α and γ above satisfy the following relations:

$$\alpha^7 = \gamma^2 = (\alpha\gamma)^3 = [\alpha, \gamma]^4 = 1,$$

which is a well known presentation for the simple group $L_2(7)$ of order 168. Since the group is simple and the generating matrices non-trivial, we must have $\langle \alpha, \gamma \rangle \cong L_2(7)$ and, since this group attains the Hurwitz bound, we must have

$$\langle \alpha, \beta, \gamma \rangle = \langle \alpha, \gamma \rangle \cong L_2(7).$$

So, by seeking the group preserving the quartic $F = 0$, Klein, and then Baker, had constructed a 3-dimensional, irreducible, complex representation of the group $L_2(7)$. Turning the process around, they had shown that the group $L_2(7)$ possesses an *invariant* of degree 4, namely $F = 0$.

Associated with a curve $F = 0$ is its *Hessian*, defined as follows:

$$\mathcal{H}(x, y, z) = \begin{vmatrix} \dfrac{\partial^2 F}{\partial x^2} & \dfrac{\partial^2 F}{\partial x \partial y} & \dfrac{\partial^2 F}{\partial x \partial z} \\[2ex] \dfrac{\partial^2 F}{\partial y \partial x} & \dfrac{\partial^2 F}{\partial y^2} & \dfrac{\partial^2 F}{\partial y \partial z} \\[2ex] \dfrac{\partial^2 F}{\partial z \partial x} & \dfrac{\partial^2 F}{\partial z \partial y} & \dfrac{\partial^2 F}{\partial z^2} \end{vmatrix} = 0.$$

It may be shown, as in, for example, ref. [49], p. 100, that a curve meets its Hessian at *points of inflexion* and *multiple points*. Since our curve possesses no multiple points, all points of intersection of $F = 0$ and $\mathcal{H} = 0$ are points of inflexion. Now, each entry in the determinant defining $\mathcal{H}$ has degree 2, and so the degree of $\mathcal{H}$ is 6. Thus the Klein quartic possesses $4 \times 6 = 24$ points of inflexion, and these correspond to the faces of the Klein map κ.

The group $L_2(7)$ possesses two further invariants of degrees 14 and 21, giving rise, respectively, to $4 \times 14 = 56$ and $4 \times 21 = 84$ special points of the curve. The first type are *points of contact of bitangents*, that is tangents which touch the curve at two distinct points. The quartic possesses 28 bitangents; indeed, the *group of the 28 bitangents*, which is isomorphic to the symplectic group $Sp_6(2)$, is one of the most important and most studied classical groups. We shall discuss it in some detail in Chapter 5. These points correspond to the vertices of the Klein map.

The second type are known as *sextactic points*. Recall that five points in the plane determine a unique conic, and so, given a higher degree differentiable curve such as our quartic, for any point on the curve there is a conic making 5-point contact with the curve at that point. Consider the analogy with tangents: any two distinct points in the plane determine a straight line, and so at any point on a differentiable curve there is a unique line making 2-point contact at that point. Of course, some of those lines will make 3-point contact with the curve; these are the points of inflexion. Analogously, at some points the associated conic will make 6-point contact with the curve; these are the sextactic points. They correspond to the edges of the Klein map.

It is of interest to ask for a geometric description of the triples of points of inflexion, which correspond to triples of faces at maximal distance from one another in the Klein map. To furnish such a description, first note that α can only fix three points in $\mathbb{C}^3$, namely $(1, 0, 0)$, $(0, 1, 0)$ and $(0, 0, 1)$, and so these are three points of inflexion. Now, the tangent at a point of inflexion, as mentioned above, makes 3-point contact at that point. But a line should intersect a quartic curve four times, and so this tangent must cut $F = 0$ once more. A transformation which preserves the curve and fixes the given point of inflexion must fix the tangent at that point of inflexion. It must, therefore, fix the fourth point of contact which the tangent makes with the curve. This says that the tangent at $(1, 0, 0)$ must cut the curve again at a point which is fixed by α, namely at $(0, 1, 0)$ or $(0, 0, 1)$. Whichever of these two it is, we can repeat the process, arriving back where we started after three moves. Thus we obtain *triangles of inflexion*, which, in the correspondence with 7-cycles, are the sets $\{\lambda, \lambda^2, \lambda^4\}$.

2.3 The algebraic approach

We see from the ATLAS [25], p. 96, that $M \cong M_{24}$ contains a class of maximal subgroups isomorphic to $L_2(7)$ and acting transitively on the 24 points. Let L be such a subgroup. Then the stabilizer of a point in L must be cyclic of order 7, and so a subgroup of L isomorphic to S_4 must act regularly. Now, L contains two classes of such subgroups, corresponding to the points and lines of the projective plane of order 2. In one case these subgroups are self-centralizing within M_{24}; however, if $H \cong S_4$ is in the other class, we find that H centralizes an involution in M_{24}. By the transitivity of H, such an involution must have cycle shape 2^{12}. Under conjugation by L, such an element will have seven images, which we may label $\{t_0, t_1, \ldots, t_6\}$, corresponding, without loss of generality, to the points of the plane in which $\{1, 2, 4\}$ and its translates modulo 7 are the lines. Now, the normalizer of the subgroup $\langle t_0, t_1, \ldots, t_6 \rangle$ properly contains the maximal subgroup L; thus,

$$\langle t_0, t_1, \ldots, t_6 \rangle = M.$$

Suppose now we start with a group $L \cong L_3(2)$ acting transitively on 24 letters and thus embedded in the symmetric group $\Sigma \cong S_{24}$. In order to produce the configuration which we know exists within M_{24}, we must produce involutions in Σ which commute with subgroups of L isomorphic to S_4. As above, such a subgroup $H \cong S_4$ must act regularly on the 24 points, and thus (as in Section 1.3) we have $K = C_\Sigma(H) \cong S_4$. From the previous paragraph we realize that at least one of the involutions of K together with

L will generate a copy of M_{24}, and, of course, K contains six transpositions and a further three involutions in its Klein fourgroup. As mentioned above, the normalizer in Σ of L has the following form:

$$3 \times \mathrm{PGL}_2(7),$$

for, as in Section 1.1, an element centralizing L must be regular and preserve the eight blocks of imprimitivity of size 3; so the centralizer has order at most 3. But $\mathrm{PGL}_2(7)$ certainly does have a transitive action on 24 letters, since it contains dihedral subgroups of order 14; and its derived subgroup remains transitive. Moreover, the blocks of imprimitivity of length 3 have a sense attached to them, as the unique (maximal) subgroup containing such a D_{14} is Frobenius of order 42. Thus, there is a unique cyclic subgroup of order 3, generated by a regular element ζ, commuting with this $\mathrm{PGL}_2(7)$. Now, the outer elements of our $\mathrm{PGL}_2(7)$ interchange the two classes of subgroups isomorphic to S_4, which we can choose as H. Having fixed H, the centralizing element ζ permutes the nine involutions of K by conjugation into three orbits of length 3. In fact, it is the elements which lie in the Klein fourgroup which generate copies of M_{24}. One set of transpositions gives rise to copies of the alternating group A_{24}, whilst the other gives groups of shape

$$3^7 : (2 \times 2^3 : \mathrm{L}_3(2)).$$

Thus, a copy of $\mathrm{L}_3(2)$ acting transitively on 24 letters can be extended to a copy of M_{24} on those letters in just $2 \times 3 = 6$ ways. To verify this without assuming the above statement about the three types of involution in K, we count in two ways the following set:

$$\{(L, M) \mid L \cong \mathrm{L}_3(2), M \cong M_{24}, L \leq M \leq \Sigma, L \text{ transitive}\},$$

where Σ denotes the symmetric group S_{24} acting on a given set of 24 letters. Thus,

$$\mid \Sigma : N_\Sigma(L) \mid \times d = \mid \Sigma : N_\Sigma(M) \mid \times \mid M : N_M(L) \mid,$$

and so

$$d = \frac{24!}{|M|} \times \frac{|M|}{|L|} \times \frac{6|L|}{24!} = 6,$$

where d is the number of copies of M_{24} containing a given L.

2.4   Independent proofs

We now let $\mathcal{T} = \{t_0, t_1, \ldots, t_6\}$ and define $M = \langle \mathcal{T} \rangle$ to be the subgroup of Σ_{24}, the symmetric group on 24 letters, generated by the t_i as defined in Table 2.2. From here we deduce the well known properties of M_{24}. The element $\hat{a} = (0\ 18\ 3\ 20\ 8\ 14\ 15)(6\ 12\ 5\ 9\ 19\ 21\ 1)(11\ 10\ 4\ 2\ 16\ 7\ 13)$ cycles the seven generators by conjugation as in the M_{12} case. Thus $t_i^{\hat{a}} = t_{i+1}$ for $i = 0, \ldots, 6$ and the subscripts are read modulo 7. We first show Lemma 2.1.

LEMMA **2.1** The group $M = \langle t_0, t_1, \ldots, t_6 \rangle$ contains L, the group of all rotational symmetries of the Klein map as labelled in Figure 2.3.

Proof We simply show that $1 \neq (t_i t_j)^3 \in L$ for all $i \neq j$. Then, since L acts doubly transitively on the set $\mathcal{T}$ by conjugation, these elements must form a complete conjugacy class of L. Since L is simple, we see that these elements must generate L. In fact, we have

$$(t_3 t_5)^3 = (\infty \ 0)(1 \ 12)(2 \ 3)(4 \ 9)(5 \ 20)(6 \ 17)$$
$$(7 \ 19)(8 \ 21)(10 \ 13)(11 \ 22)(14 \ 16)(15 \ 18),$$

the involutory symmetry of κ about the edge separating faces ∞ and 0. $\square$

In order to prove our assertions about the group M generated by $\mathcal{T}$, it will prove useful to know the cycle-shapes of some of its elements.

LEMMA **2.2** The group $M = \langle t_0, t_1, \ldots, t_6 \rangle$ possesses elements of cycle-shapes 3^8, $1^2.11^2$, 1.23 and $1^8.2^8$.

Proof It is readily checked that the element $t_i t_j$ for $i \neq j$ has cycle-shape 6^4, and that $t_i t_j t_k$, for i, j, k distinct, has cycle-shape $2^4.4^4$ if $\{i, j, k\}$ is a line, and $1^2.11^2$ if not. Moreover, the element $\hat{a}t_0$ has cycle-shape $1.3.5.15$. Now, Lemma 2.1 states that $L \subset M$ and, since L acts transitively on Ω, so does M. The point-stabilizer in M contains elements of cycle-shapes 1.11^2 and $3.5.15$ and, since any orbit of this point-stabilizer must have length a sum of cycle lengths in each case, we see that M must act doubly transitively. But this implies that M contains elements of cycle-shape 1.23. Elements of cycle-shape 3^8 and $1^8.2^8$ follow by squaring elements of shape 6^4 and $2^4.4^4$, respectively. $\square$

The desired properties of our group M are best deduced by introducing a new structure. Recall that $P(\Omega)$, the power set of a set Ω, can be regarded as a vector space over the field of order 2 of dimension the cardinality of Ω. In this space the sum of two vectors – which are, of course, subsets of Ω – is their symmetric difference. Thus,

$$X + Y = (X \setminus Y) \cup (Y \setminus X) \quad \text{for } X, Y \in P(\Omega).$$

The set of 1-element subsets of Ω forms a natural basis for the space $P(\Omega)$. Such a space has a bilinear form defined on it by

$$(X, Y) = \begin{cases} 0 \text{ if } |X \cap Y| \in 2\mathbb{Z} \\ 1 \text{ if } |X \cap Y| \in 2\mathbb{Z} + 1. \end{cases}$$

Thus, two subsets of Ω are orthogonal if, and only if, they intersect in an even number of points. For us Ω will be the set of faces of the Klein map κ, labelled with the points of the projective line $P_1(23)$ as in Figure 2.3, and so the space $P(\Omega)$ will have dimension 24; see ref. [5]. We shall define a subspace of $P(\Omega)$ of dimension 12. Firstly, note that the 24 faces fall into eight blocks

∞	0	1	2	3	4	5	6
∞	0	18	3	20	8	14	15
22	6	12	5	9	19	21	1
17	11	10	4	2	16	7	13

Figure 2.5. The eight terns.

of imprimitivity under the action of the group of rotational symmetries of the Klein map. These are the sets of faces at maximal distance from one another in the map κ; and an element, its square and its fourth power when the faces are regarded as 7-cycles. For convenience, these triples, which we refer to as *terns*, are displayed as the columns in Figure 2.5, where they are arranged so that the 7-fold symmetry $\hat{a}$ which fixes the face ∞ preserves the rows. The group L permutes the eight terns as the projective special linear group $L_2(7)$ acting on the points of the projective line $P_1(7)$, and so we label the terns accordingly in Figure 2.5. Note that a subset of Ω consisting of a face and the heptagon of neighbours around it contains one face from each tern (this is certainly true when the central face is ∞ and so is true for the 24 images of ∞ under the rotations of L), and so its complement contains two points from each tern; let u_i denote this complement in the case when the heptagon is centred on i, so that u_∞ consists of the bottom two rows of Figure 2.5. We shall consider

$$\mathcal{E} = \langle u_i \mid i \in \Omega \rangle ,$$

the subspace of $P(\Omega)$ spanned by these 24 *generating 16-ads*. Thus every member of $\mathcal{E}$ will intersect each tern evenly. We shall denote the vector consisting of the 1-element subset $\{i\}$ by v_i, and for $X \subset \Omega$ we let v_X denote

$$\sum_{i \in X} v_i .$$

In order to write the vectors of $\mathcal{E}$ (which are simply subsets of Ω) more concisely, we make use of $GF(4)$, the field of order 4, and let entries in the terns be denoted by

$$0 = \begin{array}{|c|}\hline \\ \hline \\ \hline \\ \hline \end{array}\,, \quad 1 = \begin{array}{|c|}\hline \\ \hline \times \\ \hline \times \\ \hline \end{array}\,, \quad \omega = \begin{array}{|c|}\hline \times \\ \hline \\ \hline \times \\ \hline \end{array}\,, \quad \bar{\omega} = \begin{array}{|c|}\hline \times \\ \hline \times \\ \hline \\ \hline \end{array}\,.$$

Note that $1 + \omega + \bar{\omega} = 0$, and that vectors of $\mathcal{E}$ may be written as 8-tuples of elements from $GF(4)$. Thus, for example,

$u_0 =$

∞	0	1	2	3	4	5	6
			×	×	×	×	
×	×	×		×	×		×
×	×	×	×			×	×

$= [1, 1, 1, \omega, \bar{\omega}, \bar{\omega}, \omega, 1]$.

A word of warning is necessary: the space $\mathcal{E}$ is *not* a subspace over the field $GF(4)$; nor can it be made into a subspace by a judicious relabelling of the 24 points. Were this possible, then, corresponding to multiplication by ω, there would be a permutation of cycle shape 3^8 fixing all the terns and commuting with the group L. Our element ζ and its inverse are the only elements to have these properties, and they cannot fix $\mathcal{E}$ as they would then fix the whole Golay code, defined below, and so lie in M_{24}. But the subgroup L has trivial centralizer in M_{24}. Of course, $\mathcal{E}$ is a *subgroup* of the vector space of 8-tuples over $GF(4)$.

LEMMA 2.3 The space $\mathcal{E}$ has dimension 8.

Proof Firstly, note that, for any pair of faces of κ, the number of faces joined to both of them is even. Thus, any two of our generating 16-ads are orthogonal to one another with respect to the bilinear form defined above. So, the subspace $\mathcal{E}$ is contained in its own orthogonal complement. Moreover, the subspace $\mathcal{B}$, spanned by the eight terns (the blocks of imprimitivity), clearly has dimension 8 and, since the vectors of $\mathcal{E}$ intersect the terns evenly, it is orthogonal to and disjoint from $\mathcal{E}$. So, the space $\mathcal{E} + \mathcal{B}$ has dimension dim $\mathcal{E}$ + 8 and is contained in the orthogonal complement of $\mathcal{E}$; i.e.

$$\dim \mathcal{E} + (\dim \mathcal{E} + 8) \leq 24,$$

and so $\mathcal{E}$ has dimension less than or equal to 8. In order to see that $\mathcal{E}$ has dimension at least 8, we consider the linear transformation which projects vectors of $\mathcal{E}$ onto the 8-element subset $u_{\infty} + \Omega$, the top row of the tern array. Clearly, the kernel of this transformation has dimension at least 1 since it contains the vector u_{∞}. Moreover, it is readily checked that

$$u_{16} + u_{18} = [\omega, 0, 0, \omega, 0, 0, 1, 1],$$

a vector of $\mathcal{E}$ which projects onto the 2-element subset of the top row $\{\infty, 3\}$. This subset has seven images under the action of the cyclic group of order 7 generated by $\hat{a}$ which fixes ∞ and cycles the rows of the tern array. These are clearly linearly independent (and generate the space of all even subsets of the top row), and so the image of $\mathcal{E}$ has dimension at least 7. We conclude that $\mathcal{E}$ has dimension at least $(1+7) = 8$, and so dim $\mathcal{E} = 8$, as required. $\square$

Now that we know the dimension of $\mathcal{E}$, we can readily use the above argument to prove Lemma 2.4.

LEMMA 2.4 The set $B = \{u_0, u_{18}, u_3, u_{20}, u_8, u_{14}, u_{15}, u_{22}\}$ is a basis for $\mathcal{E}$.

Proof Firstly, we give some relations which hold between the u_i. Since every face is joined to just one face of each tern, the sum

$$\sum_{i \in T} u_i,$$

for T a tern, is the zero vector. Secondly, note that the sum of the seven u_j for j joined to a fixed face, i say, is just u_i, for every face other than these eight faces is joined to none or two of them. In particular, we have

$$u_\infty = u_0 + u_{18} + u_3 + u_{20} + u_8 + u_{14} + u_{15} \in \langle B \rangle.$$

As above, we now project onto the top row of the tern array. That is to say, we define

$$\phi : X \mapsto X \cap (\Omega + u_\infty).$$

Certainly, $u_\infty = [1, 1, 1, 1, 1, 1, 1, 1] \in \text{Ker}(\phi)$; we need to show that $\text{Im}(\phi)$ consists of the 7-dimensional space of all even subspaces of the top row. But it is readily checked that

$$\phi : u_0 + u_{18} + u_8 + u_{14} = \{3, 20\},$$

an adjacent pair in the top row. It follows, taking images of this under the action of the element $\hat{a}$, that any even subset of the top line which does not contain ∞ is in $\langle B \rangle$. But u_{22} maps onto the whole of the top line, and so we obtain every even subset of the top line and dim $\langle B \rangle = 1 + 7 = 8$.

We now investigate the 256 vectors of $\mathcal{E}$ in further detail, and begin by producing a more subtle relation. Consider the following:

$$u_\infty + u_0 = [1, 1, 1, 1, 1, 1, 1, 1] + [1, 1, 1, \omega, \bar{\omega}, \bar{\omega}, \omega, 1] = [0, 0, 0, \bar{\omega}, \omega, \omega, \bar{\omega}, 0];$$

in other words, the sum of two generating 16-ads whose 'centres' are adjacent. This sum is illustrated in Figure 2.6(a), where the vertices represent faces of the Klein map κ and two vertices are joined if the corresponding faces are adjacent.

It is visible from the figure, and may readily be checked, however, that this vector is also obtained by taking

$$u_{10} + u_{13} = [\bar{\omega}, \omega, \bar{\omega}, \omega, \bar{\omega}, 1, 1, \bar{\omega}] + [\bar{\omega}, \omega, \bar{\omega}, 1, 1, \bar{\omega}, \omega, \bar{\omega}] = u_\infty + u_0,$$

and so we have the following relation:

$$u_\infty + u_0 + u_{10} + u_{13} = 0.$$

Now observe that our set B of eight generating 16-ads is preserved by $\hat{a}$, a symmetry of order 7 fixing face ∞. The new relation shows that

$$u_3 + u_{18} = u_{22} + u_{21},$$

and so $u_{21} \in \langle B \rangle$ as do all the u_i with face i adjacent to face 22. But we also have

$$u_\infty + u_{22} + u_{17} = 0 \text{ and } u_1 + u_9 + u_{17} + u_{10} = 0,$$

and so $u_{10} \in \langle B \rangle$, as do all the u_i for face i adjacent to face 17. Thus, $\langle B \rangle$ contains all the u_i, and

$$\langle B \rangle = \mathcal{E}.$$

$\square$

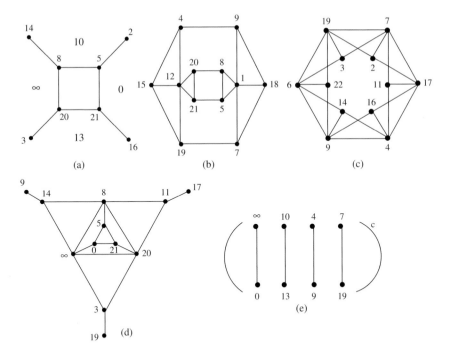

Figure 2.6. The vectors of $\mathcal{E}$.

Since we wish to understand $\mathcal{E}$ in some detail, we choose to exhibit a representative of each of the orbits of L acting on the vectors of $\mathcal{E}$. Note that all elements of L act regularly on the faces of κ, except elements of order 7 (and the identity element), and so it is straightforward to read off the order of the stabilizer of a given vector. Representatives of the orbits are given in Figure 2.6, and Table 2.3 gives the length of each orbit; thus we see that $\mathcal{E}$ has orbits of lengths $(1 + 24 + 42 + 84 + 28 + 56 + 21) = 256$ under the action of L. Note that Figure 2.6(e) is the complement of the four parallel lines shown.

Note that, since the 24 generating vectors of $\mathcal{E}$ are 16-ads intersecting one another evenly, every vector in $\mathcal{E}$ will be a subset containing a multiple of four faces. In Table 2.3 we list the cardinality of subsets corresponding to vectors in the various orbits, and term 8-element subsets *octads*, 12-element subsets *dodecads*, and 16-element subsets *16-ads*. In the column headed 'Terns' we record the manner in which vectors of each orbit intersect the eight terns. There is one further piece of information about these vectors which we need to record before we are able to proceed. Observe that the octad $u_\infty + u_0$ intersects each of the terns labelled **2, 3, 4** and **5** in two points and is disjoint from the others. Now, when $L_2(7)$ acts on the 8-point projective line, it preserves two Steiner systems $S(3, 4, 8)$, each of which contains 14 subsets of size 4. The remaining $\binom{8}{4} - 28 = 42$ 4-element subsets, of which $\{\mathbf{2, 3, 4, 5}\}$ is an example, fall in a single orbit.

Table 2.3. Vectors of $\mathcal{E}$ and their intersection with the terns

Number	Type	Terns	Example	Stabilizer in L
1	ϕ	0^8	**0**	L
24	16-ads	2^8	u_∞	$\langle \alpha \rangle \cong C_7$
42	octads	$0^4.2^4$	(a) $u_\infty + u_0 = u_{10} + u_{13}$	$\langle \beta \rangle \cong C_4$
84	dodecads	$0^2.2^6$	(b) $u_{12} + u_1$	$\langle \beta^2 \rangle \cong C_2$
28	dodecads	$0^2.2^6$	(c) $u_6 + u_{17} = u_{19} + u_4 = u_7 + u_9$ $u_{19} + u_7 + u_{17} =$	$\langle \gamma, \beta^2 \rangle \cong S_3$
56	dodecads	$0^2.2^6$	(d) $u_9 + u_6 + u_{19} =$ $u_{17} + u_4 + u_9$	$\langle \gamma \rangle \cong C_3$
21	16-ads	2^8	(e) $u_\infty + u_0 + u_4 + u_9 =$ $u_{13} + u_{10} + u_{19} + u_7$	$\langle \beta, \delta \rangle \cong D_8$

$\alpha = (0\ 18\ 3\ 20\ 8\ 14\ 15)(6\ 12\ 5\ 9\ 19\ 21\ 1)(11\ 10\ 4\ 2\ 16\ 7\ 13)$
$\beta = (\infty\ 13\ 0\ 10)(1\ 11\ 12\ 22)(2\ 14\ 3\ 16)(4\ 19\ 9\ 7)(5\ 8\ 20\ 21)(6\ 18\ 17\ 15)$
$\gamma = (\infty\ 20\ 8)(0\ 21\ 5)(10\ 18\ 12)(1\ 15\ 13)(2\ 16\ 22)(3\ 11\ 14)(9\ 19\ 17)(4\ 6\ 7)$
$\delta = (\infty\ 4)(0\ 9)(10\ 19)(7\ 13)(14\ 15)(16\ 18)(8\ 12)(1\ 21)(11\ 20)(3\ 17)(5\ 22)(2\ 6)$

Thus the 42 octads in $\mathcal{E}$ correspond one-to-one with these *non-special tetrads*.

The two Steiner systems referred to above consist of one of $\{\infty, \mathbf{1}, \mathbf{2}, \mathbf{4}\}$ or $\{\infty, \mathbf{3}, \mathbf{6}, \mathbf{5}\}$ and its images under $L_2(7)$. We choose the first of these and, for convenience, list its tetrads in Table 2.4. We now use this Steiner system to define $\mathcal{D}$, a 4-dimensional subspace of $P(\Omega)$ which intersects $\mathcal{E}$ trivially. For each of the 14 tetrads of the Steiner system, $\mathcal{D}$ contains the dodecad consisting of the union of the four associated terns. These 14 vectors, together with the whole set Ω and the empty set ϕ, form a 4-dimensional subspace of $\mathcal{B}$. Of course, $\mathcal{E} \cap \mathcal{D} = \mathbf{0}$, since every vector of $\mathcal{E}$ intersects each tern evenly, whilst every non-trivial vector of $\mathcal{D}$ intersects at least four of the terns oddly. Moreover, for similar reasons, every vector of $\mathcal{D}$ intersects every vector of $\mathcal{E}$ evenly, and vectors of $\mathcal{D}$ intersect one another evenly. We now define the following:

$$\mathcal{C} = \mathcal{E} + \mathcal{D}.$$

Certainly, $\mathcal{C}$ will have dimension 12 and, as above, every vector of $\mathcal{C}$ will correspond to a subset of Ω of cardinality congruent to 0 modulo 4. Now adding a non-trivial vector of $\mathcal{D}$ to a vector of $\mathcal{E}$ consists of complementing it on four (or all eight) of the terns. Clearly the only way we could obtain a subset of cardinality 4 in this way would be by adding a dodecad of $\mathcal{D}$ to an octad of $\mathcal{E}$ lying in the *same* four terns. But octads of the 42-orbit lie in non-special tetrads, whilst dodecads of $\mathcal{D}$ are unions of special tetrads. So $\mathcal{C}$ contains no 4-element subsets, and its smallest subsets have eight elements in them. Since all vectors of $\mathcal{C}$ corrrespond to subsets of Ω of cardinality

Table 2.4. The tetrads of
the Steiner system $S(3, 4, 8)$

∞124	0365
∞235	1406
∞346	2510
∞450	3621
∞561	4032
∞602	5143
∞013	6254

congruent to 0 modulo 4, since cardinality 4 is impossible, and since the whole set Ω is an element of $\mathcal{C}$, we can immediately conclude the following:

(a) $\mathcal{C}$ consists of the empty set, the whole set Ω, octads, dodecads and 16-ads; and

(b) the number of octads in $\mathcal{C}$ is equal to the number of 16-ads.

We now count the number of octads in $\mathcal{C}$.

LEMMA **2.5** The vector space $\mathcal{C}$ contains 759 octads; it thus consists of the empty set, the whole set Ω, 759 octads, 759 16-ads and 2576 dodecads.

Proof Every element of $\mathcal{C}$ can be written uniquely as $\mathbf{e} + \mathbf{d}$, with $\mathbf{e} \in \mathcal{E}$ and $\mathbf{d} \in \mathcal{D}$. If $\mathbf{e}$ is the empty set, we clearly cannot obtain an octad for any $\mathbf{d} \in \mathcal{D}$, and, if $\mathbf{e}$ is a 16-ad, we can only obtain an octad by adding the whole set Ω to it. If $\mathbf{e}$ is an octad, lying as it does in a non-special tetrad of terns, we can obtain an octad by adding the empty set or a dodecad of $\mathcal{D}$ whose special tetrad has three terns in common with this non-special tetrad. In a Steiner system $S(3,4,8)$, every subset of three points is contained in a unique special tetrad, so there are four dodecads of $\mathcal{D}$ which can be added to a given octad of $\mathcal{E}$ to give a further octad. Finally, if $\mathbf{e}$ is a dodecad in $\mathcal{E}$, then we obtain an octad by adding a dodecad of $\mathcal{D}$ whose special tetrad of terns is contained in the set of six terns in which $\mathbf{e}$ lies. But, since the complement of a special tetrad is itself a special tetrad, the number of special tetrads contained in a fixed set of size 6 is equal to the number of special tetrads containing a given pair of points, namely three. Thus, the number of octads is $24 + 42 \times (1+4) + (84 + 28 + 56) \times 3 + 21 = 759$. By the remark before this Lemma, the number of 16-ads in $\mathcal{C}$ is also 759, and so the number of dodecads is $(2^{12} - 2 - 2 \times 759) = 2576$, as required. □

The space $\mathcal{C}$ is the celebrated *binary Golay code*. Since the subsets corresponding to any two vectors of $\mathcal{C}$ intersect evenly, and since the dimension of $\mathcal{C}$ is half the dimension of $P(\Omega)$, we see that, with respect to this bilinear form, $\mathcal{C}$ is its own orthogonal complement; such a code is said to be *self-dual*. Note that, in these circumstances, a subset of Ω is in $\mathcal{C}$ if, and only if, it intersects every member of $\mathcal{C}$ evenly. Indeed, a subset of Ω is in $\mathcal{C}$ if, and only if, it intersects each member of a basis for $\mathcal{C}$ evenly. We shall use this fact to prove the following theorem.

THEOREM 2.1 The Golay code $\mathcal{C}$ as defined above is preserved by the group M.

Proof The code $\mathcal{C}$ is certainly preserved by the group L and, since the seven generators in $\mathcal{T}$ are conjugate to one another under the action of L, it will suffice to show that $\mathcal{C}$ is preserved by t_0. From the preceding remarks, we need only show that the image under t_0 of each member of a basis of $\mathcal{C}$ intersects each member of that basis evenly. As basis we take v_Ω, the eight octads $u_i + v_\Omega$ for $u_i \in B$ of Lemma 2.4, and the three dodecads corresponding to the special tetrads of terns $\{\infty, \mathbf{1}, \mathbf{2}, \mathbf{4}\}$, $\{\infty, \mathbf{2}, \mathbf{3}, \mathbf{5}\}$ and $\{\infty, \mathbf{3}, \mathbf{4}, \mathbf{6}\}$. For now only, we shall denote the dodecad corresponding to the four terns $\{\mathbf{i}, \mathbf{j}, \mathbf{k}, \mathbf{l}\}$ by $\mathbf{d}(i, j, k, l)$. Then, for example,

$$t_0 : (v_\Omega + u_0) \mapsto \{21, 14, 4, 5, 2, 0, 13, 7\}.$$

As stated, all that is required is to check that this 8-element subset of Ω intersects each of our basis vectors evenly, which is readily done. However, since our decomposition of the space $\mathcal{C}$ lends itself to writing vectors as a sum of u_is and $\mathbf{d}(i, j, k, l)$, and for the sake of completeness on the printed page, we choose to write each of these images as linear combinations of our generating vectors. For instance, this image appears as

∞	0	1	2	3	4	5	6
	×					×	
			×			×	
			×	×		×	×

,

which intersects terns **0**, **3**, **5** and **6** oddly. Adding the dodecad $\mathbf{d}(0, 3, 5, 6)$ to it results in the 8-element set:

∞	0	1	2	3	4	5	6
			×			×	
	×		×	×		×	
	×		×				

$= [0, 1, 0, 1, \bar{\omega}, 0, 0, \bar{\omega}] = u_8 + u_{14} \in \mathcal{E},$

as the member of the orbit of length 42; see Figure 2.6(a). Similarly, we have

$$
\begin{aligned}
t_0: \quad (v_\Omega + u_{18}) &\mapsto u_5 + u_{21} + \mathbf{d}(1,3,4,5), \\
(v_\Omega + u_3) &\mapsto u_8 + u_{11} + \mathbf{d}(\infty,1,2,4), \\
(v_\Omega + u_{20}) &\mapsto u_1 + u_5 + \mathbf{d}(1,3,4,5), \\
(v_\Omega + u_8) &\mapsto u_0 + u_2 + \mathbf{d}(1,3,4,5), \\
(v_\Omega + u_{14}) &\mapsto u_0 + u_{18} + \mathbf{d}(\infty,2,3,5), \\
(v_\Omega + u_{15}) &\mapsto u_5 + u_{10} + \mathbf{d}(\infty,0,2,6), \\
(v_\Omega + u_{22}) &\mapsto u_8 + u_{14} + \mathbf{d}(0,3,5,6), \\
\mathbf{d}(\infty,1,2,4) &\mapsto u_0 + u_{21} + u_{10} + u_{17} + \mathbf{d}(\infty,1,2,4), \\
\mathbf{d}(\infty,2,3,5) &\mapsto u_\infty + u_{14} + u_{11} + u_{12} + \mathbf{d}(\infty,2,3,5), \\
\mathbf{d}(\infty,3,4,6) &\mapsto u_{18} + u_{20} + \mathbf{d}(0,1,2,5).
\end{aligned}
$$

But permutations of Ω must map $P(\Omega)$ onto itself, and so t_0 maps $\mathcal{C}$ onto $\mathcal{C}$. □

We have shown that the permutation t_0 preserves the Golay code $\mathcal{C}$ by showing that the image of each vector in a basis of $\mathcal{C}$ intersects every vector of that basis evenly, and so lies in $\mathcal{C}$. To make life slightly easier for ourselves, we chose one of our basis vectors to be the whole set Ω, which is certainly preserved by any permutation. In fact, in ref. [30], the author showed that it is possible to produce a set of eight octads with the property that a permutation of Ω preserves $\mathcal{C}$ if, and only if, it maps these eight octads to vectors in $\mathcal{C}$. We can extend this set of vectors to a basis by adjoining the whole set Ω and three further octads, so a permutation preserves $\mathcal{C}$ if, and only if, the image of each of the eight octads intersects each of the eleven octads evenly. The reason that this holds is that the subspace spanned by the eight octads extends to a copy of the Golay code in precisely one way. The test is minimal in the sense that any set of seven octads is contained in *no* copy of the Golay code, or in more than one. The set of eight octads is spatially unique in that the subspace it spans has to belong to a uniquely defined type under the action of M_{24}. We are now in a position to observe the following.

THEOREM 2.2 The 759 octads of $\mathcal{C}$ form a Steiner system $S(5,8,24)$.

Proof No two distinct octads of $\mathcal{C}$ can intersect in more than four points, as their sum would then contain fewer than eight points. Thus, no subset of five points of Ω can be contained in more than one octad. So the total number of 5-element subsets of Ω contained in an octad is $759 \times \binom{8}{5}$. Since this number is precisely $\binom{24}{5}$, we deduce that every 5-element subset of Ω is contained in precisely one octad. □

So we see that the group M preserves the Golay code $\mathcal{C}$, and in so doing preserves a Steiner system $S(5,8,24)$. We conclude with some results about the manner in which M acts on these two combinatorial structures.

LEMMA 2.6 M acts transitively on the 759 octads of $\mathcal{C}$.

Proof We have already seen in Lemma 2.2 that the group M contains elements of cycle-shapes 1.23, $1^2.11^2$ and 3^8, none of which could fix an octad. So any orbit of M on octads must have length divisible by $3 \times 11 \times 23 = 759$, which is the total number of octads in $\mathcal{C}$. Thus M acts transitively on the octads of the Steiner system. □

We now need a lemma which tells us that a permutation fixing a Steiner system $S(5, 8, 24)$ cannot fix too many points.

LEMMA 2.7 A permutation of Ω which preserves a Steiner system $S(5, 8, 24)$ fixes every point of an octad, and a point not in this octad must be the identity.

Proof Let π be the permutation, let $U = \{a_1, a_2, \ldots, a_8\}$ be the octad which is fixed pointwise by π, and let b be the fixed point not in U. Let $x \neq b$ be any other point not in U. We shall show that x must be fixed by π and so π is the identity. Consider firstly the octad containing $\{a_1, a_2, a_3, b, x\}$. It must contain a further point of U as all octads intersect evenly, so, without loss of generality, let this be a_4. It can contain no further members of U, and so $V = \{a_1, a_2, a_3, a_4, b, x, y, z\}$ is an octad, where y and z are not in U. Now, the octad containing $\{a_1, a_2, a_5, b, x\}$ must also contain a further member of U which cannot be a_3 or a_4 and so may be taken to be a_6. Thus, $W = \{a_1, a_2, a_5, a_6, b, x, v, w\}$ is an octad, where v and w are distinct from y and z and are not in U. But π fixes five members of each of V and W, and so fixes them both. Thus, $\{x, y, z\}^\pi = \{x, y, z\}$; $\{x, v, w\}^\pi = \{x, v, w\}$; and $x^\pi \in \{x, y, z\} \cap \{x, v, w\} = \{x\}$. Thus, $x^\pi = x$, and π is the identity permutation. □

COROLLARY 2.1 No element which preserves a Steiner system $S(5, 8, 24)$ can act as a transposition on an octad and fix an element outside that octad.

Proof Let the element π interchange a_1 and a_2 in the proof of Lemma 2.7, and fix b and all other points of U. Then the same argument shows that π must fix every other point outside U, and so must be a transposition of S_{24}. But there are octads which contain a_1 but not a_2, and such an octad would be taken by π to an octad having seven points in common with itself: a clear contradiction. □

What then is the action of the stabilizer in M of an octad on the points of that octad?

THEOREM 2.3 The stabilizer in M of an octad of the Steiner system $S(5, 8, 24)$ acts as the alternating group A_8 on the points of the octad, and has shape $2^4 : A_8$. Thus, M has order $2^4 \times 8!/2 \times 759 = 244\,823\,040$.

Proof Since 759 is not congruent to 0 modulo seven, we know that elements of order 7 in L must fix octads and act with cycle-shape 1.7 on the points of the octad (indeed, we know that our element $\hat{a}$ fixes the octad $u_\infty + \Omega$). Moreover, we have seen that the element $\hat{a}t_0$ has cycle-shape 1.3.5.15. The five points in the 5-cycle of this element determine an octad which must be fixed by this element. Transitivity on octads ensures that any octad thus has permutations of cycle-shapes 1.7 and 3.5 acting on its eight points. But the subgroup of the symmetric group S_8 generated by any two permutations of these shapes must be doubly transitive and so primitive, and, since it contains 3-cycles (the fifth power of $\hat{a}t_0$), it must be the alternating group A_8. But we know that elements which fix every point of an octad must act regularly on the remaining 16 points, since the only element which fixes a further point is the identity; so the maximal order of the subgroup of M fixing every point of an octad is 16. But we know that M contains elements of cycle shape $1^8.2^8$; indeed, the element

$$(t_0 t_1 t_3)^2 = (\infty\ 11)(0\ 3)(2\ 13)(5\ 16)(7\ 15)(8\ 22)(9\ 10)(12\ 14)$$

is one such. The fixed point set of this element is given by

$$\{1, 4, 6, 17, 18, 19, 20, 21\} = [\bar{\omega}, \omega, 1, \bar{\omega}, 1, \omega, \omega, \omega]^c = u_1 + u_4 + u_{17} + u_{18} + \Omega,$$

an octad of $\mathcal{C}$ (the superscript 'c' denotes complementation). Now, an element of cycle shape 1.3.5.15 fixing this octad must conjugate the involution into 15 distinct involutions, all with the same fixed point set. But, together with the identity, this is the maximum number of elements which can fix every point of an octad, and so they form an elementary abelian group of order 2^4. Suppose the stabilizer of an octad acts as S_8 on the points of that octad. An element σ of M which acts $1^6.2$ on the points of an octad must square to the identity or an element of shape $1^8.2^8$. But, in the first case, in order for σ to be even it would have to fix points outside the octad, which is impossible by the Corollary to Lemma 2.7. The second case would imply that σ has cycle shape $1^6.2.4^4$, which is odd. Thus, the stabilizer of an octad has shape $2^4 : A_8$, and the result follows by the transitivity of M on octads. $\square$

We conclude this section by describing the orbits of the group L on the 759 octads of $\mathcal{C} = \mathcal{E} + \mathcal{D}$. The 24 octads of the form $u_i + v_\Omega$ certainly form an orbit, as do the 42 octads of form $u_i + u_j$, where i is joined to j, such as $u_\infty + u_0$. Now, as was remarked above, octads of $\mathcal{E}$ have two points in each of four terns which form a *non-special* tetrad. In order to obtain a further octad by adding a vector of $\mathcal{D}$ to such an octad, that vector must contain three of those four terns. There are thus four possible vectors to add, and the stabilizer of our original vector, which is isomorphic to C_4, acts transitively on them. We thus get an orbit of $42 \times 4 = 168$ octads. The dodecads of $\mathcal{E}$ lie in six terns, and so to obtain an octad we must add one of the three dodecads of $\mathcal{D}$ which lie entirely in those terns. Now, L has three orbits on the dodecads of $\mathcal{E}$, as shown in Figure 2.6, of lengths 84, 28 and 56, respectively. In the latter two cases, the subgroup of L fixing such a dodecad acts transitively on the three possible vectors of $\mathcal{D}$. However, the stabilizer of a dodecad in the 84-orbit is cyclic of order 2; in fact, if the dodecad is $u_{12} + u_1$, this group is generated by β^2, which acts on the eight terns as $(\infty\ 0)(1\ 6)(2\ 3)(4\ 5)$, and so the three possible vectors of $\mathcal{D}$ fall into two orbits consisting of $\{\mathbf{d}(1,2,3,6)\}$ and $\{\mathbf{d}(2,4,5,6), \mathbf{d}(1,3,4,5)\}$. Lastly, the complements of the 16-ads in the 21-orbit of $\mathcal{E}$ are clearly in one orbit. There are thus eight orbits of lengths $24 + 42 + 42 \times 4 + 84 \times 1 + 84 \times 2 + 28 \times 3 + 56 \times 3 + 21 = 759$. In Figure 2.7, we display a representative of each of these orbits as follows: $u_\infty + v_\Omega$ (24); $u_\infty + u_0$ (42); $u_\infty + u_0 + \mathbf{d}(\infty, 2, 3, 5)$ (168); $u_1 + u_{12} + \mathbf{d}(1, 2, 3, 6)$ (84);

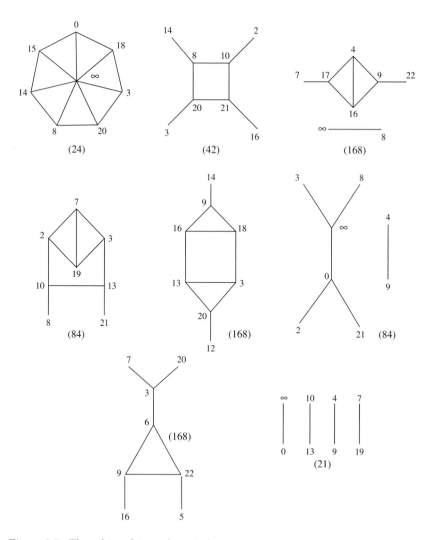

Figure 2.7. The orbits of L on the octads.

$u_1 + u_{12} + \mathbf{d}(2, 4, 5, 6)$ (168); $u_\infty + u_2 + \mathbf{d}(\infty, 0, 4, 5)$ (84); $u_0 + u_3 + u_{18} + \mathbf{d}(\infty, 0, 4, 5)$ (168); $u_\infty + u_0 + u_4 + u_9 + v_\Omega$ (21), and we show how the eight orbits of the subgroup L on octads correspond to faces of the Klein map κ. The vertices in the graphs of the figure correspond to faces of the map as labelled, and they are joined by an edge if those faces are adjacent.

This decomposition of the Golay code $\mathcal{C}$ was described in the author's Ph.D. thesis [28], where the action of the maximal subgroup L on the various combinatorial objects associated with the code was analysed in some detail. Although it gives a convenient notation for the vectors of the code, it does not compare with the so-called Miracle Octad Generator (MOG) and Hexacode (see Conway and Sloane [26] and Curtis [29]) for actually working with the code and thus with the group M_{24} and related groups. Our aim here has been to derive the group and many of its main properties directly from $L_2(7)$, the second smallest non-abelian simple group, and the Klein map on which it acts.

This link between the Klein map κ and the Mathieu group M_{24} is a source of great delight to the author. Both objects were found in the 1870s, but no connection between them was known. Indeed, the class of maximal subgroups of M_{24} isomorphic to the simple group of order 168 (often known, especially to geometers, as the Klein group; see Baker [8]) remained undiscovered until the 1960s. That generators for the group can be read off so easily from the map is immensely pleasing.

Conclusions to Part I

In Chapters 1 and 2 we have used the two smallest non-abelian simple groups, namely the projective special linear groups $L_2(5)$ and $L_2(7)$, to produce generators for the Mathieu groups M_{12} and M_{24}, probably the most remarkable finite simple groups of all. In each case, it was the high degree of symmetry possessed by the set of generators which enabled us to write them down so easily, either combinatorially, geometrically or in a more abstract algebraic manner. The number of generators in the two cases, five for M_{12} and seven for M_{24}, exploits the following exceptional isomorphisms:

$$L_2(5) \cong A_5 \text{ and } L_2(7) \cong L_3(2).$$

Indeed, it was shown by Galois that the projective special linear group $L_2(p)$, $p \geq 5$, cannot act non-trivially on fewer than $p+1$ letters except in the cases $p = 5, 7$ and 11. The above isomorphisms explain how the first two actions are possible; $L_2(11)$ acts on 11 letters since members of the generic maximal class of subgroups isomorphic to A_5 have index 11. So, in a sense, it is the exceptional actions of $L_2(5)$ and $L_2(7)$ which have given rise to the Mathieu groups. In Chapter 5, we shall see how the exceptional action of $L_2(11)$ may be exploited to produce the smallest Janko group J_1, the first sporadic simple group to be discovered after the Mathieu groups.

It is natural to ask how we might generalize these observations and constructions so as to obtain other simple groups. Firstly, we ask ourselves what it means to say that a group G can be generated by a set of seven involutions – as happens here in the M_{24} case. This is simply stating that G is a homomorphic image of F, a free product of seven copies of the cyclic group of order 2. Thus, if we have

$$F = \langle \tau_1, \tau_2, \ldots, \tau_7 \mid \tau_i^2 = 1 \rangle \text{ and } G = \langle t_1, t_2, \ldots, t_7 \rangle,$$

then the map $\phi : F \mapsto G$, where $\phi : \tau_i \mapsto t_i$ for $i \in \{1, 2, \ldots, 7\}$, defines such a homomorphism. The kernel of ϕ will be the normal closure $\langle w_1(\tau_1, \ldots, \tau_7), w_2(\tau_1, \ldots, \tau_7), \ldots \rangle^F$, where $\{w_1, w_2, \ldots\}$ is a set of defining relations for the group G. We should then have

$$G \cong \langle \tau_1, \tau_2, \ldots, \tau_7 \mid \tau_1^2 = \cdots = \tau_7^2 = 1, w_1 = w_2 = \cdots = 1 \rangle,$$

as a presentation for G.

In our case, however, we are saying much more than that $G \cong M_{24}$ is generated by some set of seven involutions, which would be a very weak requirement. We are asserting that M_{24} is generated by a set of seven involutions which possesses all the symmetries of $L_3(2)$ acting on the points of the 7-point projective plane, in the sense that these symmetries are realized by conjugation within G. That is to say that there exists a set of involutions $\mathcal{T} = \{t_0, t_1, \ldots, t_6\}$ in G, such that $\langle \mathcal{T} \rangle = G$, and there exists a surjective homomorphism

$$\psi : N_G(\mathcal{T}) = \{\pi \in G \mid \mathcal{T}^\pi = \mathcal{T}\} \mapsto L \cong L_3(2).$$

So, if $\alpha = (0\ 1\ 2\ 3\ 4\ 5\ 6)$ and $\gamma = (2\ 6)(4\ 5)$ are permutations of $\{0, 1, \ldots, 6\}$ such that $\langle \alpha, \gamma \rangle \cong L_3(2)$, then, after suitable relabelling of the t_i if necessary, there exists an element $a \in G$ such that $t_i^a = t_{\alpha(i)} = t_{i+1}$ and an element $c \in G$ such that $t_i^c = t_{\gamma(i)}$ for $i \in \{0, 1, \ldots, 6\}$, where subscripts are read modulo 7. Note that, since $\langle \mathcal{T} \rangle = G$, it must be possible to find words w_a and w_c in the t_i such that

$$a = w_a(t_0, t_1, \ldots, t_6) \text{ and } c = w_c(t_0, t_1, \ldots, t_6).$$

Let us, for convenience, denote F, the free product of seven copies of the cyclic group of order 2, by $2^{\star 7}$. Thus,

$$F = \langle \tau_1, \tau_2, \ldots, \tau_7 \mid \tau_i^2 = 1 \rangle \cong \underbrace{C_2 \star C_2 \star \ldots \star C_2}_{\text{seven times}} \cong 2^{\star 7}.$$

Certainly, permutations of $\{\tau_1, \tau_2, \ldots, \tau_7\}$, the set of involutory generators of F, induce automorphisms of F, and so given a subgroup $N \leqslant S_7$ we can form a semi-direct product of shape

$$2^{\star 7} : N,$$

where the automorphisms act by conjugation. In other words,

$$\tau_i^\pi = \pi^{-1} \tau_i \pi = \tau_{\pi(i)} \text{ for } \pi \in N.$$

Since all the relations we have used implicitly to define this semi-direct product plainly hold in G, and since $\langle \mathcal{T} \rangle = G$, we can assert that G is a homomorphic image of

$$P = 2^{\star 7} : L_3(2).$$

Moreover, since every element of the semi-direct product P can be written as πw, where $\pi \in N \cong L_3(2)$ and w is a word in the generators $\{\tau_1, \tau_2, \ldots, \tau_7\}$, any homomorphic image of P can be defined by setting certain such expressions equal to the identity. Thus,

$$\frac{2^{\star 7} : L_3(2)}{\pi_1 w_1, \pi_2 w_2, \ldots} \cong M_{24}$$

for a suitable choice of elements $\pi_i w_i$.

It turns out that this approach is particularly revealing and that many simple groups, both sporadic and classical, have surprisingly simple definitions of this type. Indeed, in the smaller cases, the identification can be readily proved by hand, furnishing as a by-product a concise representation for the elements of the group.

Part II

Involutory symmetric generators

3

The (involutory) progenitor

In this chapter, we introduce a family of infinite groups, the members of which include among their homomorphic images all the non-abelian simple groups. A member of this family will be called a *progenitor*, its progeny being the set of finite images it possesses. It will turn out that the elements of these groups take a particularly simple form and are very easy to work with.

3.1 Free products of cyclic groups of order 2

We shall start by considering a group generated by two elements of order 2, with no further relation holding between them. Thus,

$$G = \langle a, b \mid a^2 = b^2 = 1 \rangle.$$

Note that the element $x = ab$ has infinite order and, since $\langle ab, a \rangle = \langle a, b \rangle$, we have

$$G = \langle x, a \mid a^2 = 1, x^a = x^{-1} \rangle.$$

For this reason, we often refer to G as an *infinite dihedral group*; we may write its elements as follows:

$$G = \{1, a, b, ab, ba, aba, \dots\},$$

where elements of odd length in a and b are involutions, whilst elements of even length have infinite order. Multiplication of elements of G is achieved by juxtaposition followed by cancellation of any adjacent repetitions, and inversion by reversing the word in a and b. It is intuitively clear from the symmetrical manner in which the group G was defined that interchanging a and b gives rise to an automorphism of G, and we shall verify this assertion in a more general context. We call G, which is generated by two cyclic subgroups of order 2 with no relation between them, a *free product* of these groups, and write

$$G \cong \langle a \rangle \star \langle b \rangle \cong C_2 \star C_2.$$

For convenience, we denote this free product by $2^{\star 2}$.

We can readily extend these ideas to n generators and define a free product of n copies of the cyclic group of order 2 as follows:

$$E \cong 2^{\star n} = \langle \tau_1, \tau_2, \ldots, \tau_n \mid \tau_1^2 = \tau_2^2 = \cdots = \tau_n^2 = 1 \rangle$$

$$= \langle \tau_1 \rangle \star \langle \tau_2 \rangle \star \cdots \star \langle \tau_n \rangle \cong \underbrace{C_2 \star C_2 \star \cdots \star C_2}_{n \text{ times}}.$$

So, E consists of all finite products of the elements τ_i without adjacent repetitions, i.e.

$$E = \left\{ \tau_{k_1} \tau_{k_2} \cdots \tau_{k_r} \mid r \in \mathbb{N}, k_i \neq k_{i+1}, \text{ for } i = 1, 2, \ldots, r-1 \right\},$$

where $\mathbb{N}$ denotes the natural numbers, together with zero, and the empty word with $r = 0$ is interpreted as the identity 1. We shall call such finite products *reduced words in the τ_i*.

Suppose now that $\pi \in S_n$, the symmetric group on n letters. Then π induces an automorphism of the group E given by

$$\hat{\pi} : \tau_i \mapsto \tau_{(i)\pi}.$$

Now, $\hat{\pi}$ certainly maps E onto E and, if $u = \tau_{k_1} \tau_{k_2} \cdots \tau_{k_r}$ and $v = \tau_{m_1} \tau_{m_2} \cdots \tau_{m_s}$, then

$$
\begin{aligned}
(uv)^{\hat{\pi}} &= \left(\tau_{k_1} \tau_{k_2} \cdots \tau_{k_r} \cdot \tau_{m_1} \tau_{m_2} \cdots \tau_{m_s} \right)^{\hat{\pi}} \\
&= \left(\tau_{k_1} \tau_{k_2} \cdots \tau_{k_{r-d+1}} \cdot \tau_{m_d} \tau_{m_{d+1}} \cdots \tau_{m_s} \right)^{\hat{\pi}} \\
&= \tau_{(k_1)\pi} \tau_{(k_2)\pi} \cdots \tau_{(k_{r-d+1})\pi} \cdot \tau_{(m_d)\pi} \tau_{(m_{d+1})\pi} \cdots \tau_{(m_s)\pi} \\
&= \tau_{(k_1)\pi} \tau_{(k_2)\pi} \cdots \tau_{(k_r)\pi} \cdot \tau_{(m_1)\pi} \tau_{(m_2)\pi} \cdots \tau_{(m_s)\pi} \\
&= u^{\hat{\pi}} v^{\hat{\pi}},
\end{aligned}
$$

where $k_r = m_1, \ldots, k_{r-d+2} = m_{d-1}$ and $k_{r-d+1} \neq m_d$. So, if N is a subgroup of the symmetric group S_n, then the map $\pi \mapsto \hat{\pi}$ is an injection of N into the group of automorphisms of E.

3.2   Semi-direct products and the progenitor P

We are now in a position to form the infinite semi-direct product, which we call the progenitor; we use the following standard construction.

LEMMA **3.1** Let K be a group, and let $A \leq \operatorname{Aut} K$, be a subgroup of the automorphism group of K. Then the Cartesian product $A \times K$ becomes a group under the binary operation $\circ$ defined by $(a, x) \circ (b, y) = (ab, x^b y)$, for $a, b \in A$ and $x, y \in K$.

Firstly, a note of warning: I have used the notation $A \times K$ because the set on which the product is defined *is* the Cartesian product of A and K. However, the *group* $(A \times K, \circ)$, which is not, in general, the direct product of A and K, is often denoted by $A \ltimes K$. Here we shall adhere to the notation used in the ATLAS [25], namely $K{:}A$. From the definition, it is clear that what is really needed is an action of A on K or, in other words, a homomorphism $\alpha : A \mapsto \operatorname{Aut} K$ when the definition of $\circ$ takes the following more general form:

$$(a, x) \circ (b, y) = (ab, x^{\alpha(b)} y).$$

The resulting group is then more properly denoted by $A \ltimes_\alpha K$. This will clearly reduce to the direct product if α maps every element of A to the trivial automorphism of K. In the case when the homomorphism α is an injection from A into the automorphism group of K, we may identify A with its image in $\operatorname{Aut} K$, as we have done here.

Proof of Lemma 3.1 Certainly $A \times K$ is closed under $\circ$. Moreover, we see that

$$[(a, x) \circ (b, y)] \circ (c, z) = (ab, x^b y) \circ (c, z) = (abc, x^{bc} y^c z)$$

and

$$(a, x) \circ [(b, y) \circ (c, z)] = (a, x) \circ (bc, y^c z) = (abc, x^{bc} y^c z),$$

for all $a, b, c \in A$ and for all $x, y, z \in K$. So $(A \times K, \circ)$ is associative. Now, if ι is the identity automorphism of K and e is the identity element of K, it is readily checked that (ι, e) is the identity of $A \times K$ and that

$$(a, x)^{-1} = (a^{-1}, (x^{-1})^{a^{-1}}).$$

$\square$

A group constructed from the Cartesian product of two groups A and K in this manner is called a *semi-direct product* and, following the notation used in the ATLAS [25], we denote such a semi-direct product of the groups K and A by

$$K{:}A.$$

Such a semi-direct product is also referred to as a *split extension* of the group K by the group A.

Note that $K{:}A$ possesses two subgroups of particular interest:

$$\hat{A} = \{(a, e) \mid a \in A\} \text{ and } \hat{K} = \{(\iota, x) \mid x \in K\}.$$

It is easy to see that $\hat{A}$ and $\hat{K}$ are isomorphic copies of A and K, respectively, with $\hat{K} \cap \hat{A} = \langle I_{K{:}A} \rangle$, where $I_{K{:}A}$ denotes the identity of the group $K{:}A$. Moreover, $(a, e) \circ (\iota, x) = (a, x)$ and so $K{:}A = \hat{A}\hat{K}$. In a moment, we shall see that the subgroup K is normal, and so, although the colon notation reverses the order of the semi-direct product in a slightly unfortunate manner, we have $\hat{A}\hat{K} = \hat{K}\hat{A}$. Indeed, note that

$$(a, x)^{-1} \circ (\iota, y) \circ (a, x) = (a^{-1}, (x^{-1})^{a^{-1}}) \circ (a, y^a x) = (\iota, x^{-1} y^a x) \in \hat{K},$$

and so $\hat{K} \triangleleft (K{:}A)$. So, $K{:}A$ is a product of a normal subgroup isomorphic to K and a subgroup isomorphic to A which intersect trivially. In practice, we shall identify $\hat{K}$ with K and $\hat{A}$ with A, and write the element (a, x) simply as ax. So the product formula becomes

$$(a, e) \circ (\iota, x) \circ (b, e) \circ (\iota, y) = (ab, e) \circ (\iota, x^b y) = (a, e) \circ (b, e) \circ (\iota, x^b) \circ (\iota, y),$$

which may be written as follows:

$$\hat{a}\,\hat{x} \cdot \hat{b}\,\hat{y} = \hat{a}\,\hat{b} \cdot \hat{x}^b\,\hat{y},$$

and dropping the 'hats', or equivalently identifying A with $\hat{A}$ and K with $\hat{K}$, we obtain the simple form of the product formula:

$$ax \cdot by = ab \cdot x^b y.$$

Putting $a = b^{-1}$ and $y = e$, we see that we can interpret

$$x^b \text{ as } b^{-1}xb, \text{ the conjugate of } x \text{ by } b.$$

In this way, we have essentially passed from the *external* semi-direct product (in which A and K are only related by the fact that there is an action of A on K) to the *internal* semi-direct product (in which A and K are subgroups of some larger group, A is contained in the normalizer of K and A and K intersect trivially).

Now take $K \cong 2^{\star n} = \langle t_1, t_2, \ldots, t_n \mid t_1^2 = t_2^2 = \cdots = t_n^2 = 1 \rangle$, and let $N \le \mathrm{S}_n$ be a transitive subgroup of the symmetric group S_n acting on the set $\Lambda = \{1, 2, \ldots, n\}$. Then we have the following definition.

DEFINITION **3.1** A *progenitor* is a semi-direct product of the following form:

$$P \cong 2^{\star n} : N = \{\pi w \mid \pi \in N, w \text{ a reduced word in the } t_i\},$$

where $2^{\star n}$ denotes a free product of n copies of the cyclic group of order 2 generated by involutions t_i for $i = 1, \ldots, n$; and N is a transitive permutation group of degree n which acts on the free product by permuting the involutory generators.

We refer to the subgroup N as the *control subgroup* and to the involutory generators of the free product as the *symmetric generators*. Note that this definition should more precisely be of an *involutory progenitor* in that the symmetric generators are of order 2. A more general definition, in which the cyclic subgroups have other orders, is given in Chapter 4, but it is probably advisable to become familiar with this simpler version first. Of course, we have

$$\pi u \cdot \sigma v = \pi \sigma u^{\sigma} v \tag{3.1}$$

and

$$\pi^{-1} t_i \pi = t_i^{\pi} = t_{(i)\pi}. \tag{3.2}$$

The ease with which we can work in such a group P, and in its homomorphic images, is illustrated in Example 3.1 and Exercise 3.2(4).

EXERCISE 3.1

(1) Let K be an abelian group. Show that $\alpha : K \mapsto K$ defined by

$$x^{\alpha} = x^{-1} \text{ for all } x \in K$$

is an automorphism of K, which has order 2 unless K is an elementary abelian group of order 2^n, in which case it is trivial. In the non-trivial case, construct a semi-direct product of K by the cyclic group of order 2. Such a group of shape $K{:}2$ is referred to as a *generalized dihedral group*. If K is itself cyclic of order $n > 2$, then $K : 2 \cong D_{2n}$, the dihedral group of order $2n$.

(2) Show that there are four non-isomorphic semi-direct products consisting of a cyclic group of order 7 extended by a cyclic group of order 6. (In ATLAS notation, the symbol $7{:}6$ is generally reserved for the semi-direct product of a cyclic group of order 7 by a cyclic group of order 6 in which all the elements of order 7 are conjugate, that is to say a *Frobenius* group of order 42.) How would the other three semi-direct products be denoted in an unambiguous manner?

(3) Let $K = \{e, (0\ 1)(2\ 3), (0\ 2)(3\ 1), (0\ 3)(1\ 2)\} \le G \cong S_4$ be the Klein fourgroup and let $S_3 \cong A = \langle (1\ 2\ 3), (2\ 3) \rangle \le G$. Show that $K{:}A \cong G$, where x^a for $x \in K$, $a \in A$ denotes the conjugate of x by a.

(4) Let K be a finite group and let $A = \mathrm{Aut}\ K$ be the automorphism group of K. The semi-direct product $K{:}A$ is called the *holomorph* of K. Let $K \cong 2^n$, an elementary abelian group of order 2^n, then $A = \mathrm{Aut}\ K \cong L_n(2)$, the general linear group in n dimensions over the field of order 2. Show that

$$AF_n = K{:}A = 2^n : L_n(2) \cong \left\{ \left(\begin{array}{c|c} 1 & u \\ \hline 0 & X \end{array} \right) \ \middle| \ \begin{array}{c} X \in L_n(2) \\ u \in F^n \end{array} \right\} \le L_{n+1}(2),$$

where F denotes the field of order 2. Exhibit AF_3 as permutations on eight letters.

3.3   The Cayley graph of *P* over *N*

For H and K subgroups of the group G, we define a relation on G as follows:

$$x \sim y \iff \exists\, h \in H \text{ and } k \in K \text{ such that } y = hxk.$$

Then it is readily checked that $\sim$ is an equivalence relation and that the equivalence classes are sets of the following form:

$$HxK = \{hxk \mid h \in H, k \in K\} = \bigcup_{k \in K} Hxk = \bigcup_{h \in H} hxK.$$

Such a subset of G, which from the above is clearly both a union of right cosets of H and a union of left cosets of K, is called a *double coset*. Indeed, if we consider the group G acting by right multiplication on the right cosets of H in G, then the double cosets of form HxK correspond to the orbits of K in this action. In particular, if $H = K$ then the number of double cosets of form HxH gives the *rank* of this action, where the rank of a transitive permutation group is defined as the number of orbits of the point-stabilizer. So a rank 2 permutation group is doubly transitive and has double coset decomposition of form $G = H \cup HxH$, where H is the stabilizer of a point. Unlike single cosets, however, double cosets of two finite subgroups H and K do not, in general, all contain the same number of elements. In fact, we have the following lemma.

LEMMA 3.2 If H and K are finite subgroups of the group G and x is an element of G, then $|HxK| = |H|.|K|/|H^x \cap K|$.

   Proof We shall count the number of (single) right cosets of H in HxK. We have

$$\begin{aligned}
Hxk_1 \neq Hxk_2 &\iff Hxk_1 k_2^{-1} x^{-1} \neq H \\
&\iff k_1 k_2^{-1} \notin x^{-1}Hx \cap K = H^x \cap K \\
&\iff (H^x \cap K)k_1 \neq (H^x \cap K)k_2.
\end{aligned}$$

So, the number of single cosets of H in HxK is equal to the number of single cosets of $H^x \cap K$ in K, and we have

$$|HxK| = |H|.|K : H^x \cap K| = |H|.|K|/|H^x \cap K|.$$

$\square$

Let us now consider the double cosets of form NxN in the progenitor $P \cong 2^{*n} : N$. We first note that we can write $x = \pi w$ for some $\pi \in N$ and w a reduced word in the t_i. So $NxN = N\pi wN = NwN = [w]$, say. If we let $\mathcal{T} = \{t_1, t_2, \ldots, t_n\}$ and $u, v \in \langle \mathcal{T} \rangle$, then $[u] = [v]$ if, and only if, $Nu = Nv\sigma$ for some $\sigma \in N$. But

$$Nv\sigma = N\sigma^{-1}v\sigma = Nv^\sigma = Nu,$$

if, and only if, $v^\sigma u^{-1} \in \langle \mathcal{T} \rangle \cap N = \langle 1 \rangle$, which happens if, and only if, $v^\sigma = u$, since there is no non-trivial relation between the t_i. In other words, the double cosets are in one-to-one correspondence with the orbits of N on finite sequences of integers from $\Lambda = \{1, 2, \ldots, n\}$ without adjacent repetitions. An example will make this clear.

Example 3.1 Let $P \cong 2^{\star 3} : S_3$.

Thus P is a semi-direct product of a free product of three copies of the cyclic group of order 2, and the symmetric group on three letters which permutes the three *symmetric generators* by conjugation. Multiplication in P is straightforward; for example,

$$(1\ 2\ 3)t_2 t_3.(1\ 3)t_1 t_2 t_1 = (1\ 2\ 3)(1\ 3)t_2^{(1\ 3)} t_3^{(1\ 3)} t_1 t_2 t_1$$
$$= (1\ 2)t_{2^{(1\ 3)}} t_{3^{(1\ 3)}} t_1 t_2 t_1 = (1\ 2)t_2 t_1 t_1 t_2 t_1 = (1\ 2)t_1.$$

Inversion of elements is also immediate for

$$[(1\ 2\ 3)t_2 t_3]^{-1} = t_3 t_2(3\ 2\ 1) = (3\ 2\ 1)t_2 t_1.$$

The decomposition into double cosets takes the following form:

$$P = N \cup Nt_1 N \cup Nt_1 t_2 N \cup Nt_1 t_2 t_1 N \cup Nt_1 t_2 t_3 N \cup \cdots,$$

which corresponds in the order given to the sequences without adjacent repetition:

$$[\ ],\ [1],\ [1,2],\ [1,2,1],\ [1,2,3],\ldots$$

Moreover, it is easy to see which single cosets a given double coset contains; for example,

$$Nt_1 t_2 N = \bigcup_{\pi \in N} Nt_1 t_2 \pi = \bigcup_{\pi \in N} Nt_{(1)\pi} t_{(2)\pi} = \bigcup_{i \neq j} Nt_i t_j$$

is the union of six distinct single cosets of N since, as above, $Nt_i t_j = Nt_k t_l$ if, and only if, $i = k$ and $j = l$.

In general, we have

$$NwN = \bigcup_{\pi \in N} Nw^\pi.$$

Let us now define

$$N^{(w)} = \{\pi \in N \mid Nw\pi = Nw\}, \tag{3.3}$$

as the *coset stabilizing subgroup* of the coset Nw. Then, as in the proof of Lemma 3.2, we see that

$$\begin{aligned} N^{(w)} &= \{\pi \in N \mid Nw\pi = Nw\} = \{\pi \in N \mid Nw\pi w^{-1} = N\} \\ &= \{\pi \in N \mid w\pi w^{-1} \in N\} = \{\pi \in N \mid \pi \in N^w\} \\ &= N \cap N^w. \end{aligned}$$

and so the number of single cosets in NwN is given by $|N : N^{(w)}|$. It is worth noting that the double coset $Nw^{-1}N$ consists precisely of all the inverses of

elements in NwN, and so the two double cosets contain the same number
of single cosets.

In Example 3.1, $N \cong S_3$, and so if $w = t_1$ we have $N^{(w)} = N_1$, the stabilizer
in S_3 of 1 which, of course, has order 2. Thus the double coset Nt_1N contains
$6/2 = 3$ distinct single cosets, which are clearly Nt_1, Nt_2 and Nt_3. However, if
$w = t_1t_2$, as in the example, we have $N^{(w)} = N_{12}$, the stabilizer in S_3 of 1 and
2, which is the trivial group. So, Nt_1t_2N contains six distinct single cosets as
calculated. Indeed, all other double cosets contain six single cosets, except
for the trivial double coset $N = NeN$, which visibly contains just one.

Returning to the general case, we now draw a graph Γ_C, the *Cayley graph*
of $P \cong 2^{*n} : N$ over N, whose vertices are the (single) cosets of N in P. As we
have seen, every coset can be written as Nw, for w a reduced word in the
t_i. We now join the vertex Nw to the n vertices Nwt_i, for $n \in \{1, 2, \ldots, n\}$.
Clearly every vertex of this graph will have valence n. Moreover, the graph
will be a tree since a circuit would correspond to a word in the t_i lying
in the subgroup N, which can only happen if that word reduces to the
identity element. Suppose that u is the reduced word corresponding to a
circuit in the graph Γ_C. Then we have $Nwu = Nw$ for some vertex Nw, and
so $wuw^{-1} \in N \cap \langle \mathcal{T} \rangle = \langle 1_P \rangle$. Thus, we conclude that $u = 1$, which would
contradict the hypothesis that there is no non-trivial relation between the
t_i unless u is the trivial word.

In Figure 3.1 we illustrate part of the *Cayley graph* for the above example
of $P \cong 2^{*3} : S_3$. Note that in the Cayley graph Γ_C of the progenitor, each
vertex is joined to n distinct other vertices, and so possesses neither loops
nor multiple edges, since either $Nwt_i = Nw$ or $Nwt_i = Nwt_j$, for $i \neq j$, would
imply a non-trivial relation between the t_i. However, when we move to
homomorphic images of the progenitor (as we shall be doing shortly), the
situation will change dramatically. For instance, if we factor out a relator
$\pi t_1t_2t_1t_2t_1$, say, then we not only introduce circuits into the Cayley graph, in
this case of length 5, but also we see that $t_1t_2t_1t_2t_1$ is in N; thus, $Nt_1t_2t_1t_2t_1 =$
N, and so, multiplying both sides by t_1t_2, we see that the cosets $Nt_1t_2t_1$ and
Nt_1t_2 are equal. That is to say the coset Nt_1t_2 is joined to itself by the edge
corresponding to the generator t_1, and so the Cayley graph contains loops.
This will become clear when we investigate some particular examples in the
following sections.

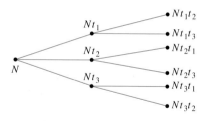

Figure 3.1. The Cayley graph of $P \cong 2^{*3} : S_3$.

EXERCISE **3.2**

(1) Let $G \cong S_6$ acting naturally on $\Lambda = \{1, 2, \ldots, 6\}$, and let $H, K_1 \leq G$ with $H = G_1$, the stabilizer in G of 1, and $K_1 = G_{\{1,2\}} \cong 2 \times S_4$, the stabilizer in G of the unordered pair $\{1, 2\}$. Work out $|H \cap K_1|$ and $|H \cap K_1^{(1\ 3)}|$, and hence work out the double coset decomposition of G over H and K_1.

Now let $h = 12.34.56$ be a partition of Λ into three pairs, and let $K_2 = G_h$ denote the stabilizer in G of h. Show that $K_2 \cong K_1 \cong 2 \times S_4$. Obtain the double coset decomposition of G over H and K_2.

What do these two double coset decompositions tell you about the various permutation actions described? How many double cosets would there be of the form $K_1 x K_2$?

(2) Define $L_3(2)$, the *general linear group of dimension 3 over the field of order 2*, to be the group of all non-singular 3×3 matrices over the field $\mathbb{Z}_2$. Work out the order of $G \cong L_3(2)$. Let

$$H = \left\{ \left(\begin{array}{c|c} 1 & 0 \\ \hline v^t & A \end{array} \right) \;\middle|\; A \in L_2(2), v \in \mathbb{Z}_2^2 \right\},$$

$$K = \left\{ \left(\begin{array}{c|c} 1 & u \\ \hline 0 & B \end{array} \right) \;\middle|\; B \in L_2(2), u \in \mathbb{Z}_2^2 \right\}.$$

Show that H is the stabilizer of a 1-dimensional subspace when G acts by right multiplication on row vectors, and that K is the stabilizer of a 2-dimensional subspace. How many double cosets of the form HxK would you expect there to be? Justify your answer. Work out $|H \cap K^x|$ for each double coset HxK.

(3) Let $G = \langle a, b \mid a^4 = b^3 = (ab)^2 = 1 \rangle$. Show that $G \cong S_4$ as follows. Firstly, show that S_4 is an image of G by exhibiting permutations on four letters which satisfy the presentation. Secondly, let $H = \langle a \rangle \cong C_4$ and draw the Cayley diagram whose vertices are the right cosets of H in G and in which Hw is joined to Hwa by a directed edge with a single arrowhead, whilst Hw is joined to Hwb by a directed edge with a double arrowhead. Once your diagram has closed you should have six vertices and you will be able to read off the way in which a and b permute these six cosets, thus obtaining a permutation represention of S_4 of degree 6. [Hint: You should make sure that each of the relations holds at every vertex.]

(4) (a) Let $P = 2^{*n} : N$ be a progenitor with symmetric generators $\mathcal{T} = \{t_1, t_2, \ldots, t_n\}$. For each of (i), (ii) and (iii) below, find conditions on the permutation $\pi \in N$ for the given element to be an involution: (i) πt_i, (ii) $\pi t_i t_j$, and (iii) $\pi t_i t_j t_k$. Note that you have effectively found all involutions in NwN for w a word in the t_i of length less than or equal to 3.

(b) Let $P = 2^{*5} : N$, where $N \cong A_5$ and the symmetric generators are $\mathcal{T} = \{t_1, t_2, \ldots, t_5\}$. Put each of the following into canonical form πw: (i) $((1\,2)(3\,4)t_1)^4$; (ii) $((2\,3\,4)t_2)^4$; (iii) $(1\,2\,3\,4\,5)t_3(1\,3)(4\,5)t_1 t_4$.

3.4   The regular graph preserved by P

It is important to note that the Cayley graph is not preserved by the group P in the sense that N is joined to Nt_1 in Γ_C but Nt_2 is not joined to $Nt_1 t_2$ when $n \geq 2$. However, we may define different edges on the same vertex set to form the *transitive graph* Γ_T by joining Nw to $Nt_i w$ for $i \in \{1, 2, \ldots, n\}$. The progenitor certainly does preserve this graph, almost by definition, as, if u is any word in the t_i, then Nwu is joined to $N(t_i w)u = Nt_i(wu)$. Of course, P acts transitively on the vertices of Γ_T, which are simply cosets of a subgroup, and the property of preserving incidence is carried over to homomorphic images.

In Figure 3.2, we illustrate part of the transitive graph for the same example of $P \cong 2^{*3} : S_3$.

3.5   Homomorphic images of P

In order to obtain a homomorphic image of the infinite group P, we must factor out a normal subgroup K, say. Now, every element of P can be written uniquely as πw, where $\pi \in N$ and w is a reduced word in the t_i. We choose a set of elements $\{\pi_1 w_1, \pi_2 w_2, \ldots\}$ such that K is the smallest normal subgroup of P containing them. That is to say,

$$K = \langle \pi_1 w_1, \pi_2 w_2, \ldots \rangle^P,$$

the *normal closure* of $\{\pi_1 w_1, \pi_2 w_2, \ldots\}$. In other words, we factor out the relators $\{\pi_1 w_1, \pi_2 w_2, \ldots\}$ to obtain

$$G \cong \frac{2^{*n} : N}{\pi_1 w_1, \pi_2 w_2, \ldots}.$$

An example will make this process clear.

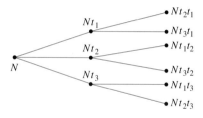

Figure 3.2. The transitive graph of $P \cong 2^{*3} : S_3$.

EXAMPLE **3.2** Let

$$G \cong \frac{2^{\star 4} : S_4}{(3\ 4)t_1 t_2 t_1 t_2}.$$

The element of $(3\ 4)t_1 t_2 t_1 t_2$ and all its conjugates have been set equal to the identity, or, equivalently, the word $(t_1 t_2)^2$ in $\langle \mathcal{T} \rangle$ is set equal to the permutation $(3\ 4)$ in the control subgroup N. So, if $\{i, j, k, l\} = \{1, 2, 3, 4\}$, then the identity

$$(i\ j) = (t_k t_l)^2$$

holds in G. In particular, this means that

$$Nt_1 t_2 t_1 t_2 = N, \text{ and so } Nt_1 t_2 = Nt_2 t_1 \text{ and } Nt_1 t_2 t_1 = Nt_2. \tag{3.4}$$

If we now investigate the double cosets of form NwN, we find we have the trivial double coset N and $Nt_1 N$, which appears to contain the four distinct cosets Nt_i as in the infinite progenitor. But the double coset $Nt_1 t_2 N$ contains at most six single cosets now, since $Nt_i t_j = Nt_j t_i$, as in equation (3.4). Furthermore, we have $Nt_1 t_2 t_1 N = Nt_1 N$ by equation (3.4), so the next double coset to consider is $Nt_1 t_2 t_3 N$. At this stage we make more subtle use of our relation, for

$$\begin{aligned}
Nt_1 t_2 t_3 &= Nt_1 \cdot t_2 t_3 t_2 t_3 \cdot t_3 t_2 \\
&= Nt_1 (1\ 4) t_3 t_2 \\
&= Nt_1^{(1\ 4)} t_3 t_2 \\
&= Nt_4 t_3 t_2 \\
&= Nt_1 t_2 t_3 (1\ 4)(2\ 3),
\end{aligned}$$

where we have forced in the word $t_2 t_3 t_2 t_3$ and replaced it by the permutation $(1\ 4)$. But we also have $Nt_1 t_2 = Nt_2 t_1 = Nt_1 t_2 (1\ 2)$. Thus, if $w = t_1 t_2 t_3$, then we have $N^{(w)} \geq \langle (1\ 2), (1\ 4)(2\ 3) \rangle \cong D_8$, the dihedral group of order 8. So, the maximum number of single cosets contained in $Nt_1 t_2 t_3 N$ is given by

$$|N : N^{(w)}| = 24/8 = 3.$$

If there *are* three cosets in this double coset, they can be labelled by the three ways in which the set $\{1, 2, 3, 4\}$ can be partitioned into two pairs, with the coset $Nt_1 t_2 t_3$ corresponding to 12.34. The coset corresponding to the partition *ij.kl* would have eight 'names' as follows:

$$\begin{aligned}
Nt_i t_j t_k &= Nt_i t_j t_l = Nt_j t_i t_k = Nt_j t_i t_l \\
&= Nt_k t_l t_i = Nt_k t_l t_j = Nt_l t_k t_i = Nt_l t_k t_j.
\end{aligned}$$

Observe that, since the coset stabilizing subgroup $N^{(w)}$ acts transitively on the set of four symmetric generators, all cosets joined to this coset in the Cayley graph lie in the double coset $Nt_1 t_2 N$. Indeed, unless there is further collapsing of which we are not yet aware, the Cayley graph is now complete and may be repesented diagramatically by the *collapsed Cayley graph* shown in Figure 3.3.

The reader will observe that this is a regular bipartite graph of valence 4 on $(7+7)$ vertices. The associated transitive graph will be shown in

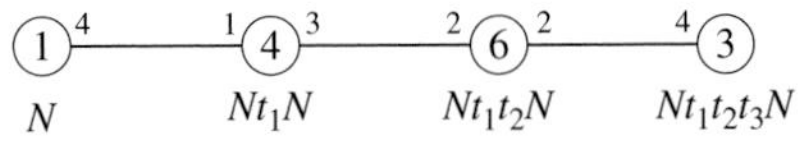

Figure 3.3. Collapsed Cayley graph of $PGL_2(7)$ over S_4.

Chapter 4 to be that of the points and lines of the projective plane of order 2, with points joined to the lines which do *not* pass through them. Once we have verified that there is no further collapse, we see that $|G| = 24 \times 14 = 2 \times 168$, and we have constructed the group $L_3(2):2$. The isomorphism $L_3(2):2 \cong PGL_2(7)$ follows easily from this construction, and will be verified in Chapter 4.

Examples of this nature inspire the following definition and are the subject matter of this book.

DEFINITION 3.2 A *symmetric presentation* of a group G is a definition of G of the form

$$G \cong \frac{2^{\star n} : N}{\pi_1 w_1,\ \pi_2 w_2, \ldots},$$

where $2^{\star n}$ denotes a free product of n copies of the cyclic group of order 2, N is a transitive permutation group of degree n which permutes the n generators of the cyclic groups by conjugation, thus defining a semi-direct product, and the relators $\pi_1 w_1, \pi_2 w_2, \ldots$ have been factored out.

Now, the progenitor $2^{\star n} : N$ is clearly an infinite group (if $n \geq 2$), and we are primarily interested in finite images G, so we certainly do not wish to restrict ourselves to faithful images. However, we do want images which contain an isomorphic copy of the control subgroup N and which have a set of n distinct involutions which are permuted under conjugation by this copy of N in the prescribed manner. This leads us to Definition 3.3.

DEFINITION 3.3 A *true* image of a progenitor $P = 2^{\star n} : N$ is a homomorphic image G in which the image of N is an isomorphic copy of N and in which the image of the set of symmetric generators $\mathcal{T} = \{t_1, t_2, \ldots, t_n\}$ consists of n distinct elements of order 2.

EXERCISE 3.3

(1) Let $G \cong S_{n+1}$ and let $t_i = (i\ n+1) \in G$ for $i = 1, 2, \ldots, n$. Show, by induction or otherwise, that $\langle t_1, t_2, \ldots, t_n \rangle = G$. Let G_{n+1} denote the subgroup of G stabilizing $n+1$. Then if $\pi \in G_{n+1}$ observe that $t_i^\pi = t_{i\pi}$ and so G is a homomorphic image of the progenitor $2^{\star n} : N$, where $N \cong S_n$.

Show, moreover, that $t_i t_j t_i = (i\ j) \in N$ and so G is an image of

$$S = \frac{2^{\star n} : S_n}{(1\ 2)t_1 t_2 t_1}$$

for $n \geq 2$. Count the elements in $Nt_1 N$ and hence show that $S \cong G$.

(2) Show that $\langle(1\ 4)(2\ 5), (2\ 4)(3\ 5), (3\ 4)(1\ 5)\rangle \cong A_5$. Let

$$N = \langle(1\ 2\ 3), (2\ 3)(4\ 5)\rangle \leq G \cong A_5$$

and show that N permutes the three involutory generators of G as the symmetric group S_3. Deduce that A_5 is an image of the progenitor $2^{\star 3} : S_3$.

(3) The 1-dimensional subspaces of a 3-dimensional vector space over $\mathbb{Z}_2$ may be labelled by the elements of $\mathbb{Z}_7$, the integers modulo 7, when the 2-dimensional subspaces correspond to the triple $\{1, 2, 4\}$ and its translates modulo 7. These are the *points* and *lines* of the 7-point projective plane. The group $L_3(2)$ may be shown to be the group of all permutations of the seven points which preserve the lines. Write down t_0, a permutation of the symmetric group S_7 of cycle shape 1.2^3 which fixes all three lines passing through the point 0. Hence show that the symmetric group S_7 is a homomorphic image of the progenitor $2^{\star 7} : L_3(2)$, and verify that in S_7 the additional relation $((0\ 1\ 2\ 3\ 4\ 5\ 6)t_0)^7 = 1$ holds. [Note: In fact,

$$\frac{2^{\star 7} : L_3(2)}{((0\ 1\ 2\ 3\ 4\ 5\ 6)t_0)^7} \cong 3^{\cdot} S_7,$$

the triple cover of the symmetric group S_7.]

(4) Show that the elements $\beta : x \mapsto 3x \equiv (\infty)(0)(1\ 3\ 9\ 5\ 4)(2\ 6\ 7\ 10\ 8)$ and $\delta : x \mapsto (2x-2)/(x-2) \equiv (\infty\ 2)(0\ 1)(3\ 4)(9\ 7)(5\ 10)(6\ 8)$ of the projective general linear group $G \cong \text{PGL}_2(11)$ satisfy the following presentation:

$$\langle b, d \mid b^5 = d^2 = (bd)^3 = 1 \rangle$$

and thus generate a subgroup isomorphic to A_5. Show that the element $t_\infty : x \mapsto -x \equiv (\infty)(0)(1\ 10)(2\ 9)(3\ 8)(4\ 7)(5\ 6)$ has just six images under conjugation by $N = \langle \beta, \delta \rangle$, that $\langle t_\infty^N \rangle = G$, and deduce that G is an image of

$$2^{\star 6} : L_2(5).$$

Show further that if we let $t_0 = t_\infty^\delta$ then $\beta^2 t_0$ has order 4, so that G is an image of

$$\frac{2^{\star 6} : L_2(5)}{((0\ 1\ 2\ 3\ 4)t_0)^4}.$$

Show* that this is in fact an isomorphism. (An asterisk, here and elsewhere, denotes a more difficult exercise.)

3.6   The lemma

Now, every element of the progenitor P has the form πw, where π is a permutation in N and w is a word in the symmetric generators. Thus, every relation by which we might factor P in order to get an interesting homomorphic image has the form $\pi w = 1$. Of course, such a relation is essentially telling us how to write certain permutations of N in terms of the symmetric generators, thus converting outer automorphisms of the free product of cyclic groups into inner automorphisms of its homomorphic image. It is natural to ask what form these relations should take. In particular, we might ask whether it is possible to write any permutations of N in terms of just two symmetric generators. There is a simple but very revealing answer to this question.

LEMMA 3.3

$$N \cap \langle t_i, t_j \rangle \leq C_N(N_{ij}),$$

where N_{ij} denotes the stabilizer in N of the two points i and j.

   Proof   If $\pi \in N$ and $\pi = w(t_i, t_j)$, a word in the two symmetric generators t_i and t_j, and if $\sigma \in N_{ij}$, then

$$\pi^\sigma = w(t_i, t_j)^\sigma = w(t_i^\sigma, t_j^\sigma) = w(t_{i^\sigma}, t_{j^\sigma}) = w(t_i, t_j) = \pi,$$

and so π commutes with every element of N_{ij}, as required. $\square$

Of course, the reader will observe that this lemma and its proof extend immediately to any number of generators, and if a permutation of N can be written in terms of a set of r symmetric generators, then it must lie in the centralizer of the stabilizer of the corresponding r letters, i.e.

$$N \cap \langle t_{i_1}, t_{i_2}, \ldots, t_{i_r} \rangle \leq C_N(N_{i_1 i_2 \cdots i_r}).$$

This easy lemma will prove remarkably powerful in what follows and, even at this stage, shows us why the relation in Example 3.2 is a sensible thing to factor by. We had $N \cong S_4$ and so $N_{12} = \langle (3\ 4) \rangle \cong C_2$, and so $C_N(N_{12}) = \langle (1\ 2), (3\ 4) \rangle$. Since $\langle t_1, t_2 \rangle \cong D_{2k}$, a dihedral group of order $2k$, if $(t_1 t_2)^2$ is an involution then $k = 4$ and $(t_1 t_2)^2$ commutes with t_1 and t_2. So, if $(t_1 t_2)^2 \in N$ it can only be $(3\ 4)$ or the identity, which points us towards the relation $(3\ 4) = t_1 t_2 t_1 t_2$.

   Note that this lemma gives us a straightforward 'recipe' for constructing groups: take a transitive permutation group N of degree n and form the progenitor $P = 2^{\star n} : N$. For each 2-point stabilizer N_{ij}, find its centralizer in N and attempt to write elements of this centralizer as words in t_i and t_j.

   Naturally we should like to choose the relators by which we factor the progenitor to be as short and as easily understood as possible. The following lemmas show that in many circumstances relators of form πw with $l(w) \leq 3$, except for the form $\pi t_i t_j t_i$, are of little interest.

LEMMA 3.4 Let $G = \langle \mathcal{T} \rangle$, where $\mathcal{T} = \{t_0, t_1, \ldots, t_{n-1}\} \subseteq G$ is a set of involutions in G with $N = N_G(\mathcal{T})$ acting primitively on $\mathcal{T}$ by conjugation. (Thus G is a homomorphic image of the progenitor $2^{\star n} : N$.) Suppose $t_0 t_i \in N$, $t_0 \notin N$ for some $i \neq 0$; then $|G| = 2|N|$.

Proof Firstly, note that N^0 is maximal in N, and so $\langle N^0, N^i \rangle = N$ unless $N^0 = N^i$. If the latter, then subsets of $\Lambda = \{0, 1, 2, \ldots, n-1\}$ consisting of members with the same stabilizer clearly form blocks of imprimitivity, and so for N primitive there must be just one block, the point-stabilizer is the identity and N acts regularly. A regular action can only be primitive if the group has no proper subgroup, which can only happen if $N \cong C_p = \langle x \rangle$, say, with p prime. But in this case, without loss of generality, $t_0 t_i = x$, and so $G = \langle N, t_0 \rangle = \langle x, t_0 \rangle \cong D_{2n}$. Otherwise,

$$
\begin{aligned}
N t_0 t_i = N &\Rightarrow 0i \sim \star \\
&\Rightarrow 0 \sim i \\
&\Rightarrow N^{(0)} \geq \langle N^0, N^i \rangle = N \\
&\Rightarrow 0 \sim j \text{ for all } j \\
&\Rightarrow G = N \cup N t_0. \qquad \square
\end{aligned}
$$

LEMMA 3.5 Suppose, for G as above, we have $t_i t_j t_k \in N$ for some $i \neq j \neq k$; then *either* $G = N$ *or* (i) $N^{ij} \leq N^k$ and (ii) $N^{kj} \leq N^i$.

Proof We have

$$
\begin{aligned}
t_i t_j t_k \in N &\Leftrightarrow ijk \sim \star \\
&\Leftrightarrow ij \sim k \text{ and } kj \sim i.
\end{aligned}
$$

If $ij \sim k$ and $N^{ij} \not\leq N^k$, then $\exists \pi \in N$ such that $i^\pi = i$, $j^\pi = j$ but $k^\pi = k' \neq k$, so

$$
N t_i t_j \pi = N t_i t_j = N t_k = N t_k \pi = N t_{k'},
$$

i.e. $k' \sim k$. But then, by Lemma 3.4, $k \sim l$ for all l. In particular, $k \sim j \sim ij$, and so $i \sim \star$. That is to say, $t_i \in N$ and $G = N$. Similarly if $N^{kj} \not\leq N^i$. $\qquad \square$

Note The case $t_i t_j t_i \in N$ frequently does occur and gives rise to many interesting groups, such as those in the Suzuki chain.

3.7 Further properties of the progenitor

Example 3.2 shows that interesting groups do emerge as images of progenitors. The extent to which this is true is made clear by the following lemma.

LEMMA 3.6 Any finite non-abelian simple group is an image of a progenitor of form $P = 2^{\star n} : N$, where N is a transitive subgroup of the symmetric group S_n.

Proof Let G be a finite non-abelian simple group; let M be a maximal subgroup of G; and let x be an involution of G which is not in M. (Certainly, G has even order, and if all the involutions of G lay in M then M would contain a non-trivial normal subgroup of G.) Now, $\langle M, x \rangle = G$, since M is maximal and x does not lie in M. But the subgroup $\langle x^M \rangle$ is normalized by M and by x, which lies in it, and so is normal in G. The simplicity of G forces $\langle x^M \rangle = G$, and so if $|x^M| = n$ then G is an image of $2^{\star n}\!:\!M$. It remains to show that M acts faithfully on the n elements of x^M by conjugation. But any element of M which commutes with every element of the generating set x^M is central in the non-abelian simple group G, and is thus trivial. $\square$

Note Of course, G is trivially an image of $2^{\star n}\!:\!G$, where $n = |x^G|$ and the set of symmetric generators is a complete conjugacy class of G, but our proof provides a much richer source of pre-images than this.

Although we are mainly interested in the homomorphic images of a particular progenitor P, it is worth noting that P itself is likely to have very interesting properties which are inherited by all its images. Firstly, recall the following definition.

DEFINITION 3.4 The *commutator* or *derived subgroup* G' of the group G is the subgroup of G generated by all commutators of elements of G. Thus

$$G' = \langle [x, y] = x^{-1} y^{-1} xy \mid x, y \in G \rangle.$$

If G is a group, then G/G' is obtained by factoring out the smallest subgroup containing all the commutators. (Note that this subgroup is certainly normal since the set of commutators is normal.) If G is given by generators and relations, which is to say

$$G := \langle X \mid R \rangle,$$

where X is a set of generators and R is a set of words in those generators which are to be set equal to the identity, then this factoring is accomplished by supplementing the set of relations R by the set of commutators of generators, i.e. by $R' = \{[x_i, x_j] \mid x_i, x_j \in X\}$.

DEFINITION 3.5 A group is said to be *perfect* if it is equal to its derived, or commutator, subgroup.

EXERCISE 3.4 Note that an asterisk on the Exercise number denotes a harder example.

(1) Show that the commutator subgroup of a group G is the smallest normal subgroup K such that G/K is abelian.

(2) Show that if G is perfect and K is a maximal normal subgroup of G, then G/K is simple. Deduce that if the perfect group G has a non-trivial finite homomorphic image, then it has a finite non-abelian simple image.

(3) Show that the alternating group A_5 is generated by an element x of order 5 and an element y of order 3, whose product xy has order 2. Show that any group that is generated by such elements is perfect by taking the presentation

$$\langle x, y \mid x^5 = y^3 = (xy)^2 = 1 \rangle, \tag{3.5}$$

adding the relation $[x, y] = 1$, and showing that the resulting group is trivial. (In fact, equation (3.5) is a presentation of A_5; you might like to show this by constructing a Cayley diagram on the cosets of the subgroup $H = \langle x \rangle$ and showing that it has just 12 vertices.)

(4*) Show that the general linear group $G \cong L_3(2)$ satisfies the following presentation:

$$\langle x, y \mid x^7 = y^3 = (xy)^2 = [x, y]^4 = 1 \rangle. \tag{3.6}$$

[Hint: It may help to write the elements of G as permutations of the seven non-trivial vectors in the 3-dimensional vector space on which G naturally acts.] As in the above example, adjoin the relator $[x, y]$ and show that the resulting quotient group is trivial. (In fact, equation (3.6) is a presentation for $L_3(2)$, as we shall show later.)

THEOREM 3.1 A group G which is generated by a perfect subgroup of itself together with an involution is either perfect or contains a perfect subgroup to index 2.

Proof Let $N \leq G$ with $N' = N$, and let $t \in G$ be an involution of G such that $G = \langle N, t \rangle$. Let $\mathcal{T} = t^N$ be the set of conjugates of t under conjugation by N. Then, every element of G has the following form:

$$\pi_1 t \pi_2 t \cdots t \pi_r = \pi_1 \pi_2 \cdots \pi_r t^{\pi_2 \cdots \pi_r} \cdots t^{\pi_r} = \pi w,$$

where $\pi_1, \ldots \pi_r, \pi \in N$, and w is a word in the elements of $\mathcal{T}$. Now, the subset

$$K = \{\pi w \mid \pi \in N, w \text{ a word of even length in the elements of } \mathcal{T}\}$$

is easily seen to be a subgroup, and, since $G = K \cup Kt$, it has index 1 or 2 in G. If t_i and t_j are two elements of $\mathcal{T}$ then, since N acts transitively on $\mathcal{T}$, there exists a $\sigma \in N$ such that $t_i{}^\sigma = t_j$. Thus $[t_i, \sigma] = t_i t_i{}^\sigma = t_i t_j$ and $K \leq G'$. Moreover, $N = N'' \leq G'' \lhd G$, so $[t_i, \sigma] = t_i t_j = (\sigma^{-1})^{t_i} \sigma \in G''$ and $G' = G'' = K$. □

Note that in this case $G = G'$ if, and only if, some product of oddly many elements of $\mathcal{T}$ is in N. Of course, if the control subgroup N is perfect, then the progenitor $P = 2^{\star n} : N$ satisfies the conditions of the theorem; so P contains a perfect subgroup to index 2, since no non-trivial product of the symmetric generators lies in the control subgroup. The property of possessing a perfect subgroup to index 1 or 2 is inherited by any homomorphic image of P. If

P/L is any image of P, then $(P/L)' = P'L/L$ and $(P/L)'' = P''L/L$. So, we have

$$|P/L : (P/L)'| = |P/L : P'L/L| = |P : P'L| \leq |P : P'| = 2$$

and $(P/L)'' = P''L/L = P'L/L = (P/L)'$. In particular, if N is perfect and the word w has odd length, then the group

$$G = \frac{2^{\star n} : N}{\pi w}$$

is perfect. We have thus proved the following corollary.

COROLLARY 3.1 Let $P = 2^{\star n} : N$ be a progenitor in which the control subgroup N is perfect. Then any homomorphic image of P is either perfect or possesses a perfect subgroup to index 2. If w is a word in the symmetric generators of *odd* length, then the image

$$G = \frac{2^{\star n} : N}{\pi w}$$

is perfect.

3.8   Coxeter diagrams and Y-diagrams

A convenient way of exhibiting diagrammatically presentations of groups generated by involutions, which by the *Odd Order Theorem* includes all finite simple groups, is afforded by the so-called *Coxeter diagrams*. These are simple connected graphs in which the vertices represent involutions and the edges are labelled with the order of the product of the two involutions at its endpoints. Commuting vertices are not joined, and by convention an edge is left unlabelled if its endpoints have product of order 3. Of course, a presentation displayed in this manner may be supplemented by various additional relations, and Soicher [81] and others have obtained Coxeter presentations for many interesting groups.

We say that a Coxeter diagram possesses a *tail* if it has a vertex of valence 1 joined to a vertex of valence 2, as in Figure 3.4. Such a presentation corresponds to an image of a progenitor in a natural manner. Let the symmetric generator t correspond to the vertex labelled a_r and let $N = \langle \Gamma_0, a_{r-1} \rangle$; then, if the index of $\langle \Gamma_0 \rangle$ in N is n, the progenitor $2^{\star n} : N$ maps onto $G = \langle \Gamma_0, a_{r-1}, a_r \rangle$. This holds because $t = a_r$ commutes with $\langle \Gamma_0 \rangle$ and thus has (at most) n images under conjugation by N; the group G is certainly

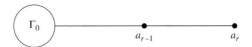

Figure 3.4. A Coxeter diagram with a tail.

generated by N and t. The element t will have strictly fewer images than n under conjugation by N only if the centralizer of t in N strictly contains $\langle \Gamma_0 \rangle$. In this case, the n symmetric generators will fall into blocks of imprimitivity under the action of N, and generators in the same block will be set equal to one another.

An easy but important example of a Coxeter presentation (with a tail) is as follows.

THEOREM **3.2**

$$\cong \frac{2^{\star r} : S_r}{(1\ 2) = t_1 t_2 t_1} \cong S_{r+1}.$$

Proof Firstly, note that the transpositions $a_i = (i\ i+1)$ for $i = 1, \ldots, r$ defined as permutations on the set $\Lambda = \{1, 2, \ldots, r, r+1\}$ satisfy the relations of the Coxeter diagram in the statement of the theorem, and generate the symmetric group S_{r+1}. Thus S_{r+1} is certainly a homomorphic image of the group defined by this presentation, and so $G = \langle a_1, \ldots, a_r \rangle$ has order greater than or equal to $(r+1)!$. We proceed by induction on r; noting that $\langle a_1 \rangle = C_2 \cong S_2$, we assume that $\langle a_1, \ldots, a_k \rangle \cong S_{k+1}$ for all $k < r$. Put $t = a_r$ and $N = \langle a_1, \ldots, a_{r-1} \rangle$, which by induction is isomorphic to S_r. Note that the extension by an additional generator cannot cause any collapse as $\langle a_1, \ldots, a_r \rangle$ maps onto S_{r+1}. But then the centralizer in N of $t = a_r$ is precisely $\langle a_1, \ldots, a_{r-2} \rangle \cong S_{r-1}$, since t commutes with this maximal subgroup of N and does not commute with the whole of N. So, under conjugation by N, t has r images which we may label $t_1, t_2, \ldots, t_r = t$. Now, a_{r-1} acts as a transposition in N, since it commutes with $\langle a_1, a_2, \ldots, a_{r-3} \rangle \cong S_{r-2}$, and so, without loss of generality, a_{r-1} conjugates t_r and t_{r-1} into one another and fixes the other t_i, i.e. $a_{r-1} \sim (r-1\ r)$. Moreover,

$$1 = (a_{r-1} a_r)^3 = (a_{r-1} t_r)^3 = a_{r-1}^3 t_r^{a_{r-1}^2} t_r^{a_{r-1}} t_r = (r-1\ r) t_r t_{r-1} t_r.$$

Conjugating this relation by permutations in N, we see that $(i\ j) = t_i t_j t_i$ for all $i \neq j$, and the group G is an image of

$$\frac{2^{\star r} : S_r}{(1\ 2) = t_1 t_2 t_1}.$$

But, since $t_i t_j t_i \in N$ for all $i \neq j$, we have $N t_i t_j = N t_i$ for all $i \neq j$, and so $G = N \cup N t_1 N$ is a double coset decomposition of G. Since

$$N t_1 N = \bigcup_{i=1}^{i=r} N t_i,$$

we see that $|G| = |N|(1 + r)$. Thus $|G| = (r+1)!$ and $G \cong S_{r+1}$, as required. $\qquad \square$

3.9   Introduction to MAGMA and GAP

Perhaps the most appealing feature of this approach is the ease with which
we are able to construct by hand surprisingly large groups. However, it is
often useful, and sometimes (dare one say) reassuring, to use the amazingly
effective coset enumeration programmes which now exist. The basic Todd–
Coxeter coset enumeration algorithm has been implemented and refined
by many different computational algebraists, but the contributions made
by teams working on the CAYLEY package (now MAGMA) in Sydney and
the GAP package in Aachen and now St Andrews deserve special mention. Researchers in algebra and many other branches of mathematics have
benefited enormously from their efforts.

In order to produce a presentation for the progenitor $2^{\star n} : N$, we first
need a presentation for the control subgroup N. Thus,

$$\langle\, X \mid R \,\rangle \cong N, \text{ where } X = \{x, y, \dots\}.$$

We now need a set of words $\{c_1(x, y, \dots), c_2(x, y, \dots), \dots\}$ in the generators X such that $\langle c_1, c_2, \dots \rangle \cong N_1$, the point-stabilizer in the permutation
group N. A further generator t will correspond to the symmetric genera-
tor t_1 and so have order 2. By requiring that t commutes with $c_1, c_2, \dots,$
we ensure that t commutes with the point-stabilizer N_1 and thus has just
$|N : N_1| = n$ images under conjugation by N. So, $t^N = \{t_1, t_2, \dots, t_n\}$ and N
permutes the elements of t^N in the required manner. Thus we have

$$\langle\, X, t \mid R,\ [c_1, t] = [c_2, t] = \cdots = 1 = t^2 \,\rangle \cong 2^{\star n} : N.$$

In our previous example, we have $P \cong 2^{\star 4} : S_4$, so we start with the classical
presentation for $N \cong S_4$ given by

$$\langle x, y \mid x^2 = y^3 = (xy)^4 = 1\rangle \cong S_4,$$

where we have, for instance, $x = (1\ 2)$, $y = (2\ 3\ 4)$ and $xy = (1\ 3\ 4\ 2)$. We
readily see that $x^{yx} = (2\ 3)$, and so $\langle y, x^{yx}\rangle = N_1$, the stabilizer in N of 1.
Thus we have

$$\langle\, x, y, t \mid x^2 = y^3 = (xy)^4 = t^2 = [y, t] = [x^{yx}, t] = 1\rangle \cong 2^{\star 4} : S_4.$$

Of course, here we have $t_1 = t, t_2 = t^x, t_3 = t^{xy}, t_4 = t^{xy^2}$. In the language of
MAGMA, we write

```
P<x,y,t> :=
Group<x,y,t | x^2=y^3=(x*y)^4=t^2=(y,t)=(x^(y*x),t)=1>;
```

Note that commutators are defined by *round* brackets rather than square
brackets. We can then define N to be the subgroup of P defined by x and
y by writing

```
N:=sub<P|x,y>;
```

where P is, of course, an infinite group, and so N has infinite index in P. However, if we factor P by the relator used in Example 3.2 we should get finite index. The additional relator is given as $(3\ 4)(t_1 t_2)^2$, and, since $[t, x] = t\ t^x = t_1 t_2$, this may be written $x^{yxy}[t, x]^2$. Thus we obtain the following MAGMA code:

```
> G<x,y,t>:=Group<x,y,t|x^2=y^3=(x*y)^4=t^2=(y,t)
> =(x^(y*x),t)=x^(y*x*y)*(t,x)^2=1>;
> N:=sub<G|x,y>;
> Index(G,N);
14
> f,gp,k:=CosetAction(G,N);
> f;
Mapping from: GrpFP: G to GrpPerm: gp
> gp;
Permutation group gp acting on a set of cardinality 14
    (2, 3)(7, 9)(8, 10)(12, 13)
    (3, 4, 6)(5, 7, 8)(9, 11, 10)(12, 14, 13)
    (1, 2)(3, 5)(4, 7)(6, 8)(9, 12)(10, 13)(11, 14)
> Order(gp);
336
```

The instruction `CosetAction(G,N)` returns the action of the group G on the 14 cosets of the subgroup N, `f` maps a word in the generators of G onto a permutation of the numbers $\{1, 2, \ldots, 14\}$, `gp` is the image of G under this homomorphism, and `k` is the kernel of the mapping. There are many functions available in the MAGMA package and the reader is referred to the *Handbook of Magma Functions* [19]; however, as a taste of what can be done, we include the following which is self-explanatory:

```
> oo:=Orbits(DerivedGroup(gp));
> oo;
[
   GSet{ 1, 5, 7, 8, 9, 10, 11 },
   GSet{ 2, 3, 4, 6, 12, 13, 14 }
]
> IsSimple(DerivedGroup(gp));
true
> Order(DerivedGroup(gp));
168
```

So we see that the group G defined by our presentation and constructed as permutations on 14 letters contains a simple subgroup to index 2 which acts with two orbits of length 7. Thus MAGMA confirms the hand calculations in Section 3.5.

EXERCISE **3.5**

(1) Using the presentation $A_5 \cong \langle x, y \mid x^5 = y^3 = (xy)^2 = 1 \rangle$ mentioned in previous exercises, obtain words in x and y which generate a point-stabilizer in A_5 and hence write down a presentation for the progenitor $P = 2^{\star 5} : A_5$.

(2) As we shall prove later, the alternating group A_5 is isomorphic to the projective special linear group $L_2(5)$ and thus has an action on six letters. Using the presentation in the previous exercise, obtain words in x and y which generate a subgroup of $L_2(5)$ of index 6 and hence write down a presentation for the progenitor $P = 2^{\star 6} : L_2(5)$.

(3) Another important isomorphism that we shall prove later is as follows:

$$L_3(2) \cong L_2(7),$$

which are naturally permutation groups of degrees 7 and 8, respectively. We have seen the presentation

$$L_3(2) \cong L_2(7) \cong \langle x, y \mid x^7 = y^3 = (xy)^2 = [x, y]^4 = 1 \rangle;$$

use this to obtain presentations for the progenitors $2^{\star 7} : L_3(2)$ and $2^{\star 8} : L_2(7)$.

(4) Each of the four control subgroups in Examples 3.1, 3.2 and 3.3 acts doubly transitively on the symmetric generators. Apply Lemma 3.3 in each case to see which elements of N could be written as words in two of the symmetric generators without causing collapse.

3.10   Algorithm for double coset enumeration

In order to find the index of N in G in the above example, MAGMA has used a version of the famous Todd–Coxeter [84] coset enumeration algorithm, which has proved immensely successful at finding the index of a subgroup of a finitely presented group, given a presentation of that group and words generating the subgroup. It has been implemented, along with many refinements and modifications (see Havas [43]), on both GAP [41] and MAGMA [19], and can handle very large indices with impressive ease.

When we performed the calculation by hand, however, we enumerated the *double cosets* of form NxN, and the way in which the group was defined seemed to lend itself to this approach. After all, even performing the above single coset enumeration to find just 14 cosets would be quite laborious by hand, whereas the double coset enumeration takes just a few minutes. Moreover, as we have pointed out, most finite groups, including all finite simple groups, can be defined in this way. In order to handle larger examples, it was decided to design a computer algorithm for performing double coset enumerations on symmetrically presented groups.

Double coset enumeration has both advantages and disadvantages when compared with single coset enumeration. On the plus side, there will in general be far fewer double cosets than single cosets. Moreover, a double coset decomposition of a group with respect to two subgroups H and K, say, contains a great deal of information about how each of these subgroups acts on the cosets of the other; for instance, the number of double cosets equals the number of orbits of each subgroup on the cosets of the other. In our case we choose $H = K = N$, and so the number of cosets gives the rank of G acting on the cosets of N. On the negative side, the fact that double cosets have different sizes requires additional algebraic information to be stored, probably in the form of coset stabilizing subgroups, in a way that was not needed in the case of single coset enumeration.

In 1984, Conway [6] suggested an algorithm for double coset enumeration, but this has proved difficult to implement. Linton [67] successfully computerized an algorithm before 1989, but required that one of the two subgroups used had to be quite small. His GAP share package [68] has now relaxed the restrictions on the small subgroup.

A first computer program implementing the above process for symmetrically presented groups was written in 1998 by Sayed [73]; it worked well for relatively small groups, but could not be made to cope with large index or rank. Since then, the author has made the process more algorithmic and closer to Todd–Coxeter enumeration. John Bray has refined this version and implemented it very effectively within the MAGMA package [15]. Indeed, as will be described below when we deal with the largest Janko group J_4, our most modern version of the program has proved able to find the index of a subgroup for which, with the amount of space available to us, we were unable to make single coset enumeration complete.

3.10.1   Input and stored data

In order to define our group as follows:

$$G = \frac{2^{\star n} : N}{\pi_1 v_1, \pi_2 v_2, \ldots},$$

we need a permutation group N of degree n and a sequence of relators which tell us how elements of N can be written in terms of the n symmetric generators. In practice, the relators are fed in as a sequence:

$$\texttt{rr} := [<u_1, \pi_1>, <u_2, \pi_2>, \ldots, <u_s, \pi_s>],$$

where the π_i are elements of N and the u_i are sequences of integers from $\Lambda = \{1, \ldots, n\}$ without adjacent repetitions. The sequence u_i gives the reverse of the sequence of subscripts of the t_j in the ith relator; therefore, the elements $<u_i, \pi_i>$ of $\texttt{rr}$ really correspond to the relations $\pi_i = v_i^{-1}$. Thus in Example 3.2 we would have

$$\texttt{rr} := [<[1, 2, 1, 2], (3\ 4)>].$$

For each double coset we must store a canonical representative which could be the lexicographically earliest as in the example or, more usefully in our implementation, the first member of that double coset the machine finds. As in **rr**, these *canonical double coset representatives* (CDCRs) are stored as sequences of integers from Λ; for each CDCR w we need its coset stabilizing subgroup $N^{(w)}$, so we set up

$$\mathtt{ss} := [w_1, w_2, \dots] \quad \text{and} \quad \mathtt{gg} := [N^{(w_1)}, N^{(w_2)}, \dots].$$

In order to store enough information to recover Γ_{C}, we need to find and save all n Cayley joins from each of our canonical representatives. The other edges could then be obtained by the action of the control subgroup N. In practice we do not have to apply all n of the symmetric generators to the CDCR w, just one from each orbit under the action of the coset stabilizing subgroup $N^{(w)}$. When we apply the generator t_k on the right to the CDCR w_i, we must end up in some double coset with CDCR w_j, say; that is to say,

$$w_i t_k \in N w_j N.$$

So, there exists a permutation π_{ik} in N such that

$$N w_i t_k = N w_j \pi_{ik}^{-1} \quad \text{and so} \quad N w_i t_k \pi_{ik} = N w_j.$$

We store this information as a triple $<k, j, \pi_{ik}>$ in the ith entry of the *joins* sequence

$$\mathtt{jo} := [[<1, 2, e>], [<,,>, <,,>, \dots, <,,>], \dots],$$

where the number of triples in the sequence $\mathtt{jo}[i]$ is the number of orbits of $N^{(w_i)}$ on Λ, and e stands for the identity of N. For Example 3.2, the completed sequences are thus given by

$$\mathtt{ss} = [[\,], [1], [1, 2], [1, 2, 3]];$$

$$\mathtt{gg} = [N, \langle(2\ 3\ 4), (3\ 4)\rangle, \langle(1\ 2), (3\ 4)\rangle, \langle(1\ 2), (1\ 3)(2\ 4)\rangle];$$

$$\mathtt{jo} = [[<1, 2, e>], [<1, 1, e>, <2, 3, e>], [<1, 2, (1\ 2)>, <3, 4, e>],$$
$$[<1, 3, (1\ 3)(2\ 4)>]].$$

So, $\mathtt{jo}[3]$ contains two triples, since the coset stabilizing subgroup $\mathtt{gg}[3]$ which is isomorphic to V_4 has two orbits $\{1, 2\}$ and $\{3, 4\}$ on $\Lambda = \{1, 2, 3, 4\}$. Now, $N t_1 t_2 t_1 = N t_2 = N t_1 (1\ 2)$ and so $\pi_{31} = (1\ 2)$, whilst $N t_1 t_2 t_3$ is our CDCR for the double coset $N t_1 t_2 t_3 N$, and so $\pi_{33} = e$, the identity of N.

Of course, as we proceed we shall discover identifications between cosets which were hitherto considered different, and so the coset stabilizing subgroups will tend to increase in size; this will lead in turn to fewer orbits on the symmetric generators and to fewer triples in terms of the sequence $\mathtt{jo}[i]$. At a stage when we do not yet know to which double coset $N w_i t_k$ belongs, we store the triple $<k, 0, e>$ in $\mathtt{jo}[i]$. We shall consider the process to have terminated when no zeros remain; at this point we shall say that the table has *closed* and we shall know to which double (and indeed single) coset each symmetric generator maps each CDCR.

3.10.2   The process

Our objective, then, is to construct the sequences $\mathtt{ss}$, $\mathtt{gg}$ and $\mathtt{jo}$. In order to do this we form a table for each additional relation $\pi_i = u_i$, rather as we do for Todd–Coxeter single coset enumeration. If the length of u_i is r, say $u_i = [k_1, k_2, \ldots, k_r]$, then our table will have $r + 1$ columns and the spaces between the columns will correspond to $t_{k_1}, t_{k_2}, \ldots, t_{k_r}$, respectively. We place a right N-coset representative w in the first column, stored in the usual way as a sequence; then wt_{k_1} in the second; $wt_{k_1}t_{k_2}$ in the third; and so on. In the $(r + 1)$th column we store the sequence w^{π_i}. When this results in the identification of certain cosets, we record corresponding information in our sequences $\mathtt{gg}$ and $\mathtt{jo}$.

To see this in action in an easy case, consider the relation of Example 3.2, which takes the form $\pi_1 = (3\ 4) = [1, 2, 1, 2] = u_1$. Our table will begin as shown in Table 3.1, in which the symbol $\star$ is used to denote the coset N. From this we deduce that coset $[1, 2, 1]$ is the same as coset $[2]$, and, working from right to left, this implies that cosets $[2, 1]$ and $[1, 2]$ are the same. Note that $[1]$ and so on are actually words in the symmetric generators; our convention of allowing w to stand for the coset Nw as well is a mild abuse of notation. The first coset other than $\star$ to appear is $[1]$, which of course also stands for Nt_1, and so this is a CDCR. Since $[1, 2]$ is the first coset representative of length 2 to occur, it becomes a CDCR. Multiplication of it by t_1 yields $[1, 2, 1]$, which equals $[2]$, and needs to be multiplied by $(1\ 2)$ to yield the CDCR $[1]$; thus we obtain the triple $<1, 2, (1\ 2)>$ in the term of $\mathtt{jo}$ corresponding to $[1, 2]$. The fact that $[1, 2]$ and $[2, 1]$ give the same coset means that the corresponding coset stabilizing subgroup contains the element $(1\ 2)$ as well as the more obvious $(3\ 4)$. So, at this stage, we have the double cosets

$$N \sim [\,], \quad Nt_1 N \sim [1] \quad \text{and} \quad Nt_1 t_2 N \sim [1, 2],$$

which are shown here with their CDCRs; so $w_1 = 1$, the empty word, $w_2 = t_1$ and $w_3 = t_1 t_2$. The sequence $\mathtt{jo}$ now appears as follows:

$$\mathtt{jo}[1] = [<1, 2, e>],$$

$$\mathtt{jo}[2] = [<1, 1, e>, <2, 3, e>],$$

$$\mathtt{jo}[3] = [<1, 2, (1\ 2)>, <3, 0, e>],$$

where the triple $<3, 0, e>$ tells us that we do not yet know where the generator t_3 takes the CDCR $[1, 2]$. Having gleaned as much information as we can from this line of the table, and recorded the information, we continue as in Table 3.2.

The fourth line of Table 3.2 ends in '4' because the word of symmetric generators corresponding to our relation, namely $[1, 2, 1, 2]$, equals the permutation $(3\ 4)$ in our control subgroup. So, we have a new double coset with CDCR $[3, 1, 2]$; we see that it is the same as $[4, 2, 1]$, and so the coset stabilizing subgroup for this CDCR not only contains $(3\ 1)$, but also

Table 3.1. First relation table for
Example 3.2

		(3 4)		
	1	2	1	2
★	1	12	121	★
		= 21	= 2	

Table 3.2. Second relation table for
Example 3.2

		(3 4)		
	1	2	1	2
★	1	12	2	★
1	★	2	21	1
2	21	1	★	2
3	31	312	3121	4
		= 421	= 42	

$(3\ 4)(1\ 2)$. So $N^{(w)} \geq \langle (3\ 4)(1\ 2), (3\ 1) \rangle \cong D_8$, when w is the CDCR $[3, 1, 2]$.
The joins sequence then terminates with

$$\mathsf{jo}[3] = [<1, 2, (1\ 2)>, <3, 4, (1\ 3\ 2)>],$$

$$\mathsf{jo}[4] = [<1, 3, (1\ 4)>].$$

The sequence of CDCRs is given by

$$\mathsf{ss} = [[\ ], [1], [1, 2], [3, 1, 2]],$$

which is different from what we had earlier, and the coset stabilizing
sequence is given by

$$\mathsf{gg} = [N \cong S_4, \langle (2\ 3\ 4), (3\ 4) \rangle \cong S_3, \langle (1\ 2), (3\ 4) \rangle$$

$$\cong V_4, \langle (1\ 3), (1\ 2)(3\ 4) \rangle \cong D_8].$$

Thus the total number of cosets of N in G is (at most) given by

$$\frac{24}{|S_4|} + \frac{24}{|S_3|} + \frac{24}{|V_4|} + \frac{24}{|D_8|} = 1 + 4 + 6 + 3 = 14.$$

3.10.3   A more ambitious example

We now illustrate the process with an example which has two additional
relations. Our input consists of the permutation group $N \cong S_4$ acting on

$\Lambda = \{1, 2, 3, 4\}$ as before, together with the relations

$$\mathbf{rr} := [<[1, 2, 1, 2, 1], (1\ 2)(3\ 4)>, <[1, 2, 3, 4, 1, 3, 4, 2], (4\ 3\ 2)>].$$

This in fact is equivalent to the following.

EXAMPLE 3.3 Consider

$$G = \frac{2^{\star 4} : S_4}{[(1\ 2)(3\ 4)t_1]^5, [(2\ 3\ 4)t_1 t_2]^5},$$

in which the second relation has been shortened by substituting the first into it. Before attempting the enumeration, we may note that we are only told how to write *even* permutations of N in terms of the symmetric generators, and so G cannot be generated by the t_i and must have a subgroup of index 2 which, by Theorem 3.1, is perfect.

In Tables 3.3 and 3.4, which correspond to the relations $(1\ 2)(3\ 4) = t_1 t_2 t_1 t_2 t_1$ and $(4\ 3\ 2) = t_1 t_2 t_3 t_4 t_1 t_3 t_4 t_2$, respectively, the roman numerical superscripts indicate the point at which a new CDCR is defined; thus, the CDCR $\mathsf{ss}[4]$ is defined where the superscript (iv) appears. Superscript letters indicate where a new mapping from one coset to another first appears, or where a hitherto unknown identification of cosets is revealed. Of course, these discoveries must then be conjugated by permutations of N so that they refer to the CDCR.

Thus we find that

$$<1, 3, e> \in \mathsf{jo}[3]^{(a)},$$

$$<1, 4, (3\ 4)> \in \mathsf{jo}[4]^{(b)},$$

$$<1, 6, (3\ 4)> \in \mathsf{jo}[6]^{(d)}.$$

Using the information gleaned from the tables, we are able to complete the data sequences as follows:

$$\mathsf{ss} = \left[[\,]^{(i)}, [1]^{(ii)}, [1, 2]^{(iii)}, [3, 1, 2]^{(iv)}, [1, 3, 2, 1]^{(v)}, [3, 4, 1, 2]^{(vi)} \right],$$

$$\mathsf{gg} = [N \cong S_4, \langle (2\ 3\ 4), (3\ 4) \rangle \cong S_3, \langle (3\ 4) \rangle \cong C_2, \langle e \rangle \cong 1,$$
$$\langle (1\ 2)(3\ 4)^{(c)}, (2\ 3\ 1\ 4)^{(f)} \rangle \cong C_4, \langle (2\ 3\ 4)^{(e)} \rangle \cong C_3]$$

and

$$\mathsf{jo} = [[<1, 2, e>],$$
$$[<1, 1, e>, <2, 3, e>],$$
$$[<1, 3, e>, <2, 2, e>, <3, 4, (1\ 3\ 2)>],$$
$$[<1, 4, (3\ 4)>, <2, 3, (1\ 2\ 3)>, <3, 5, (1\ 3)>, <4, 6, (1\ 4\ 2)>]$$
$$[<1, 4, (1\ 3)>],$$
$$[<1, 6, (3\ 4)>, <2, 4, (4\ 1\ 2)>]].$$

Table 3.3. First relation table for Example 3.3

	(1 2)(3 4)				
	1	2	1	2	1
★[i]	1[ii]	12[iii]	121[a] = 12	1	★
1	★	2	21	21	2
2	21	21	2	★	1
3	31	312[iv]	3121 = 412[b]	41	4
12	12	1	★	2	21
13	13	132	1321[v] = 2412[c]	241	24
23	231	2312 = 1421	142	14	14
34	341	3412[vi]	34121 = 4312[d]	431	43

Table 3.4. Second relation table for Example 3.3

	(4 3 2)							
	1	2	3	4	1	3	4	2
★	1	12	123	1234 = 2431[e]	243	24	2	★
1	★	2	23	234	2341 = 1243	124	12	1
2	21	21	213	2134 = 4231	423	42	42	4
3	31	312	3123 = 4314[f]	431	43	4	★	2

Thus we see that the maximum possible index of N in G is given by

$$\frac{24}{|S_4|} + \frac{24}{|S_3|} + \frac{24}{|C_2|} + \frac{24}{1} + \frac{24}{|C_4|} + \frac{24}{|C_3|} = 1 + 4 + 12 + 24 + 6 + 8 = 55.$$

The collapsed Cayley graph follows from the information contained in these data sequences and has the form given in Figure 3.5. It follows that the

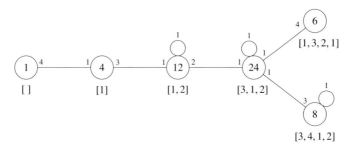

Figure 3.5. Collapsed Cayley graph for $PGL_2(11)$.

maximal order of the image group G is $24 \times 55 = 2 \times 660$. That this is indeed the order of G can be confirmed by regarding G as a permutation group on the 55 symbols we have produced. Certainly the actions of the control group N and the symmetric generator t_1 are well defined, so it remains to show that (i) t_1 has just four images under conjugation by N and (ii) that the additional relations hold within the symmetric group S_{55}.

In a case like this, however, we can guess that the image is in fact the projective general linear group $PGL_2(11)$. To verify that this is indeed the case we define the following:

$$x : \tau \mapsto \frac{2\tau+1}{-\tau+2} \equiv (\infty\ 9\ 2)(0\ 6\ 5)(1\ 3\ 4)(X\ 7\ 8) \equiv (t_1)(t_2\ t_3\ t_4),$$

$$y : \tau \mapsto \frac{-2}{\tau} \equiv (\infty\ 0)(3)(8)(1\ 9)(2\ X)(4\ 5)(6\ 7) \equiv (t_1\ t_2)(t_3)(t_4),$$

$$t_1 : \tau \mapsto \frac{-1}{\tau} \equiv (\infty\ 0)(1\ X)(2\ 5)(3\ 7)(4\ 8)(6\ 9),$$

as linear fractional permutations of the projective line $PG_1(11)$. It is readily checked that xy has order 4, and so $N = \langle x, y \rangle \cong S_4$. The permutation t_1 has just four images under conjugation by N, namely

$$t_1 = (\infty\ 0)(1\ X)(2\ 5)(3\ 7)(4\ 8)(6\ 9),$$
$$t_2 = (\infty\ 0)(9\ 2)(X\ 4)(3\ 6)(5\ 8)(7\ 1),$$
$$t_3 = (9\ 6)(2\ \infty)(7\ 1)(4\ 5)(0\ X)(8\ 3),$$
$$t_4 = (2\ 5)(\infty\ 9)(8\ 3)(1\ 0)(6\ 7)(X\ 4),$$

and the action of x and y on these symmetric generators which is shown above in the final column of their definitions. It is now a simple matter to check that $(xy)^2 t_1$ and $x t_2 t_1$ both have order 5. Now, N is maximal in $PGL_2(11)$ and $t_1 \notin N$, and so $PGL_2(11)$ is an image of G. Thus,

$$|G| \le 2 \times 660 = |PGL_2(11)| \le |G|,$$

and isomorphism follows.

These two examples are sufficiently small for us to exhibit Tables 3.2, 3.3 and 3.4 in detail, together with explanations as to how the entries are used in the enumeration. This is clearly not practical for larger cases. However, when we come to more demanding examples in later chapters we shall on occasions append the computer input and output, both as a confirmation of our hand calculations and as a demonstration of the ease with which the program handles them.

3.11   Systematic approach

We conclude this chapter by exhibiting in a systematic manner the wealth of groups which occur as homomorphic images of progenitors with rather small and familiar control subgroups. The results given here are taken directly from Curtis, Hammas and Bray [36], which builds on Hammas [42]. Our symmetric generators will be involutions, so we seek homomorphic images of the progenitor:

$$P = 2^{\star n} : N,$$

where N is a transitive permutation group on n letters. We shall limit ourselves to the six cases $N \cong$ (i) S_3, (ii) A_4, (iii) S_4, (iv) A_5, (v) $L_2(5)$ and (vi) $L_3(2)$ acting on 3, 4, 4, 5, 6 and 7 points, respectively. Here A_n and S_n denote, respectively, the alternating and symmetric groups on n letters, and $L_n(q)$ denotes the projective special linear group in n dimensions over the field of order q in its permutation action on the 1-dimensional subspaces. Our presentations for these progenitors will, in practice, take the standard form described above, namely

$$\langle x, y, t \mid \langle x, y \rangle \cong N, [t, N^0] = 1 = t^2 \rangle \cong 2^{\star n} : N,$$

where the set of relations includes a standard presentation for N in terms of the generators x and y [27] and N^0 denotes the point-stabilizer in N in its action on n letters–generators for which are given as words in x and y. In Table 3.5 we give an explicit presentation for each of our progenitors, together with the action of the generators of N on the symmetric generators.

   Now, as stated above, every element of the progenitor is uniquely express-ible as a permutation of N followed by a word in the symmetric generators (provided cancellation of adjacent repetitions is carried out). Double cosets $[w] = NwN$ are then in one-to-one correspondence with the orbits of N on ordered k-tuples without adjacent repetitions. The Cayley graph for the progenitor over N is a tree, since any circuit would imply a non-trivial relation between the symmetric generators. For example, the part of the (infinite) Cayley graph of the progenitor in (vi) of Table 3.5 over $N \cong L_3(2)$, for $l(w) \le 3$, is shown in Figure 3.6.

3.11.1   Additional relations

We now seek finite homomorphic images of such a progenitor, and must consider what additional relation or relations to factor by. As stated above,

Table 3.5. Presentations of the progenitors

(i)	$\langle x, y, t \mid x^3 = y^2 = (xy)^2 = 1 = t^2 = [t, y] \rangle \cong 2^{*3} : S_3$ $x \sim (0, 1, 2), \ y \sim (1, 2)$
(ii)	$\langle x, y, t \mid x^3 = y^3 = (xy)^2 = 1 = t^2 = [t, x] \rangle \cong 2^{*4} : A_4$ $x \sim (1, 2, 3), \ y \sim (0, 1, 2)$
(iii)	$\langle x, y, t \mid x^4 = y^2 = (xy)^3 = 1 = t^2 = [t, y] = [t^x, y] \rangle \cong 2^{*4} : S_4$ $x \sim (0, 1, 2, 3), \ y \sim (2, 3)$
(iv)	$\langle x, y, t \mid x^5 = y^3 = (xy)^2 = 1 = [t, y] = [t, y^{x^2}] \rangle \cong 2^{*5} : A_5$ $x \sim (0, 1, 2, 3, 4), \ y \sim (4, 2, 1)$
(v)	$\langle x, y, t \mid x^5 = y^3 = (xy)^2 = 1 = t^2 = [t, x] = [t^{yx^2}, xy] \rangle \cong 2^{*6} : L_2(5)$ $x \sim (0, 1, 2, 3, 4), \ y \sim (\infty, 0, 1)(2, 4, 3)$

(vi) $\langle x, y, t \mid x^7 = y^2 = (xy)^3 = [x, y]^4 = 1 = t^2 = [xy, t^{x^4}] = [y, t^{x^3}] \rangle \cong 2^{*7} : L_3(2)$
$x \sim (0, 1, 2, 3, 4, 5, 6), \ y \sim (2, 6)(4, 5)$

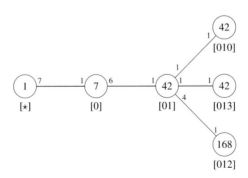

Figure 3.6. Part of the collapsed Cayley diagram for $P = 2^{*7} : L_3(2)$.

this is equivalent to asking how to write permutations of N in terms of the symmetric generators $\mathcal{T}$, and Lemma 3.3 tells us that

$$\langle t_i, t_j \rangle \cap N \le C_N(N^{ij}),$$

where N^{ij} stands for the stabilizer in N of the points i and j.

EXAMPLE 3.4 As an example let us consider the progenitor (iv) of Table 3.5, namely $2^{*4} : S_4$, when $N^{01} = \langle (2, 3) \rangle$ and so

$$\langle t_0, t_1 \rangle \cap N \le C_N(N^{01}) = \langle (0, 1), (2, 3) \rangle.$$

So, the only non-trivial elements of the control subgroup S_4 which we can write in terms of the symmetric generators t_0 and t_1 (without causing collapse) are $(0, 1), (2, 3)$ and $(0, 1)(2, 3)$.

For involutory symmetric generators, we have $\langle t_0, t_1 \rangle \cong D_{2k}$, a dihedral group of order $2k$. If k is even, this group has centre of order 2 generated by $(t_0 t_1)^{k/2}$; if k is odd, then the element $t_0 t_1 \cdots t_0$ of length k interchanges t_0 and t_1 by conjugation. Thus, in the first case, we have the following possibility:

$$(t_0 t_1)^{k/2} = (2, 3);$$

and in the second:

$$[(0, 1)t_0]^k = 1 \quad \text{or} \quad [(0, 1)(2, 3)t_0]^k = 1.$$

Consider first the group defined by

$$\frac{2^{\star 4} : S_4}{(0, 1) = t_0 t_1 t_0}, \tag{3.7}$$

the smallest possibility in the odd case. Since $010 \sim \star$, we have $[010] = [\star]$ and so $[01] = [0]$. Thus there are just two double cosets: $[\star] = N$ and $[0] = Nt_0 N$, which contains the four single cosets Nt_i for $i = 1, \ldots, 4$. We have the double coset diagram shown in Figure 3.7.

The symmetric generator t_0 acts on these five single cosets by right multiplication as the transposition interchanging coset $\star$ with coset 0, since $Nt_0 . t_0 = N$, but $Nt_i \cdot t_0 = Nt_i t_0 t_i \cdot t_i = Nt_i$ for $i \neq 0$. Since $(\star, 0)(\star, 1)(\star, 0) = (0, 1)$, we see that our additional relation is satisfied by these transpositions, and we have a symmetric presentation of S_5.

The smallest possibility in the even case is given by

$$\frac{2^{\star 4} : S_4}{(2, 3) = [t_0 t_1]^2}. \tag{3.8}$$

We now have $0101 \sim \star \Rightarrow 01 \sim 10$, and so the coset stabilizing subgroup $N^{(01)} \geq \langle (0, 1), (2, 3) \rangle$. Thus the double coset $[01]$ contains at most $24/4 = 6$ single cosets. Moreover, we see that

$$012 \sim 0.1212.21 \sim 0.(0, 3).21 \sim 321,$$

and so $N^{(012)} \geq \langle (0, 3)(1, 2), (0, 1) \rangle \cong D_8$, a transitive subgroup of N. This indicates the double coset diagram shown in Figure 3.8.

As above, this gives the action of our symmetric generators as permutations on $(1 + 4 + 6 + 3) = 14$ points, and we readily verify that they satisfy the additional relation required. Alternatively, we may recognize the group we have constructed, in this case the projective general linear group $PGL_2(7)$, and seek elements in it satisfying the symmetric presentation. If

Figure 3.7. Collapsed Cayley diagram for the symmetric presentation (3.7).

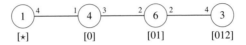

Figure 3.8. Collapsed Cayley diagram for the symmetric presentation (3.8).

we let $\Lambda = \{0, 1, 2, 3\}$, the points of the Fano plane are represented by $\star$ together with the six unordered pairs of members of Λ, and its lines by the four members of Λ together with the three partitions of Λ into two pairs. If $\lambda = \{i, j, k, l\}$, then $ij.kl$, one of the three partitions of the four letters into two pairs, represents the coset $ijk \sim ijl \sim kli \sim klj$. The three points on each line are given by

$$i = \{kl, lj, jk\} \quad \text{and} \quad ij.kl = \{\star, ij, kl\}.$$

Again, it may be verified mechanically (see ref. [57], p. 64), that

$$\frac{2^{*4}{:}S_4}{[(0, 1)(2, 3)t_0]^5} \cong (3 \times L_2(11)){:}2. \tag{3.9}$$

Indeed, using the double coset enumerator we would input the following:

```
> N:=SymmetricGroup(4);
> RR:=[<[1,2,1,2,1],N!(1,2)(3,4)>];
> HH:=[N];
> CT:=DCEnum(N,RR,HH:Print:=5,Grain:=100);
Index: 165 === Rank: 12 === Edges: 32 === Status: Early closed
```

More detailed information about the double cosets is stored in CT, and in particular we have the following:

```
> CT[7];
[ 1, 4, 12, 24, 24, 24, 12, 24, 24, 6, 8, 2 ]
> CT[4];
[
    [],
    [ 1 ],
    [ 1, 2 ],
    [ 1, 2, 3 ],
    [ 1, 2, 3, 4 ],
    [ 1, 2, 3, 4, 2 ],
    [ 1, 2, 3, 4, 2, 4 ],
    [ 1, 2, 3, 4, 2, 3 ],
    [ 1, 2, 3, 4, 2, 3, 2 ],
    [ 1, 2, 3, 4, 2, 3, 2, 4 ],
    [ 1, 2, 3, 4, 2, 3, 1 ],
    [ 1, 2, 3, 4, 2, 3, 1, 4 ]
]
```

where `CT[7]` tells us that there are 12 double cosets, each of which contains the number of single cosets given, and `CT[4]` gives us the CDCR for each of them. Thus, for example, the last double coset $Nt_1t_2t_3t_4t_2t_3t_1t_4N$ contains just two single cosets. In addition, all joins in the Cayley graph are available.

Now, in each of the six progenitors listed at the beginning of Section 3.11, the control subgroup acts doubly transitively on the symmetric generators. Applying Lemma 3.3, we see that

(i) $N \cap \langle t_0, t_1 \rangle$ is unrestricted,

(ii) $N \cap \langle t_0, t_1 \rangle$ is unrestricted,

(iii) $N \cap \langle t_0, t_1 \rangle \leq \langle (0,1), (2,3) \rangle$,

(iv) $N \cap \langle t_0, t_1 \rangle \leq \langle (2,3,4) \rangle$,

(v) $N \cap \langle t_\infty, t_0 \rangle \leq \langle (\infty, 0)(1,4), (1,4)(2,3) \rangle$,

(vi) $N \cap \langle t_0, t_1 \rangle \leq \langle (2,6)(4,5), (2,4)(5,6) \rangle$,

where the roman numerals refer to the labelling in Table 3.5 throughout. In our case, the symmetric generators are involutions, and so $\langle t_0, t_1 \rangle$ is dihedral; we thus obtain further restrictions as follows.

(i) $N \cap \langle t_0, t_1 \rangle$ is a subgroup of N normalized by $(0,1)$ and so is trivial, is $\langle (0,1) \rangle$, or is N itself. But if $\langle t_0, t_1 \rangle \geq N$, then we have $\langle t_0, t_1 \rangle = \langle t_0, N \rangle = G$, and so G is dihedral. Thus for 'interesting' G we must have $N \cap \langle t_0, t_1 \rangle \leq \langle (0,1) \rangle$. The case $(0,1) \in \langle t_0, t_1 \rangle$ is equivalent to $(0,1)t_0$ having odd order.

(ii) $N \cap \langle t_0, t_1 \rangle$ is a subgroup of N normalized by $(0,1)(2,3)$ and so is contained in $\langle (0,1)(2,3), (0,2)(1,3) \rangle$ or equals N. But $(0,2)(1,3) \in \langle t_0, t_1 \rangle \Rightarrow \langle t_0, t_1 \rangle = \langle t_0, t_1, t_2, t_3 \rangle = G$, and so G is dihedral. Thus, for 'interesting' G, we may assume that $N \cap \langle t_0, t_1 \rangle$ is trivial, or that $(0,1)(2,3)t_0$ has odd order.

(iii) As above, $(0,1)$ or $(0,1)(2,3) \in \langle t_0, t_1 \rangle$ is equivalent to $(0,1)t_0$ or $(0,1)(2,3)t_0$, respectively, having odd order. In this case, the dihedral group $\langle t_0, t_1 \rangle$ has trivial centre, and so does not contain $(2,3)$. If $(2,3) \in \langle t_0, t_1 \rangle$, then $(2,3) = (t_0 t_1)^s$ for some s. Otherwise, $N \cap \langle t_0, t_1 \rangle$ is trivial.

(iv) A dihedral group cannot have an element of order 3 in its centre and so $N \cap \langle t_0, t_1 \rangle$ is trivial.

(v)   As in (iii), the four mutually exclusive possibilities are that either $(\infty, 0)(1,4)t_\infty$ or $(\infty, 0)(2,3)t_\infty$ has odd order, that $(1,4)(2,3) = (t_\infty t_0)^s$ for some s, or that $N \cap \langle t_\infty, t_0 \rangle$ is trivial.

(vi) We have $N \cap \langle t_0, t_1 \rangle \leq Z(\langle t_0, t_1 \rangle)$, which has order 1 or 2. Since $N \cap \langle t_0, t_1 \rangle$ is normalized by the element $(0, 1)(2, 5)$, the only possibilities are that $(2, 5)(4, 6) = (t_0 t_1)^s$ or that $N \cap \langle t_0, t_1 \rangle$ is trivial.

Other useful relations by which to factor the progenitor are those of the form $(\pi u)^s$, where $\pi \in N$ and u is a word in the symmetric generators. In this context, we define the orders of elements of the form πt_i to be *the first order parameters* of the symmetric presentation; similarly the orders of elements of the form $\pi t_i t_j$ are the *second order parameters*. In Table 3.6 we give, for each progenitor, a list of elements whose orders form a complete set of first order parameters, together with other useful parameters. Note that πt_i has the same order as $\pi t_{\pi(i)}$, $t_i \pi$ and $\pi^{-1} t_i$.

3.11.2 Some generic images of $2^{\star 3} : S_3$ and other progenitors

In ref. [27], Coxeter and Moser define the following:

$$G^{p,q,r} \equiv \langle u, v, w \mid u^2 = v^2 = (uv)^2 = 1 = w^2 = (uw)^p = (vw)^q = (uvw)^r \rangle.$$

Thus, if we set $p = 3$, $v = t_0$, $v^{wu} = t_1$, $v^w = t_2$, then, by conjugation, u and w have actions $u \sim (1, 2)$, $w \sim (0, 2)$ on $\mathcal{T} = \{t_0, t_1, t_2\}$. Consideration of the presentation shows that

$$G^{3,q,r} \cong \frac{2^{\star 3} : S_3}{[(0, 1)t_0]^q, \ [(0, 1, 2)t_0]^r} \cong G^{3,r,q}, \tag{3.10}$$

i.e. that the parameters q and r may be interchanged. This may alternatively be observed by writing $s_0 = t_0(1, 2)$, $s_1 = t_1(2, 0)$ and $s_2 = t_2(0, 1)$, when we see that $\mathcal{S} = \{s_0, s_1, s_2\}$ is a second symmetric generating set permuted by the original control subgroup $N = \langle u, w \rangle \cong S_3$. But $(0, 1, 2)s_0 = (0, 2)t_0$ and $(0, 1)s_0 = (0, 2, 1)t_0$, and $(0, 1)s_0 s_2 = (0, 1)(1, 2)t_0(0, 1)t_2 = (0, 2)t_1 t_2$. Similarly, $(0, 1, 2)s_0 s_1 s_0$ has the same order as $(1, 2)t_2 t_0 t_2$, and $(0, 1)s_0 s_2 s_0$ has the same order as $(0, 1, 2)t_1 t_2 t_1$. Thus, referring to Table 3.6, the automorphism of the progenitor mapping t_i to s_i while fixing N interchanges the roles of parameters a and b, and parameters d and e, while fixing parameter c. For example, the homomorphic image of $2^{\star 3} : S_3$ defined by parameters (**7 8** 6 4 8) is isomorphic to that defined by (**8 7** 6 8 4); although, as in this case, the subgroups generated by corresponding subsets of the symmetric generators need not be isomorphic. These groups have been extensively studied and various results are given in ref. [27], p. 96, but the adjunction of our second and third order parameters enables us to obtain more factor groups.

Table 3.6. Parameters used for each of our progenitors

(i) $2^{\star 3}\!:\!S_3$	(ii) $2^{\star 4}\!:\!A_4$	(iii) $2^{\star 4}\!:\!S_4$
$[(0, 1, 2)t_0]^a$	$[(0, 1, 2)t_0]^a$	$[(0, 1, 2, 3)t_0]^a$
$[(0, 1)t_0]^b$	$[(0, 1)(2, 3)t_0]^b$	$[(0, 1, 2)t_0]^b$
$[(0, 1)t_0 t_2]^c$	$[(0, 1, 2)t_0 t_3]^c$	$[(0, 1)(2, 3)t_0]^c$
$[(0, 1, 2)t_0 t_1 t_0]^d$	$[(0, 1)(2, 3)t_0 t_2]^d$	$[(0, 1)t_0]^d$
$[(0, 1)t_0 t_2 t_0]^e$		$[(0, 1)t_0 t_2]^e$
		$(2, 3) = [t_0 t_1]^s$

(iv) $2^{\star 5}\!:\!A_5$	(v) $2^{\star 6}\!:\!L_2(5)$	(vi) $2^{\star 7}\!:\!L_3(2)$
$[(0, 1, 2, 3, 4)t_0]^a$	$[(0, 1, 2, 3, 4)t_0]^a$	$[(0, 1, 2, 3, 4, 5, 6)t_0]^a$
$[(0, 2, 4, 1, 3)t_0]^b$	$[(0, 2, 4, 1, 3)t_0]^b$	$[(1, 2, 4)(3, 6, 5)t_1]^b$
$[(0, 1, 2)t_0]^c$	$[(\infty, 0, 1)(2, 4, 3)t_\infty]^c$	$[(1, 2, 4)(3, 6, 5)t_3]^c$
$[(0, 1)(2, 3)t_0]^d$	$[(\infty, 0, 1)(2, 4, 3)t_2]^d$	$[(2, 6)(4, 5)t_2]^d$
	$[(\infty, 0)(1, 4)t_\infty]^e$	$[(1, 3)(2, 4, 6, 5)t_1]^e$
	$[(\infty, 0)(1, 4)t_1]^f$	$[(1, 3)(2, 4, 6, 5)t_2]^f$
	$(\infty, 0)(1, 4) = [t_2 t_3]^s$	$[(2, 6)(4, 5)t_0 t_2]^g$
		$[(2, 6)(4, 5)t_1 t_2]^h$
		$(2, 6)(4, 5) = [t_1 t_3]^s$

It is natural to ask which linear fractional groups are images of this progenitor. Indeed, if we write

$$(0, 1, 2) \equiv \eta \mapsto \tfrac{1}{1-\eta} \quad : \quad (\infty, 0, 1). \ldots,$$

$$(1, 2) \equiv \eta \mapsto \tfrac{1}{\eta} \quad : \quad (\infty, 0)(1)(-1). \ldots,$$

$$t_0 \equiv \eta \mapsto \tfrac{-1}{\eta} \quad : \quad (\infty, 0)(1, -1). \ldots,$$

$$t_1 \equiv \eta \mapsto \tfrac{\eta-1}{2\eta-1},$$

$$t_2 \equiv \eta \mapsto \tfrac{\eta-2}{\eta-1},$$

where η is in the projective line with $p+1$ points, we see that all these elements lie in $\mathrm{PSL}_2(p)$ for $p \equiv 1$ mod 4. But

$$(0, 1, 2)t_0 \equiv \eta \mapsto \eta - 1$$

was shown in ref. [40] to generate $\mathrm{PSL}_2(p)$ together with t_0 above. Thus $\mathrm{PSL}_2(p)$ is an image of $2^{\star 3}\!:\!S_3$ for all $p \equiv 1$ mod 4, and $\mathcal{T} = \{t_0, t_1, t_2\}$ gives a canonical set of symmetric generators; $\mathcal{T}$ is still a symmetric generating set in the case $p \equiv 3$ mod 4, but it may not possess full S_3 symmetry within the simple group.

THEOREM 3.3 The projective special linear group $\mathrm{PSL}_2(p)$, for p a prime, is a homomorphic image of $2^{\star 3}\!:\!S_3$, except for $p = 3, 7$ and 11.

Sketch of proof Let $G \cong L_2(p)$, where $p \geq 7$. We first note that a copy of S_3 in G must be contained in a (usually) maximal subgroup of shape

D_{p+1} or D_{p-1}, and that there is just one conjugacy class of such subgroups, or two classes interchanged by an outer automorphism. Let $S_3 \cong N \le G$ and let $y \in N$ have order 2. Then $C_G(y) \cong D_{p+1}$ or D_{p-1}, and so if

$$\Phi = \{t \in G \mid t^2 = 1, \ [t, y] = 1, t \notin \langle y \rangle\},$$

then $|\Phi| = (p+1)/2$ or $(p-1)/2$. We are supposing that y acts on the three symmetric generators $\{t_0, t_1, t_2\}$ as the permutation $(0)(1\ 2)$, and so we seek $t_0 \in \Phi$ and consider what the possibilities are for $H = \langle N, t \rangle$ with $t \in \Phi$. But the maximal subgroups of $PSL_2(p)$ are given in ref. [40], and we see that the only possibilities for proper subgroups H are as follows: D_{12}, when $p \equiv \pm 1$ mod 12; S_4, when $p \equiv \pm 1$ mod 8; A_5, when $p \equiv \pm 1$ mod 10. When any of these occurs, N is in just one copy of $H \cong D_{12}$ and in two copies of $H \cong S_4$ or A_5. Thus there are possibly two elements of Φ extending N to a copy of D_{12}, $2 \times 2 = 4$ elements of Φ extending N to a copy of S_4, and $2 \times 2 = 4$ elements of Φ extending N to a copy of A_5. So, if $|\Phi| > 10$, then some element of Φ together with N generates G; i.e., if $p \ge 23$, then $PSL_2(p)$ is a homomorphic image of $2^{*3}:S_3$. Consideration of the modularity condition on p for subgroups of the three types to exist in G shows that for $p = 19$, 17, 13 there are 6, 4, 4 elements of Φ, respectively, which, together with N, generate G. However, for $p = 11$ we have $|\Phi| = 6$, two elements of which generate, together with N, a copy of D_{12} and four of which, together with N, generate a copy of A_5; and for $p = 7$ we have $|\Phi| = 4$, each element of which, together with N, generates a copy of S_4. Of course, $p = 5$ works with the canonical generators given above, and the cases $p = 2$ and 3 are easy.

A more complete version of this theorem and its proof is provided by John Bray in his thesis [14]. The reader is referred to this work for further details.

THEOREM 3.4 (Bray) Each of the groups $L_2(q)$ and $PGL_2(q)$ is a homomorphic image of the progenitor $2^{*3} : S_3$ except the following:

$$L_2(3^{2m+1}), L_2(7), L_2(9) \cong A_6, L_2(11) \text{ and } PGL_2(5) \cong S_5.$$

Similar, though easier, arguments show that $PSL_2(p)$ is a homomorphic image of $2^{*4}:S_4$ if, and only if, $p \equiv \pm 1$ mod 24; and an image of $2^{*6}:L_2(5)$ if, and only if, $p \equiv \pm 1$ mod 20.

3.11.3 Tabulated results

The information given in the following tables is, for the most part, self-explanatory. Table 3.6 describes the particular parameters used for each of our progenitors, and Table 3.5 allows the reader to write down an explicit presentation for each of the groups given. Note that, in spite of our requirement that $\mathcal{T}$ be a generating set for G, we have allowed some interesting cases where this is not quite true. The notation $[n]$ is used for a soluble group of order n, whereas p^m denotes an elementary abelian p-group of rank m. Most importantly, the parameters given in bold-face type are those required to define the group; parameters given as small numerals are the values assumed in the resulting group G. Thus a particular group may occur several times, being defined by different subsets of the parameters.

As a general rule, we have attempted to identify for the reader the subgroup of G generated by subsets of the symmetric generators, although space restrictions in the tables have, on occasions, prevented our doing so. For this reason, we would mention that, in the case of the homomorphic image of $2^{\star 6}:L_2(5)$ of shape $6^{\cdot}L_3(4):2_3$, we have $\langle t_\infty, t_0, t_1 \rangle \cong PGL_2(7)$ and $\langle t_\infty, t_0, t_2 \rangle \cong 3^{\cdot}PGL_2(9)$. In the same progenitor, when the homomorphic image is $L_2(19)$, it is interesting to note that $\langle t_\infty, t_0, t_1 \rangle \cong A_5$, whereas $\langle t_\infty, t_0, t_2 \rangle$ is the whole of G.

Verification of the results given in Tables 3.7–3.12 is carried out as follows. For the smaller groups we use coset enumeration over N to find the order of G and then construct a faithful permutation representation of lower degree to find its structure in detail. Many of the smaller cases have been constructed by hand [14] using the double coset enumeration procedure described in ref. [34]. Several interesting Cayley diagrams are given in ref. [14].

For larger groups, we exploit the nesting property of the progenitors; thus, for example,

$$2^{\star 4}:S_4 \le 2^{\star 7}:L_3(2).$$

Parameters are chosen so that a subgroup generated by a subset of the symmetric generators may be identified. For example, the subgroup generated by an 'oval' $\langle t_0, t_3, t_6, t_5 \rangle$ in an image of $2^{\star 7}:L_3(2)$ is itself essentially an image of $2^{\star 4}:S_4$. The fact that

$$\langle t_0, t_3, t_6, t_5 \rangle \cong L_2(23)$$

is used in the identification of M_{24}. It is worth mentioning that the subgroup $\langle x, yxtxy \rangle$ of ref. [33] has index 24 in the group, even without factoring by the relator $[(1,3)(2,4,6,5)t_2]^{11}$, i.e. not setting the parameter $f = 11$. This relator is needed for a presentation, however, as is shown by the parameters of $2^6:S_7$ in the same table. In obtaining the Tits group $^2F_4(2)'$ as an image of $2^{\star 4}:S_4$, we introduced the parameter $e = 12$ to identify $L_3(3)$ as the subgroup generated by three symmetric generators. It may well not be needed in the presentation.

The *low index subgroups* facility on MAGMA is used to identify groups with a complicated multiplier and outer automorphism group, such as the image $6^{\cdot}L_3(4):2_3$ in the progenitor $2^{\star 6}:L_2(5)$. In this case, having found the order of the group by coset enumeration over N and established a homomorphic image isomorphic to $L_3(4):2_3$, we quote Theorem 3.1 to show that the derived subgroup has index 2 and is perfect.

Table 3.7. Some finite images of the progenitor $2^{\star 3}:S_3$

Parameters					Order of G	Shape of $\langle t_0, t_1 \rangle$	Shape of $\langle \mathcal{T} \rangle$	Shape of G
a	b	c	d	e				
5	5	3	3	3	60	D_{10}	G	A_5
7	8	6	4	8	336	D_8	$L_2(7)$	$PGL_2(7)$
8	7	6	8	4	336	D_{14}	G	$PGL_2(7)$
7	9	9	9	7	504	D_{18}	G	$L_2(8)$
8	10	5	10	8	720	D_{10}	G	$PGL_2(9)$
7	13	7	13	6	1092	D_{26}	G	$L_2(13)$
10	11	12	4	5	1320	D_{22}	G	$PGL_2(11)$
8	10	15	10	8	2160	D_{10}	G	$3 \cdot PGL_2(9)$
12	7	14	14	6	2184	D_{14}	G	$PGL_2(13)$
9	9	10	18	10	3420	D_{18}	G	$L_2(19)$
17	17	5	17	17	4080	D_{34}	G	$L_2(16)$
18	9	16	16	4	4896	D_{18}	G	$PGL_2(17)$
10	12	7	6	10	5040	D_{12}	G	S_7
18	20	5	6	20	6840	D_{20}	G	$PGL_2(19)$
13	13	5	12	12	7800	D_{26}	G	$L_2(25)$
8	13	12	12	8	11 232	D_{26}	G	$L_3(3):2$
8	11	24	24	23	12 144	D_{22}	G	$PGL_2(23)$
7	15	29	15	15	12 180	D_{30}	G	$L_2(29)$
28	13	26	4	13	19 656	D_{26}	G	$PGL_2(27)$
28	7	10	30	14	24 360	D_{14}	G	$PGL_2(29)$
32	8	32	32	5	29 760	D_8	G	$PGL_2(31)$
7	22	11	11	21	39 732	D_{22}	G	$L_2(43)$
44	11	6	14	21	79 464	D_{22}	G	$PGL_2(43)$
15	29	5	29	29	102 660	D_{58}	G	$L_2(59)$
52	52	9	4	54	148 824	D_{52}	G	$PGL_2(53)$
10	19	7	10	11	175 560	D_{38}	G	J_1
35	36	5	9	36	178 920	D_{36}	G	$L_2(71)$
8	74	9	74	74	388 944	D_{74}	G	$PGL_2(73)$
8	15	10	8	21	483 840	D_{30}	G	$2^2 \cdot L_3(4):S_3$
12	15	7	5	15	604 800	D_{30}	G	$HJ = J_2$
115	19	5	115	115	6 004 380	D_{38}	G	$L_2(229)$

Perhaps the most delightful entry in these tables is the fact that if we factor a progenitor of shape $2^{\star 5}:A_5$ by a single relation which requires a 5-cycle of A_5 times one of the symmetric generators to have order 7, we obtain J_1, the smallest of the sporadic simple groups discovered by Janko [58].

EXERCISE 3.6

(1) Show that the alternating group A_5 is an image of the progenitor $P = 2^{\star 3}:S_3$, but that the symmetric group S_5 is not.

(2) Show that the alternating group A_6 is not an image of the progenitor $P = 2^{\star 3}:S_3$.

Table 3.8. Some finite images of the progenitor $2^{*4}:A_4$

Parameters				Index of N in G	Order of G	Shape of $\langle t_0, t_1, t_2 \rangle$	Shape of $\langle \mathcal{T} \rangle$	Shape of G
a	b	c	d					
4	6	5	5	10	120	S_4	G	S_5
7	7	6	7	91	1092	G	G	$L_2(13)$
8	6	20	10	640	7680	$4^2:D_6$	G	$4\cdot2^4\cdot S_5$
8	6	10	5	320	3840	$4^2:D_6$	G	$2^{1+4}:S_5$
8	6	5	5	160	1920	$4^2:D_6$	G	$2^4:S_5$
8	6	5	5	160	1920	$4^2:D_6$	G	$2^4:S_5$
8	8	7	3	28	336	G	G	$PGL_2(7)$
9	9	6	7	63504	762048	$L_2(8)^2$	$L_2(8)^2$	$(L_2(8) \times L_2(8)):3$
10	6	12	13	1300	15600	$5^2:S_3$	G	$P\Sigma L_2(25)$
10	8	7	5	6720	80640	$PGL_2(9)$	G	$2\cdot L_3(4):2_3$
10	12	5	6	420	5040	G	G	S_7
10	12	5	6	420	5040	G	G	S_7
11	5	5	6	55	660	G	G	$L_2(11)$
11	5	5	6	55	660	G	G	$L_2(11)$
18	16	4	9	408	4896	G	G	$PGL_2(17)$
24	8	21	3	84	1008	$PGL_2(7)$	$PGL_2(7)$	$3 \times PGL_2(7)$
33	5	15	6	165	1980	$L_2(11)$	$L_2(11)$	$3 \times L_2(11)$
110	60	5	6	277200	3326400	G	G	$L_2(11) \times S_7$

The reader who is not familiar with the action of $L_2(q)$ on the projective line $P_1(q)$ is referred to Section 4.1 before tackling the following exercises (3) and (4).

(3) Verify explicitly as follows the statement in Theorem 3.3 that $L_2(7)(\cong L_3(2))$ is *not* an image of the progenitor $2^{*3}:S_3$. By Sylow's theorem, any two subgroups of order 3 in $L_2(7)$ are conjugate, so, without loss of generality, we may take $N = \langle x, y \rangle$ to be the normalizer of $\langle x \rangle$, where

$$x \equiv \eta \mapsto 2\eta; \, y \equiv \eta \mapsto -\frac{1}{\eta}.$$

Obtain the centralizer of y, which is isomorphic to D_8, and show that each of the four involutions in this subgroup (other than y) generates with x a copy of S_4 as claimed. [Hint: It may be helpful to show that the element $\eta \mapsto (\eta-1)/(\eta+1)$ squares to y.] Note: it is perhaps easier to work in $L_3(2)$ acting on the seven points of the projective plane of order 2 (see Chapter 4). Then the element x fixes a point and a line, and one can readily see that each of the four eligible involutions fixes either this point or this line and so cannot generate the whole group together with x.

Table 3.9. Some finite images of the progenitor $2^{\star 4}:S_4$

Parameters						Order of G	Shape of $\langle t_0, t_1, t_2 \rangle$	Shape of $\langle \mathcal{T} \rangle$	Shape of G
a	b	c	d	e	s				
10	**8**	**6**	6	8	—	7680	$2^2{\cdot}S_4$	$2^{1+4}:S_5$	$2^{1+4+1}:S_5$
10	**4**	6	6	4	—	240	S_4	S_5	$2 \times S_5$
5	4	6	**3**	4	—	120	S_4	G	S_5
5	4	6	3	4	—	120	S_4	G	S_5
6	24	8	8	14	—	2016	$PGL_2(7)$	$PGL_2(7)$	$S_3 \times PGL_2(7)$
24	**8**	8	8	**7**	—	43008	$PGL_2(7)$	G	$2^{1+6}_+:PGL_2(7)$
6	8	8	8	7	**2**	336	G	G	$PGL_2(7)$
8	12	12	12	13	**3**	11232	G	G	$L_3(3):2$
8	**12**	12	12	13	**3**	11232	G	G	$L_3(3):2$
8	**12**	20	20	31	**5**	744000	G	G	$L_3(5):2$
8	**12**	28	28	48	**7**	11261376	G	G	$SL_3(7):2$
11	11	12	12	11	**3**	6072	G	G	$L_2(23)$
11	**11**	12	12	11	**3**	6072	G	G	$L_2(23)$
12	**10**	12	12	21	**3**	15120	G	G	$3{\cdot}S_7$
12	10	12	12	**7**	**3**	5040	G	G	S_7
12	33	**5**	10	12	—	3960	$L_2(11)$	$L_2(11)$	$(3 \times L_2(11)):2$
12	**11**	**5**	10	12	—	1320	$L_2(11)$	$L_2(11)$	$PGL_2(11)$
14	**7**	**7**	14	12	—	2184	$L_2(13)$	$L_2(13)$	$PGL_2(13)$
30	10	10	**5**	6	—	14400	$2 \times A_5$	G	$2{\cdot}(A_5 \times A_5):2$
15	10	10	**5**	6	—	7200	$2 \times A_5$	G	$(A_5 \times A_5):2$
13	10	6	**6**	10	—	15600	$5^2:S_3$	G	$P\Sigma L_2(25)$
7	30	30	15	30	—	24360	G	G	$PGL_2(29)$
25	**8**	14	**7**	6	—	117600	$PGL_2(7)$	G	$P\Sigma L_2(49)$
82	**8**	10	**10**	10	—	1062720	$PGL_2(9)$	$L_2(81):2_2$	$2 \times L_2(81):2_2$
10	**10**	8	8	30	—	483840	$3{\cdot}PGL_2(9)$	$6{\cdot}L_3(4):2_3$	$6{\cdot}L_3(4):2_3 \times 2$
10	10	8	**8**	**5**	—	80640	$PGL_2(9)$	G	$2{\cdot}L_3(4):2_3$
20	**10**	8	**8**	**5**	—	161280	$PGL_2(9)$	G	$2^2{\cdot}L_3(4):2_3$
9	**12**	**10**	10	14	—	362880	S_7	G	$2 \times A_9$
9	**12**	10	10	**7**	—	181440	S_7	G	A_9
10	**11**	10	**10**	12	—	190080	$L_2(11)$	M_{12}	$M_{12}:2$
10	**13**	8	**8**	**12**	—	35942400	$L_3(3)$	$^2F_4(2)'$	$2 \times {}^2F_4(2)'$

(4) Verify the counting argument in Theorem 3.3 that $L_2(11)$ is *not* an image of the progenitor $2^{\star 3}:S_3$. A little care is needed here as the maximal $D_{12} \cong 2 \times S_3$ contains two non-conjugate copies of S_3; however, they are conjugate to one another in $PGL_2(11)$. You may assume the information given in the ATLAS that any maximal subgroup of $L_2(11)$ is isomorphic to one of $D_{12} \cong 2 \times S_3$, $11:5$ or A_5 (two classes). [Hint: Count the number of subgroups isomorphic to S_3 and the number of subgroups isomorphic to A_5 and deduce how many copies of A_5 a given copy of S_3 is contained in.]

Table 3.10. *Some finite images of the progenitor* $2^{\star 5}:A_5$

Parameters				Index of N in G	Order of G	Shape of $\langle t_0, t_1, t_2 \rangle$	Shape of $\langle t_0, t_1, t_2, t_3 \rangle$	Shape of G
a	b	c	d					
6	6	4	6	12	720	S_4	S_5	S_6
6	6	**4**	6	12	720	S_4	S_5	S_6
7	19	15	10	2926	175 560	G	G	J_1
8	8	14	8	1960	117 600	$2 \times L_2(7)$	G	$P\Sigma L_2(49)$
8	8	14	**8**	1960	117 600	$2 \times L_2(7)$	G	$P\Sigma L_2(49)$
12	12	**8**	8	333 312	19 998 720	$PGL_2(7)$	$2_+^{1+6}:PGL_2(7)$	$L_5(2):2$
13	13	**10**	6	78 000	4 680 000	$5^2:S_3$	$P\Sigma L_2(25)$	$S_4(5)$

Table 3.11. *Some finite images of the progenitor* $2^{\star 6}:L_2(5)$

Parameters							Order of G	Shape of $\langle t_\infty, t_0, t_1, t_2 \rangle$	Shape of $\langle \mathcal{T} \setminus \{t_4\} \rangle$	Shape of G
a	b	c	d	e	f	s				
4	10	10	12	12	12	3	1320	G	G	$PGL_2(11)$
8	8	10	10	8	8	**2**	2160	G	G	$3\cdot PGL_2(9)$
9	9	**5**	9	10	5	—	3420	G	G	$L_2(19)$
9	9	5	9	10	**5**	—	3420	G	G	$L_2(19)$
6	**6**	12	12	6	6	—	15120	$(3 \times A_5):2$	$3\cdot S_6$	$3\cdot S_7$
7	21	7	41	20	20	5	34 440	G	G	$L_2(41)$
8	8	**8**	8	6	**6**	—	190 080	G	G	$2 \times M_{12}$
6	**10**	8	10	8	8	—	241 920	G	G	$6\cdot L_3(4):2_3$

Table 3.12. *Some finite images of the progenitor* $2^{\star 7}:L_3(2)$

Parameters									Order of G	Shape of $\langle t_0, t_1, t_3 \rangle$	Shape of $\langle t_0, t_3, t_6, t_5 \rangle$	Shape of G
a	b	c	d	e	f	g	h	s				
12	6	8	8	8	6	8	7	**2**	12 096	$[2^5]$	$PGL_2(7)$	$U_3(3):2$
12	6	8	8	8	**6**	8	7	2	12 096	$[2^5]$	$PGL_2(7)$	$U_3(3):2$
6	6	10	12	6	12	12	21	3	15 120	$6^2:2$	G	$3\cdot S_7$
6	6	10	12	6	12	**4**	7	3	5040	$D_{12} \times 2$	G	S_7
8	12	12	6	12	10	4	4	—	40 320	S_4	S_5	S_8
12	12	10	12	6	12	**4**	**7**	**3**	322 560	$S_4 \times 2^2$	S_7	$2^6:S_7$
12	12	10	12	**6**	12	4	**7**	3	322 560	$S_4 \times 2^2$	S_7	$2^6:S_7$
10	12	**8**	**6**	12	10	12	8	—	5 322 240	$3:S_4$	$2^{1+4}:S_5$	$6\cdot M_{22}:2$
10	12	**8**	**6**	12	10	**4**	8	—	1 774 080	S_4	$2^{1+4}:S_5$	$2\cdot M_{22}:2$
10	12	**8**	6	12	**10**	**4**	8	—	1 774 080	S_4	$2^{1+4}:S_5$	$2\cdot M_{22}:2$
15	12	11	12	**6**	**11**	**4**	11	3	244 823 040	$S_4 \times 2^2$	$L_2(23)$	M_{24}
15	12	11	12	6	**11**	**4**	11	**3**	244 823 040	$S_4 \times 2^2$	$L_2(23)$	M_{24}

(5) The projective special linear group $L_2(q)$ for q odd will always have subgroups isomorphic to $D_{q\pm1}$, which will, in general, be maximal. If $q \cong \pm1 \pmod{24}$, then $L_2(q)$ will have subgroups isomorphic to S_4. Assuming these facts, prove that for such a q the group $L_2(q)$ is a homomorphic image of $2^{*4} : S_4$.

Let $q = 23$ and let

$$x \equiv \eta \mapsto \frac{10\eta - 1}{\eta + 10}; \; y \equiv \eta \mapsto \frac{6\eta + 9}{\eta - 6}; \; t_0 \equiv \eta \mapsto -\frac{1}{\eta}$$

be three elements of $L_2(23)$. Write these elements out explicitly as permutations of 24 letters and verify that $N = \langle x, y \rangle \cong S_4$. Verify further that $\mathcal{T} = |t_0^N| = 4$ and that N acts on $\mathcal{T}$ by conjugation as S_4 with $x \sim (0)(1\ 2\ 3)$ and $y \sim (0\ 1)$. Finally, verify that $(0\ 1\ 2\ 3)t_0$ has order 11 and $(2\ 3) = (t_0 t_1)^3$, and thus $L_2(23)$ is a homomorphic image of

$$\frac{2^{*4} : S_4}{[(0\ 1\ 2\ 3)t_0]^{11}, (2\ 3) = (t_0 t_1)^3}.$$

(Note: in fact this is an isomorphism, as stated in Table 3.9.)

(6) Similarly to the previous question, if $q \equiv \pm1 \pmod{20}$, then $L_2(q)$ contains maximal subgroups isomorphic to A_5. Prove that in this case $L_2(q)$ is a homomorphic image of $2^{*6} : L_2(5)$.

Let $q = 19$ and let

$$x \equiv \eta \mapsto \frac{4\eta - 1}{\eta + 4}; \; y \equiv \eta \mapsto \frac{\eta - 8}{\eta - 1}; \; t_\infty \equiv \eta \mapsto -\frac{1}{\eta}$$

be three elements of $L_2(19)$. Write these elements out explicitly as permutations of 20 letters and verify that $N = \langle x, y \rangle \cong A_5$. Verify further that $\mathcal{T} = |t_0^N| = 6$ and that N acts on $\mathcal{T}$ by conjugation as $L_2(5)$ with $x \sim (\infty)(0\ 1\ 2\ 3\ 4)$ and $y \sim (\infty\ 0)(1\ 4)$. Finally, verify that $xy \sim (\infty\ 0\ 4)(1\ 2\ 3)$ and that xyt_∞ has order 5, and so $L_2(19)$ is a homomorphic image of

$$\frac{2^{*6} : L_2(5)}{[(\infty\ 0\ 1)(2\ 4\ 3)t_\infty]^5}.$$

(Note: as above, Table 3.11 states that this is, in fact, an isomorphism.)

4

Classical examples

In Chapter 3 we introduced the notion of a symmetrically generated group and showed how we could enumerate double cosets so as to identify certain homomorphic images of the progenitor. In this chapter we exploit this technique to explore some familiar classical groups. In particular, we use the above techniques to construct some important combinatorial objects in a rather effortless manner and to verify a number of the exceptional isomorphisms of simple groups.

4.1   The group $\mathrm{PGL}_2(7)$

It was shown in Chapter 3 that the group G defined by

$$G \cong \frac{2^{\star 4} : \mathrm{S}_4}{(2\ 3)t_0 t_1 t_0 t_1}$$

appears to contain the subgroup $N \cong \mathrm{S}_4$ to index $(1+4+6+3) = 14$, where the four symmetric generators are labelled $\{0, 1, 2, 3\}$ here rather than $\{1, 2, 3, 4\}$ as previously. The Cayley graph of G over N appears to have the form illustrated in Figure 4.1.

This shows that $|G| \leq 24 \times 14 = 336$. In order to see that G does indeed have this order we consider its action on the 14 symbols representing cosets of N in G in Figure 4.1; in other words we embed G in the symmetric group on 14 letters. The 14 letters being permuted are labelled as follows:

$\star$	one
i	four such
ij	six such – unordered pairs
$ij.kl$	three such – the partitions of $\{0, 1, 2, 3\}$ into two pairs,

Figure 4.1. The 14-point action of G.

where $\{i, j, k, l\} = \{0, 1, 2, 3\}$. Permutations of $N \cong S_4$ act on these 14 symbols in the natural way. Moreover, the symmetric generator t_0 acts by right multiplication on the cosets to yield

$$t_0 = (\star\ 0)(1\ 01)(2\ 02)(3\ 03)(23\ 01.23)(31\ 02.31)(12\ 03.12),$$

which is preserved by permutations of 1, 2 and 3, and thus commutes with the subgroup of N fixing 0. So we see that t_0 has just four images under conjugation by N, and the permutation group we have constructed as a subgroup of S_{14} is an image of $2^{\star 4} : S_4$. But we may readily write down

$$t_0 t_1 = (\star\ 01)(0\ 1)(2\ 02.13\ 3\ 03.12)(02\ 12\ 03\ 13)(23)(01.23),$$

and so

$$(t_0 t_1)^2 = (\star)(0)(1)(2\ 3)(01)(23)(02\ 03)(12\ 13)(01.23)(02.13\ 03.12).$$

Since this element fixes the coset $\star = N$, it is a permutation in N. A glance at its action shows us that it is the transposition $(2\ 3)$ of N. Thus the subgroup, H say, of S_{14} generated by these images of N and the t_i is certainly a homomorphic image of G. But the action of t_0 on the 14 symbols shows that it fuses the four orbits of $N \cong S_4$, and so H acts transitively on the 14 letters. Furthermore, the group N fixes $\star$, and so $|G| \geq |H| \geq 14 \times 24 = 336$. This shows that G does indeed have order 336, and that the Cayley graph over N is as indicated in Figure 4.1, which is a bipartite graph, as shown in Figure 4.2.

We have chosen to show that the group G has the desired order without attempting to identify it, and this will prove a powerful technique when we wish to use this approach to construct and prove the existence of a sporadic group. It is, however, clear that the associated transitive graph is bipartite between two sets of cardinality 7 with valence 4. It is, of course, the graph of the 7-point projective plane $P_2(2)$ in which a point is joined to the four lines which do *not* pass through it.

Figure 4.2. Bipartite 14-point graph.

To complete the identification explicitly, we take

$$\mathcal{P} = \{\star, ij \mid i, j \in \{0, 1, 2, 3\}, i \neq j\}$$

and

$$\mathcal{L} = \{i, ij.kl \mid \{i, j, k, l\} = \{0, 1, 2, 3\}\}$$

as points and lines, respectively, when we see that $\star$ joins i, and ij joins $\{i, j, ik.jl, il.jk\}$, where i, j, k and l are distinct. So the three lines which *pass through* $\star$ are

$$\{01.23, 02.31, 03.12\},$$

and the three lines through ij are $\{k, l, ij.kl\}$; see Figure 4.3. In the classical construction of this plane, we take a 3-dimensional vector space V over $\mathbb{Z}_2$, the field of order 2. The points then correspond to the 1-dimensional subspaces, which in this case can be represented by the unique non-zero (row) vector each contains. The lines are the 2-dimensional subspaces, which may be denoted by homogeneous linear expressions, such as $x + y + z = 0$, or more conveniently as (column) vectors. A point $\mathbf{u} = (x, y, z)$ lies on a line $\mathbf{v}^t = (a, b, c)^t$ if, and only if,

$$\mathbf{u}\mathbf{v}^t = (x, y, z) \begin{pmatrix} a \\ b \\ c \end{pmatrix} = ax + by + cz = 0,$$

that is to say if the corresponding vectors are orthogonal. The group $\mathrm{L}_3(2)$ of all non-singular 3×3 matrices over $\mathbb{Z}_2$ has order equal to the number of ordered bases of V and is thus $(2^3 - 1) \times (2^3 - 2) \times (2^3 - 2^2) = 168$.

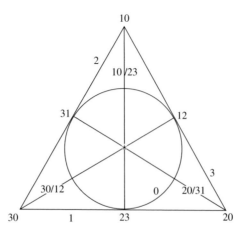

Figure 4.3. The Fano plane.

A particular matrix A permutes the points by right multiplication, and the lines by left multiplication by A^{-1}, and the equation

$$\mathbf{u}\mathbf{v}^t = \mathbf{u}A.A^{-1}\mathbf{v}^t$$

shows that such a transformation preserves incidence between points and lines. The duality between points and lines may be achieved by the *polarity* which simply transposes row vectors, representing points, into column vectors, representing lines, and vice versa. The effect of this polarity on the matrices is to map a matrix to the inverse of its transpose. This map, α say, is clearly a bijection, and we have

$$(AB)^\alpha = ((AB)^{-1})^t = (B^{-1}A^{-1})^t = (A^{-1})^t(B^{-1})^t = A^\alpha B^\alpha;$$

thus α defines an automorphism of the group $L_3(2)$. The fact that A^α has the required effect on the transposed lines is shown by the following equation:

$$\mathbf{u}AA^{-1}\mathbf{v}^t = (\mathbf{u}AA^{-1}\mathbf{v}^t)^t = \mathbf{v}(A^{-1})^tA^t\mathbf{u}^t = \mathbf{v}A^\alpha(A^\alpha)^{-1}\mathbf{u}^t.$$

In Table 4.1 we show the correspondence between the points and lines of $P_2(2)$ and the 14 cosets in our enumeration. They are arranged to display the action of the symmetric generator t_0, which visibly corresponds to the polarity α. The control subgroup $N \cong S_4$ corresponds to the $4 \times 3 \times 2$ matrices whose columns are distinct and drawn from the top four lines of Table 4.1. In particular, the permutation $(2\ 3)$ of N is represented by the following matrix:

$$(2\ 3) \sim \begin{pmatrix} 1 & 0 & 0 \\ 0 & 0 & 1 \\ 0 & 1 & 0 \end{pmatrix} = A_{23}, \text{ say, and } (0\ 1) \sim \begin{pmatrix} 1 & 0 & 0 \\ 1 & 1 & 0 \\ 1 & 0 & 1 \end{pmatrix} = A_{01}.$$

We can now check that the additional relation, namely $(t_0 t_1)^2 = (2\ 3)$, holds in $\operatorname{Aut} L_3(2)$:

$$(t_0 t_1)^2 = (t_0 t_0^{(0\ 1)})^2 = (\alpha A_{01} \alpha A_{01})^2$$

$$= [(A_{01}^{-1})^t A_{01}]^2 = [A_{01}^t A_{01}]^2$$

$$= \left[\begin{pmatrix} 1 & 1 & 1 \\ 0 & 1 & 0 \\ 0 & 0 & 1 \end{pmatrix}\begin{pmatrix} 1 & 0 & 0 \\ 1 & 1 & 0 \\ 1 & 0 & 1 \end{pmatrix}\right]^2$$

$$= \begin{pmatrix} 1 & 1 & 1 \\ 1 & 1 & 0 \\ 1 & 0 & 1 \end{pmatrix}^2 = \begin{pmatrix} 1 & 0 & 0 \\ 0 & 0 & 1 \\ 0 & 1 & 0 \end{pmatrix}$$

$$= A_{23} = (2\ 3),$$

as required.

Now, $\operatorname{PGL}_2(7)$, the projective general linear group of degree 2 over the field $\mathbb{Z}_7$, also has order $(7^2 - 1)(7^2 - 7)/(7 - 1) = 8 \times 7 \times 6 = 336$, and is well

Table 4.1. Correspondence with $P_2(2)$

$\mathcal{P}$			$\mathcal{L}$
(1 1 1)	$\star$	(1 1 1)^{t}	0
(1 0 0)	01	(1 0 0)^{t}	1
(0 1 0)	02	(0 1 0)^{t}	2
(0 0 1)	03	(0 0 1)^{t}	3
(0 1 1)	23	(0 1 1)^{t}	01/23
(1 0 1)	31	(1 0 1)^{t}	02/31
(1 1 0)	12	(1 1 0)^{t}	03/12
	$\longleftrightarrow$		

known to be isomorphic to the group $G \cong \mathrm{Aut}\, \mathrm{L}_3(2) \cong \mathrm{L}_3(2) : 2$. Indeed, it is clear how the above definition of G can be used to prove this important fact. Before doing so, we introduce some notation and terminology.

Let $\mathbb{F}$ denote a finite field of order $q = p^m$, where p is a prime number – the *characteristic* of the field – and let $V \cong \mathbb{F}^2$ be a 2-dimensional vector space over F. Then V possesses $q^2 - 1$ non-zero vectors, and therefore $q^2 - 1/q - 1 = q + 1$ 1-dimensional subspaces. These subspaces are permuted by any non-singular linear transformation acting on V. If such a 1-dimensional subspace contains the non-zero vector (x_0, y_0), then we have

$$\langle (x_0, y_0) \rangle = \{ (\mu x_0, \mu y_0) \mid \mu \in \mathbb{F} \},$$

and so the subspace is determined by the ratio of the x-coordinate to the y-coordinate of its non-zero vectors. If $y_0 \neq 0$, this is an element $\lambda \in \mathbb{F}$, and so this 1-dimensional subspace may conveniently be labelled $[\lambda]$ or simply λ. If $y_0 = 0$, we label the subspace ∞.

DEFINITION 4.1 The *projective line* $P_1(q)$ is defined to be the set $\{\infty\} \cup \mathbb{F}_q$, denoting the set of 1-dimensional subspaces of a 2-dimensional vector space over the field of order q.

The action of the matrix

$$A = \begin{pmatrix} a & c \\ b & d \end{pmatrix}$$

on these symbols representing 1-dimensional subspaces (see Exercises 4.1) is to map

$$\begin{pmatrix} a & c \\ b & d \end{pmatrix} : \infty \mapsto \frac{a}{c}; \; -\frac{d}{c} \mapsto \infty; \; \lambda \mapsto \frac{a\lambda + b}{c\lambda + d}$$

if $c \neq 0$. If $c = 0$, then it fixes ∞. With the (natural) conventions that $a\infty = \infty$ for $a \neq 0$, $\infty + a = \infty$, $1/\infty = 0$ and $a\infty/b\infty = a/b$, we see the action on the projective line is described completely by the *linear fractional map*:

$$\tau \mapsto \frac{a\tau + b}{c\tau + d}.$$

EXERCISE 4.1

(1) The function which maps the matrix

$$\begin{pmatrix} a & c \\ b & d \end{pmatrix}$$

to the linear fractional map

$$\tau \mapsto \frac{a\tau + b}{c\tau + d}$$

is a homomorphism from the general linear group $GL_2(q)$ into the symmetric group S_{q+1}. Show that the kernel of this function consists of the (non-singular) scalar matrices and so has order $q-1$. The image group is thus isomorphic to

$$GL_2(q)/Z(GL_2(q)) \cong PGL_2(q),$$

the *projective* general linear group.

(2) Let $G \cong PGL_2(q)$ and let G_∞ denote the stabilizer in G of ∞. Show that

$$G_\infty = \{\tau \mapsto a\tau + b \mid a, b \in \mathbb{F}, a \neq 0\},$$

of order $q(q-1)$. If $G_{\infty 0}$ denotes the stabilizer of both ∞ and 0, then show that

$$G_{\infty 0} = \{\tau \mapsto a\tau \mid a \neq 0\},$$

of order q. Show that $G_{\infty 0}$ acts transitively on $\mathbb{F} \setminus \{0\}$, and that $G_{\infty 0 1} = 1$, the trivial subgroup. Deduce that $G \cong PGL_2(q)$ is a *sharply* 3-transitive subgroup of S_{q+1}; that is to say G is 3-transitive and the only element fixing three points is the identity. So, $|G| = (q+1)q(q-1)$.

(3) Let k be a primitive element in the field $\mathbb{F}$, of order q; that is to say, k is a generator for the multiplicative group of $\mathbb{F}$ which is cyclic of order $q-1$. Thus $\langle k \rangle \cong C_{q-1}$. Use this to show that if q is odd then $PGL_2(q)$ possesses odd permutations and so has a normal subgroup of index 2. Show that the element $\gamma: \tau \mapsto -1/\tau$ is an even permutation if $q > 2$, no matter the parity of q, and hence, or otherwise, show that $PGL_2(2^m) \leq A_{q+1}$ for $m \geq 2$. [Hint: Consider the number of points fixed by γ; you will need to decide when -1 is a square and when a non-square.] (The normal subgroup of index 2 in the case q odd is the projective *special* linear group $PSL_2(q)$, which consists of all linear fractional maps which have $ad - bc = 1$ (or equivalently, modulo the scalar matrices, $ad - bc$ a square in $\mathbb{F}$). If q is even we have $PSL_2(2^m) = PGL_2(2^m)$ since in this case every element of $\mathbb{F}$ is a square. The group $PSL_2(q)$ is always simple for $q \geq 4$.)

(4) Define elements α, β and γ of $\mathrm{PGL}_2(q)$ as follows:

$$\alpha := \tau \mapsto \tau + 1; \, \beta := \tau \mapsto k\tau; \, \gamma := \tau \mapsto -1/\tau,$$

where k is a primitive element of multiplicative group of $\mathbb{F}$ as above. Show that $\langle \alpha, \beta, \gamma \rangle = \mathrm{PGL}_2(q)$. Work out these elements explicitly in the cases $q = 4, 5$ and 7.

Show that, for all values of q, G_∞ possesses a normal subgroup of order q.   □

Now consider $L \cong \mathrm{PGL}_2(7)$ as the set of all linear fractional maps acting on the 8-point projective line $P_1(7) = \{\infty, 0, 1, \ldots, 6\}$. That is,

$$L = \left\{ \tau \mapsto \frac{a\tau + b}{c\tau + d} \mid a, b \in \mathbb{Z}_7, ad - bc \neq 0 \right\}.$$

Now let $\lambda := \infty 365/0124$ be a partition of $P_1(7)$ into two sets of size 4, and let L_λ denote the stabilizer of λ in L. Certainly λ is preserved by

$$\gamma : \tau \mapsto -\frac{1}{\tau} \equiv (\infty\ 0)(1\ 6)(2\ 3)(4\ 5)$$

and

$$\delta : \tau \mapsto \frac{3\tau + 2}{\tau + 2} \equiv (\infty\ 3\ 5)(6)(0\ 1\ 4)(2),$$

and we readily check that $\gamma\delta$ has order 4 and so $\langle \gamma, \delta \rangle$ is an image of $\langle x, y \mid x^2 = y^3 = (xy)^4 = 1 \rangle$ containing elements of order 4; thus $N = \langle \gamma, \delta \rangle \cong S_4$. It is well known and readily proved in this case (see ref. [40]) that such a subgroup is maximal in L, and so $L_\lambda = N$. Now, the element

$$t_0 : \tau \mapsto -\tau \equiv (1\ 6)(2\ 5)(3\ 4)$$

commutes with γ and $\beta : \tau \mapsto 2\tau \equiv (1\ 2\ 4)(3\ 6\ 5)$, which together generate S_3, and so t_0 has just four images under conjugation by N, Indeed, we have

$$\begin{aligned} t_0 &\equiv (1\ 6)(2\ 5)(3\ 4); \\ t_0^\delta = t_1 &\equiv (4\ 6)(2\ \infty)(5\ 0); \\ t_1^\delta = t_2 &\equiv (0\ 6)(2\ 3)(\infty\ 1); \\ t_1^\gamma = t_3 &\equiv (5\ 1)(3\ 0)(4\ \infty). \end{aligned}$$

Now, $(t_0 t_1)^2 = (\infty\ 0)(1\ 3)(2\ 5)(4\ 6)$ clearly lies in L_λ and, by conjugation, acts $(t_0)(t_1)(t_2\ t_3)$ on the t_i. That is to say, it is the transposition $(2\ 3)$ of N. Similarly all the transpositions of $N \cong S_4$ can be written as $(t_i t_j)^2$, and so

$$\langle t_0, t_1, t_2, t_3 \rangle = \langle t_0, t_1, t_2, t_3, N \rangle = L,$$

since $t_0 \notin N$ and N is maximal. Since G and L have the same order, we have that

$$G \cong \frac{2^{\cdot 4} : S_4}{(2\ 3)t_0 t_1 t_0 t_1} \cong L_3(2) : 2 \cong \mathrm{PGL}_2(7).$$

EXERCISE 4.2

(1) Work out the elements $(0\ 1\ 2)t_0$ and $[(0\ 1\ 2), t_0]$ as permutations on the 14 cosets of N, and show that they have orders 8 and 4, respectively. Hence show that elements of G satisfy the following well known presentation:

$$\langle x, y \mid x^2 = y^3 = (xy)^8 = [x, y]^4 = 1 \rangle$$

for $\mathrm{PGL}_2(7)$. Show that $\langle (0\ 1\ 2), t_0 \rangle = G$ and deduce that G is a homomorphic image of $\mathrm{PGL}_2(7)$.

(2) Every element in the above group G of order 336 can be written as πw, where $\pi \in S_4$ and $l(w) \leq 3$. In the case when $l(w) \leq 2$, find a necessary and sufficient condition on π and w for πw to be an involution. From the Cayley diagram we see that the element $t_0 t_1 t_2$ has eight names of the form $\pi t_i t_j t_k$. Find them, and hence find all involutions in G under conjugation by $N \cong S_4$. Finally, show that these involutions fall into two conjugacy classes in G and find their class lengths.

(3) Confirm the classical presentation

$$\langle x, y \mid x^2 = y^3 = (xy)^7 = [x, y]^4 = 1 \rangle$$

for $L_2(7)$ as follows:

Step 1 Verify that $x = (0\ 1)(2\ 5)$ and $y = (1\ 2\ 4)(3\ 6\ 5)$ satisfy the presentation.

Step 2 Denote x, x^y and x^{y^2} by s_0, s_1 and s_2, respectively, y by $(0\ 1\ 2)$, and confirm that $\langle x, y \rangle = \langle s_0, s_1, s_2 \rangle$.

Step 3 Deduce from $(xy)^7 = 1$ that $(0\ 1\ 2) = s_0 s_1 s_2 s_0 s_1 s_2 s_0$ and deduce from $[x, y]^4 = 1$ that $s_i s_j$ has order 4 for $i \neq j$. Hence observe that $s_1 s_2 s_0 s_1 s_2$ has order 3 and that, since $s_1 s_2 s_0 s_1 = (s_2 s_0)^{s_1}$ has order 4, we have the product of an element of order 4 and an element of order 2 whose product has order 3. (Note that all orders are as asserted by Step 1.) Thus $H = \langle s_1 s_2 s_0 s_1, s_2 \rangle \cong S_4$.

Step 4 Construct by hand the Cayley diagram of $G = \langle s_0, s_1, s_2 \rangle$ over H, using the relations $(s_i s_j)^4 = 1$ and $s_i s_j s_k s_i s_j s_k s_i = s_j s_k s_i s_j s_k s_i s_j$, to obtain the diagram shown in Figure 4.4.

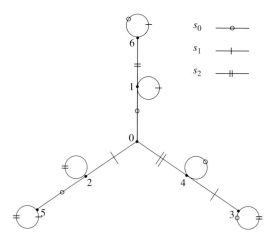

Figure 4.4. Cayley diagram of $L_3(2)$ over S_4 with respect to three symmetric generators.

4.2   Exceptional behaviour of S_n

4.2.1   An exceptional outer automorphism

The first family of finite groups a student meets is invariably, and for very good reasons, the family of symmetric groups. Not only are the basic group theoretic operations of multiplication and inversion carried out with immense ease within the symmetric group S_n, but conjugation of one element by another is almost as straightforward. The order of an element can be read off from its cycle lengths, and one can tell at a glance whether or not two elements are conjugate. Nonetheless, these groups in which we feel so comfortable working increase in size very rapidly, with even the unintimidating S_{12} having order over 400 million.

So familiar do we feel with the full symmetric group that it comes as something of a shock when particular members of the family exhibit exceptional behaviour. Moreover, S_n appears so complete in itself that the idea that it could possess an *outer* automorphism seems counter-intuitive. Thus the unease we feel at learning that the group S_6 possesses an outer automorphism seems somehow vindicated by the further knowledge that this automorphism can be used to construct the amazing Mathieu group M_{12}, whose own outer automorphism can in turn be used to construct M_{24}. These constructions were presented in the 1960s by G. Higman and J. A. Todd in postgraduate lecture courses at Oxford and Cambridge, respectively. In this section we show how symmetric generation can be used to construct the full automorphism group of A_6 in a rather revealing manner.

Background

Group theoretically, the outer automorphism of S_6 owes its existence to the fact that the projective general linear group $PGL_2(5)$ is isomorphic to S_5. Now, $PGL_2(5)$ consists of the following set of linear fractional maps:

$$PGL_2(5) = \left\{ x \mapsto \frac{ax+b}{cx+d} \;\middle|\; a, b, c, d \in \mathbb{Z}_5, ad - bc \neq 0 \right\},$$

which acts triply transitively on the six points of the projective line $P_1(5) = \{\infty, 0, \ldots, 4\}$. Thus S_6 possesses two sets of isomorphic but non-conjugate subgroups of index 6: one consists of the stabilizers of the six letters on which our S_6 acts naturally, and the other consists of six copies of $PGL_2(5)$ acting triply transitively on those letters. The outer automorphism interchanges these two sets.

Combinatorially, an outer automorphism can exist because the number of unordered pairs of six letters is equal to the number of ways in which six letters can be partitioned into three pairs; which is to say that the two conjugacy classes of odd permutations of order 2 in S_6 contain the same number of elements, namely 15. Sylvester [62] refers to the unordered pairs as *duads* and the partitions as *synthemes*. Certain collections of five synthemes, which will reveal themselves spontaneously below, he refers to as *synthematic totals* or simply *totals*; each total is stabilized within S_6 by a subgroup acting triply transitively on the six letters as $PGL_2(5)$ acts on the projective line; see Definition 4.1. If we draw a bipartite graph on $(15 + 15)$ vertices by joining each syntheme to the three duads it contains, we obtain the famous *8-cage* (a graph of valence 3 with minimal cycles of length 8); see Figure 4.5. For the interested reader we remark that this graph may be extended by adjoining six vertices corresponding to the original letters, which in Sylvester's nomenclature are *monads*, and six vertices corresponding to the as yet undefined totals. If we then join each duad to the two monads it contains, and each total to the five synthemes it contains, we obtain the bipartite graph on $(21 + 21)$ vertices indicated in Figure 4.6. This, in fact, consists of the points and lines of the projective plane of order 4, and its group of symmetries is isomorphic to the full automorphism group of the projective special linear group $L_3(4)$ with shape $L_3(4) : (S_3 \times 2)$.

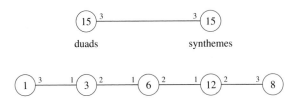

Figure 4.5. The 8-cage on duads and synthemes.

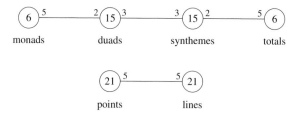

Figure 4.6. The projective plane of order 4.

Now, the above remarks can be sharpened to prove the existence of an outer automorphism of S_6, and can, with considerably more work, be used to construct it. Perhaps the best place to see the full group $S_6 : 2$, however, is in the Mathieu group M_{12}. Recall that M_{12} is a permutation group on 12 letters which preserves a Steiner system $S(5, 6, 12)$. That is to say a collection of 6-element subsets of the 12 letters, known as *special hexads*, which have the property that any five of the 12 letters lie together in precisely one of them. It turns out that up to relabelling there is only one such system, and that in it the complement of a special hexad is another special hexad. The stabilizer of such a pair of complementary hexads is isomorphic to $S_6 : 2$, and S_6 can be viewed acting simultaneously on the two hexads in non-identical fashion. This, however, requires familiarity with a much more complicated configuration than the one we are attempting to describe. In what follows we shall describe an elementary, new, direct construction which is easy to work with and facilitates the identification $S_6 : 2 \cong P\Gamma L_2(9)$, the projective general linear group in 2 dimensions over the field of order 9 extended by the field automorphism.

Automorphisms of S_n

Let G be a group, let $x \in G$ and let $x^G = \{g^{-1}xg \mid g \in G\}$ be the conjugacy class of G containing x. Then if $\alpha \in \mathrm{Aut}\, G$, the automorphism group of G, we have $(x^G)^\alpha = (x^\alpha)^G$, the conjugacy class of G containing x^α. Now, any automorphism of S_n must preserve the parity of a permutation and so must fix the class of transpositions or take it to another class of odd involutions. If we draw a graph Γ_n with vertices corresponding to the $\binom{n}{2}$ transpositions, joined if they do not commute, we soon verify that the full group of symmetries of this graph is realized by S_n. (By induction on n: if $(1,2)$ and $(1,3)$ are fixed by α then so is $(1, 2)^{(1,3)} = (2, 3)$. The subgraph of vertices joined to none of these three is then Γ_{n-3} with group of automorphisms isomorphic to S_{n-3}. From here one may readily argue that $|\mathrm{Aut}\,\Gamma_n| = \binom{n}{2} \times 2(n-2) \times (n-3)! = n!$) But the transpositions generate S_n, and so the only automorphism which fixes every transposition is the identity. Thus, all automorphisms of S_n are realized by inner automorphisms unless S_n contains a second class of odd involutions with $\binom{n}{2}$ members.

Now, the number of elements of cycle shape $2^m.1^{n-2m}$ is $n!/2^m m!(n-2m)!$ and, on simplifying, the condition that this should equal $\binom{n}{2}$ reduces to the following:

$$\frac{2^{m-1}}{(m-2)!} = \binom{n-2}{m-2}\binom{n-m}{m} \in \mathbb{Z},$$

with $m \geq 3$ odd. Clearly this forces $m = 3$ when it follows that $n = 6$. Thus the only symmetric group which could possess an outer automorphism is S_6.

The progenitor for $\mathbf{Aut\ A_6 \cong S_6 : 2}$

In this section we are interested in a particular image of the progenitor $P = 2^{*6} : \mathrm{PGL}_2(5)$. Thus the symmetric generators are labelled t_i for i in $P_1(5)$, the projective line $\infty \cup \mathbb{Z}_5$.

We immediately have, as a slight extension of the proof of Theorem 3.1, the following lemma.

LEMMA 4.1 If $P = 2^{*6} : \mathrm{PGL}_2(5)$, then $P/P' \cong V_4$, the Klein fourgroup, and $P'' = P'$.

Proof We have $N \cong \mathrm{PGL}_2(5)$, and so

$$\begin{aligned}
P' &\geq \langle N' \cong \mathrm{PSL}_2(5), [(\infty\ 0)(1\ 4), t_\infty] = t_0 t_\infty \rangle \\
&= \langle N', t_i t_j \mid i, j \in \{\infty, 0, \dots, 4\} \rangle \\
&= \{\pi w \mid \pi \text{ even}, l(w) \in 2\mathbb{Z}\} = K,
\end{aligned}$$

say, since N' is doubly transitive. But $P/K = \langle Kt_\infty, K(\infty\ 0)(1\ 2)(3\ 4) \rangle \cong V_4$. Thus, $P' \leq K \leq P'$, and so $P' = K$ and $P/P' \cong V_4$.

Similarly,

$$P'' \geq \langle N' \cong \mathrm{PSL}_2(5), [(1\ 4)(2\ 3), t_\infty t_1] = t_4 t_\infty t_\infty t_1 = t_4 t_1 \rangle = P'. \qquad \square$$

It follows that any image of P contains a perfect subgroup to index less than or equal to 4. The group we wish to study is defined to be

$$G = \frac{2^{*6} : \mathrm{PGL}_2(5)}{[(\infty\ 0\ 1\ 4\ 2\ 3)t_\infty]^4}, \tag{4.1}$$

and we may immediately note that, since the relator lies in P', we have $G/G' \cong V_4$ and $G'' = G'$. If we let $\pi = (\infty\ 0\ 1\ 4\ 2\ 3)$, we may rewrite our additional relation:

$$\begin{aligned}
(\pi t_\infty)^4 &= \pi^4 t_\infty^{\pi^3} t_\infty^{\pi^2} t_\infty^\pi t_\infty \\
&= (\infty\ 2\ 1\ )(0\ 3\ 4) t_4 t_1 t_0 t_\infty = 1.
\end{aligned}$$

In order to see that G is, in fact, finite and to find its order, we perform a manual double coset enumeration of G over N in the usual manner. Now, the number of single cosets Nx in the double coset NwN is given by the index $|N:N^{(w)}|$, where we define

$$N^{(w)} = \{\pi \in N \mid Nw\pi = Nw\} = \{\pi \in N \mid w\pi w^{-1} \in N\}$$

$$= \{\pi \in N \mid \pi \in w^{-1}Nw\} = N \cap w^{-1}Nw.$$

In particular, this means that

$$|N^{(w)}| = |N \cap w^{-1}Nw| = |wNw^{-1} \cap N| = |N^{(w^{-1})}|, \tag{4.2}$$

a fact which we shall use later. Besides the trivial double coset N, which we label $[\star]$, we have

$$[\infty] = Nt_\infty N = \bigcup_{i \in \{\infty,0,\dots,4\}} Nt_i,$$

in which the six single cosets appear to be distinct, and

$$[\infty 0] = Nt_\infty t_0 N = \bigcup_{\substack{i,j \in \{\infty,0,\dots,4\} \\ i \neq j}} Nt_i t_j.$$

Since $t_4 t_1 t_0 t_\infty \in N$, from our additional relation, we see that $Nt_4 t_1 t_0 t_\infty = N$ and so $Nt_4 t_1 = Nt_\infty t_0 = Nt_\infty t_0 (\infty\ 4\ 0\ 1)$. Thus,

$$N^{(\infty 0)} = \{\pi \in N \mid Nt_\infty t_0 \pi = Nt_\infty t_0\} \geq \langle (1\ 2\ 4\ 3), (\infty\ 4\ 0\ 1) \rangle \cong S_4,$$

acting transitively on the six letters. This says that

$$Nt_\infty t_0 = Nt_4 t_1 = Nt_3 t_2 = Nt_0 t_\infty = Nt_1 t_4 = Nt_3 t_2,$$

and so the coset $Nt_\infty t_0$ can be labelled by the syntheme $\infty 0.14.23$, and the double coset $[\infty 0]$ contains at most $|N|/24 = 5$ single cosets, which may be labelled by the special synthemes of Table 4.2. The transitivity of $N^{(\infty 0)}$ ensures that for any i, j, k we have $Nt_i t_j t_k = Nt_l t_k t_k = Nt_l$ for some l, and so $[\star]$, $[\infty]$ and $[\infty 0]$ are the only three double cosets. If we draw a graph with vertices the 12 cosets obtained, and coset Nw joined to the six cosets Nwt_i, we obtain the Cayley graph indicated in Figure 4.7 *provided there is no further collapsing.* It visibly takes the form of a complete bipartite graph $K_{6,6}$. Note that each coset in the 5-orbit has six 'names', one of which has the form ∞i for $i \in \{0, 1, \dots, 4\}$; the alternative names can be read from Table 4.2. Right multiplication by t_∞ induces a permutation on these 12 cosets, which can easily be written down using Table 4.2:

$$t_\infty : (\star\ \infty)(0\ \infty 0)(1\ \infty 1)(2\ \infty 2)(3\ \infty 3)(4\ \infty 4). \tag{4.3}$$

We know how $N \cong PGL_2(5)$ acts on these 12 cosets, however, and it is obvious that t_∞ above commutes with the stabilizer of ∞ in N. Thus the permutation group on 12 letters which we have constructed is certainly a homomorphic image of the progenitor. To see that it is in fact G, it remains to check that the additional relation holds. But

$$(\infty\ 0\ 1\ 4\ 2\ 3)t_\infty : (\star\ \infty\ \ \infty 0\ 3)(\infty 1\ 2\ \ \infty 3\ 0)(\infty 2\ 4)(\infty 4\ 1),$$

Table 4.2. The synthematic
total preserved by $\mathrm{PGL}_2(5)$

$$\infty0.14.23$$
$$\infty1.20.34$$
$$\infty2.31.40$$
$$\infty3.42.01$$
$$\infty4.03.12$$

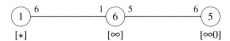

Figure 4.7. The Cayley diagram of G over the cosets of N.

which clearly has order 4. Note that it is repeated use of Table 4.2 that enables us to write this down so easily; for example,

$$(\infty\ 0\ 1\ 4\ 2\ 3)t_\infty : \infty0 \mapsto 01\infty \mapsto 3\infty\infty = 3.$$

Thus we have constructed a group G of order $120 \times 12 = 2 \times 6!$, which acts transitively on 12 letters, but imprimitively with two blocks of size 6, namely

$$\{\infty, 0, \ldots, 4\} \cup \{\star, \infty0, \ldots, \infty4\}. \tag{4.4}$$

The subgroup of index 2 fixing these two blocks has order 6! and consists of the following elements:

$$\{\pi, \pi t_i t_\infty \mid \pi \in N, i \in \{0, 1, \ldots, 4\}\}.$$

Since the permutation actions on these two sets of size 6 are clearly faithful, the block stabilizer is a permutation group on six letters of order 6! and is thus isomorphic to S_6. We observe, furthermore, that the permutation $(\infty\ 0)(1\ 2)(3\ 4)$ of N has action $(\star)(\infty0)(\infty1)(\infty2\ \infty3)(\infty4)$ on the other block. So permutations of cycle shape 2^3 on one side act as transpositions on the other. Similarly, the permutation $(\infty\ 0\ 1)(2\ 4\ 3)$ of N has action $(\star)(\infty0\ \ \infty3\ \ \infty1)(\infty2)(\infty4)$ on the other side; thus, elements of order 3 have cycle shape 3^2 on one side and 3.1^3 on the other. From these two observations it follows that 6-cycles on one side must have cycle shape 1.2.3 on the other, and the reader may like to use Table 4.2 to confirm this for the permutation $(\infty\ 0\ 1\ 4\ 2\ 3)$ of N. Thus the permutation of the 12 cosets achieved by right multiplication by the symmetric generator t_∞, say, interchanges the two blocks and so swaps the two classes of odd involutions, the two classes of elements of order 3, and the two classes of elements of order 6.

The notation for the $(6+6)$ letters permuted by Aut S_6 given in Equation (4.4) is much shorter than the somewhat cumbersome synthematic

Table 4.3. Correspondence between our notation
and the synthematic totals

Label	Total	Abbreviation
★	$\begin{cases} \infty0.14.23 \\ \infty1.20.34 \\ \infty2.31.40 \\ \infty3.42.01 \\ \infty4.03.12 \end{cases}$	$(\infty \mid 01234)$
$\infty0 \sim \infty0.14.23$	$\begin{cases} \infty0.14.32 \\ \infty1.30.24 \\ \infty3.21.40 \\ \infty2.43.01 \\ \infty4.02.13 \end{cases}$	$(\infty \mid 01324)$

totals. However, it is of interest to see how the two notations correspond
to one another. Now the coset which we label ★ clearly corresponds to the
synthematic total in Table 4.2 and, since this is preserved by our subgroup $N \cong \mathrm{PGL}_2(5)$, it possesses just six images under the action of S$_6$. In
Figure 4.6 we see that each syntheme is contained in just two totals. Our
notation ∞i for points in the second block is really an abbreviation for a
syntheme, as in Table 4.2. Thus $\infty0$ is an abbreviation for $\infty0.14.23$, which
gives its six possible names. The synthematic total corresponding to each
of the synthemes in Table 4.2 is thus the unique *other* synthematic total
containing it.

In Table 4.3 the other correspondences are obtained by conjugation by
the permutation (0 1 2 3 4). The abbreviations given in the third column
are simply a mnemonic for the synthematic totals.

EXERCISE 4.3

(1) Confirm that the permutation $(\infty\ 0\ 1\ 4\ 2\ 3) \in N \cong \mathrm{PGL}_2(5)$ acts as a
permutation of cycle shape 1.2.3 on the other six points.

(2) Confirm that $(\infty\ 0)(1\ 2)(3\ 4)t_\infty t_0 = (\infty\ 0)(1\ 2)(3\ 4)^{t_0}$ acts as
a transposition on $\{\infty, 0, 1, 2, 3, 4\}$ and has cycle shape 2^3 on
$\{\star, \infty0, \infty1, \infty2, \infty3, \infty4\}$.

(3) Work out the action of the element $(1\ 3\ 4\ 2)t_\infty t_0$ on the 12 points and
show that it is a transposition on one side and has cycle shape 2^3 on
the other. (Note that this implies that the relation $(1\ 4)(2\ 3) = (t_\infty t_0)^2$
holds in G.) Hence write down all (15+15) transpositions in canonical
form as π or $\pi t_\infty t_i$.

(4*) Perform a double coset enumeration for the group

$$G = \frac{2^{\star 6} : \mathrm{PGL}_2(5)}{(t_\infty t_0)^2 = (1\ 4)(2\ 3)}.$$

The isomorphism $S_6 : 2 = \mathrm{P\Gamma L}_2(9)$

The group $\mathrm{PGL}_2(9)$ acts triply transitively on the ten points of the projective line $P_1(9) = \{\infty, 0, \pm 1, \pm i, \ \pm 1 \pm i \mid i^2 = -1, \ 1+1+1 = 0\}$ with order $10 \times 9 \times 8$. Thus the group $\mathrm{P\Gamma L}_2(9)$, obtained by extending this group by the *field automorphism* which interchanges i and $-$i, has order $2 \times 6! = |\mathrm{Aut}\, A_6|$.

We now prove the following theorem.

THEOREM 4.1 $S_6 : 2 \cong \mathrm{P\Gamma L}_2(9)$.

Proof Note that all non-trivial normal subgroups of the group $G \cong A_6 . 2^2$ defined in Equation (4.1) contain A_6, and so any homomorphic image of G containing elements of order 5 must be an isomorphic copy. Thus, if we construct such a homomorphic image of G in $\mathrm{P\Gamma L}_2(9)$, the fact that the two groups $\mathrm{Aut}\, A_6$ and $\mathrm{P\Gamma L}_2(9)$ have the same order will suffice to prove that they are isomorphic. To achieve this we take

$$\eta \equiv x \mapsto \frac{(1-i)x + 1 + i}{x + 1 - i} : (\infty, \ 1-i, \ -1, \ 1, \ -1+i)(0, \ i, \ -1-i, \ 1+i, \ -i)$$

and let σ denote the field automorphism, which acts by 'complex conjugation'. Then it is easily checked that η and σ satisfy the following presentation:

$$\langle \eta, \sigma \mid \eta^5 = \sigma^2 = (\eta\sigma)^4 = [\eta, \sigma]^3 = 1 \rangle,$$

which is a classical presentation for $S_5 \cong \mathrm{PGL}_2(5)$; see ref. [27]. The element

$$t_\infty \equiv x \mapsto \frac{1+i}{x}$$

commutes with a subgroup of $N = \langle \eta, \sigma \rangle$ of order 20, which is generated by η and

$$\nu = (\sigma\eta)^{\eta\sigma\eta} \equiv \sigma\left(x \mapsto \frac{i}{x}\right) : (\infty, \ 0)(1, \ i, \ -1, \ -i)(1+i, \ 1-i, \ -1-i, \ -1+i).$$

Thus t_∞ has just $120/20 = 6$ images under conjugation by N. For convenience, we display these symmetric generators:

$$
\begin{aligned}
t_\infty \ & (\infty, \ 0)(1, \ 1+i)(-1, \ -1-i)(i, \ 1-i)(-i, \ -1+i); \\
t_0 \ & (\infty, \ 0)(1, \ 1-i)(-1, \ -1+i)(-i, \ 1+i)(i, \ -1-i); \\
t_1 \ & (\infty, \ 1)(0, \ -i)(-1, \ -1+i)(i, \ 1-i)(1+i, \ -1-i); \\
t_2 \ & (\infty, \ 1)(0, \ i)(-1, \ -1-i)(-i, \ 1+i)(1-vi, \ -1+i); \\
t_3 \ & (\infty, \ -1)(0, \ -i)(1, \ 1+i)(i, \ -1-i)(1-i, \ -1+i); \\
t_4 \ & (\infty, \ -1)(0, \ i)(1, \ 1-i)(-i, \ -1+i)(1+i, \ -1-i).
\end{aligned}
$$

It is readily checked that η, ν and ρ permute these elements as $(\infty)(0\ 1\ 2\ 3\ 4), (1\ 2\ 4\ 3)$ and $(\infty\ 0)(1\ 2)(3\ 4)$, respectively. It remains to check the additional relation of Equation (4.1). But $\sigma\eta^2$ acts as $(\infty\ 2\ 3\ 1\ 4\ 0)$, and so we must simply observe that

$$\sigma\eta^2 t_\infty : (\infty,\ -1-\mathrm{i},\ \mathrm{i},\ 1-\mathrm{i})(0,\ -1,\ -\mathrm{i},\ 1)(1+\mathrm{i})(-1+\mathrm{i})$$

has order 4. □

Now, the ten objects permuted by the group S_6 may be thought of as the partitions of the letters $\Gamma = \{\infty, 0, \ldots, 4\}$ into two threes, thus there exists a correspondence between these partitions and the ten points of the projective line $P_1(9)$. For completeness this correspondence is indicated in Figure 4.8. Moreover, this set of ten points is well known to have a Steiner system $S(3, 4, 10)$ defined on it. That is to say, there is a set of special subsets of size 4 such that any three of the points belong to just one of them. In the case of the projective line these *special tetrads* correspond to the 30 images of the subset $\{\infty, 0, 1, -1\}$ under the action of $\mathrm{P\Gamma L}_2(9)$. In fact, these are precisely the quadruples of points whose cross-ratio[1] is -1.

In the case of the ten partitions of six letters into two threes, the special tetrads correspond to the 15 duads and the 15 synthemes; the special tetrad corresponding to the duad xy corresponds to the four partitions in which x and y appear 'on the same side', and the special tetrad corresponding to the syntheme $uv.wx.yz$ corresponds to the four partitions in which each of the duads uv, wx and yz has a point on either side. That is to say,

$$\infty 0 \sim \left\{ \frac{\infty 01}{234}, \frac{\infty 02}{134}, \frac{\infty 03}{124}, \frac{\infty 04}{123} \right\};$$

$$\infty 0.14.23 \sim \left\{ \frac{\infty 12}{034}, \frac{\infty 13}{024}, \frac{\infty 42}{013}, \frac{\infty 43}{012} \right\}.$$

We end this section with some remarks. Firstly, note that implicit within our construction is a family of presentations which have been proved by hand to define $\mathrm{Aut}\,A_6$. Suppose x and y generate $N \cong \mathrm{PGL}_2(5)$ and t, a symmetric generator, is made to commute with a subgroup of $N = \langle x, y \rangle$ of index 6; suppose furthermore that $u = u(x, y)$ is a word in x and y which

---

[1] Recall that the *cross-ratio* (ab, cd) of four numbers a, b, c, d is defined as follows:

$$\lambda = \frac{(a-c)(b-d)}{(a-d)(b-c)};$$

permutations of the four numbers map λ into one of

$$\left\{ \lambda, \frac{1}{1-\lambda}, 1-\frac{1}{\lambda}, \frac{1}{\lambda}, 1-\lambda, \frac{\lambda}{\lambda-1} \right\}.$$

If $\lambda = -1$ we say that (ab, cd) is a *harmonic ratio*, and in this case permutation of the four numbers results in just the three cross-ratios $\{-1, 1/2, 2\}$. Thus, if the numbers lie in a field of characteristic 3, these three cross-ratios coincide and the property of being in harmonic ratio is independent of the order in which the numbers are taken. It is readily verified that cross-ratio is preserved by linear fractional maps.

has order 6 in N. Factoring by the relator $(ut)^4$ must define $\mathrm{Aut}\,A_6$. Thus we have proved, for example, that

$$\langle x, y, t \mid x^5 = y^2 = (xy)^4 = [x, y]^3 = t^2 = [t, x] = [t, (yx)^{(xy)}] = (yx^2 t)^4 = 1 \rangle$$

is a presentation for $\mathrm{Aut}\,A_6$, as may be confirmed computationally using for instance the coset enumerator in the MAGMA package [19]. Secondly, note that every element of $\mathrm{Aut}\,A_6$ can be written in the form πw with $\pi \in N$ and w a word in the t_i of length less than, or equal to, 2. We have

$$
\begin{aligned}
\{\pi t_i t_\infty \mid \pi \in \mathrm{PSL}_2(5), i \in \{\infty, 0, \dots, 4\}\} &\cong \mathrm{PSL}_2(9) \cong A_6; \\
\{\pi t_i t_\infty \mid \pi \in \mathrm{PGL}_2(5), i \in \{\infty, 0, \dots, 4\}\} &\cong \mathrm{P\Sigma L}_2(9) \cong S_6; \\
\{\pi t_i, \pi t_i t_\infty \mid \pi \in \mathrm{PSL}_2(5), i \in \{\infty, 0, \dots, 4\}\} &\cong \mathrm{PGL}_2(9);
\end{aligned}
$$

$$
\left\{ \pi t_i t_\infty, \rho t_j \middle| \begin{array}{l} \pi \in \mathrm{PSL}_2(5), i \in \{\infty, 0, \dots, 4\} \\ \rho \in \mathrm{PGL}_2(5) \setminus \mathrm{PSL}_2(5)\} \end{array} \right\} \cong M_{10}.
$$

Embedding **Aut A$_6$** in the Mathieu groups

As was mentioned above, perhaps the best place to see the outer automorphism of S_6 is in the Mathieu group M_{12}, and ultimately in the Mathieu group M_{24}. Now we have seen $G = \mathrm{Aut}\,A_6$ acting on 12 letters, with two blocks of imprimitivity of size 6, and on ten letters as $\mathrm{P\Gamma L}_2(9)$; plainly G can also act on two letters. For the convenience of the reader who is familiar with the Mathieu groups we show in Figure 4.9 how these orbits may be displayed in the Miracle Octad Generator. Each of the three diagrams represents the same 4×6 array. The first diagram shows the 24 points permuted by M_{24} labelled by the projective line $P_1(23)$; see Definition 4.1. The quadratic residues modulo 23 together with 0, which lie in the symmetric difference of the even columns and the top row, form a *dodecad* whose stabilizer is M_{12}; its complement, which consists of the non-residues together with ∞, is also a dodecad. When the 24 points are partitioned into two complementary dodecads in this way, a pair in one dodecad completes to an octad of the Steiner system $S(5, 8, 24)$, whose remaining six points lie in the complementary dodecad in precisely two ways, and the resulting two *hexads* in the complementary dodecad are disjoint from one another. Accordingly, in the second diagram we show how fixing two points of the first dodecad, namely the two $\star$'s, has the effect of partitioning the other dodecad into two

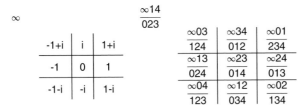

Figure 4.8. Correspondence with the projective line $P_1(9)$.

0	∞	1	11	2	22
19	3	20	4	10	18
15	6	14	16	17	8
5	9	21	13	7	12

★	●	★	●	○	●
○	◯	○	◯	●	◯
○	◯	○	◯	●	◯
○	◯	○	◯	●	◯

A	★	B	∞2	1+i	∞4
∞	-1-i	2	-1	∞1	∞
0	1	4	1-i	∞0	-1+i
1	-i	3	0	∞3	i

Figure 4.9. The embedding Aut $A_6 \leq M_{12} \leq M_{24}$.

sixes, the ●'s and the ○'s. The stabilizer of this pair of points in M_{12} is our group Aut $A_6 \cong P\Gamma L_2(9) \cong M_{10} : 2$. The final diagram shows the fixed pair of points labelled A and B, the remaining ten points labelled $P_1(9)$, and the two sixes labelled $P_1(5)$ and with $\{\star, \infty 0, \infty 1, \infty 2, \infty 3, \infty 4\}$, respectively. In order to avoid confusion between the two occurences of the symbols $\infty, 0$ and 1, the points of $P_1(9)$ are shown in a smaller font size than the two 6-orbits.

As we have seen, our control subgroup isomorphic to $PGL_2(5)$ can be generated by an element of order 5, which may be taken to be $t \mapsto t+1$ in $PGL_2(5)$, and to be

$$\eta \equiv x \mapsto \frac{(1-i)x+1+i}{x+1-i}$$

in $P\Gamma L_2(9)$, together with an element σ of order 2 which acts as $t \mapsto 2/t$ in $PGL_2(5)$ and as the field automorphism in $P\Gamma L_2(9)$. Thus

$\eta:$ $(\infty)(0\ 1\ 2\ 3\ 4)(\star)(\infty 0\ \infty 1\ \infty 2\ \infty 3\ \infty 4)$
   $(A)(B)(\infty\ 1-i\ -1\ 1\ -1+i)(0\ i\ -1-i\ 1+i\ -i);$

$\sigma:$ $(\infty\ 0)(1\ 2)(3\ 4)(\star)(\infty 0)(\infty 1)(\infty 2\ \infty 3)(\infty 4)$
   $(A\ B)(\infty)(0)(1)(-1)(i\ -i)(1+i\ 1-i)(-1+i\ -1-i);$

$t_\infty:$ $(\infty\ \star)(0\ \infty 0)(1\ \infty 1)(2\ \infty 2)(3\ \infty 3)(4\ \infty 4)$
   $(A\ B)(\infty\ 0)(1\ 1+i)(-1\ -1-i)(i\ 1-i)(-i\ -1+i).$

4.2.2 The exceptional Schur multipliers of S_n

A *central extension* of a group G is defined to be a group H such that there exists a homomorphism ρ from H onto G with Ker $\rho \leq Z(H)$, the centre of H. In other words, H is such that if you factor out a subgroup of its centre the resulting quotient is isomorphic to G. A central extension H is said to be *universal* if, for any other central extension K, there exists a homomorphism from H onto K. It turns out that G possesses a universal central extension if, and only if, G is perfect, and in this case the universal central extension $\tilde{G}$ is called the *universal covering group*; the kernel of the homomorphism from $\tilde{G}$ onto G is called the *Schur multiplier* of G. Finite perfect groups have finite universal covering groups and finite multipliers.

The Schur multiplier of the alternating group A_n, $n \geq 5$, has order 2 except for $n = 6$ and $n = 7$ when it has order 6. For a non-homological proof of this fact, the reader is referred to Aschbacher (ref. [3], p. 170). The double cover of the alternating group A_n, which we write as $2 \cdot A_n$, extends to a group $2 \cdot S_n$ in two different ways depending on whether the transpositions of S_n 'lift' to elements of order 2 or 4 in the pre-image. These are usually denoted by $2 \cdot S_n^+$ and $2 \cdot S_n^-$, respectively. Even here S_6 is exceptional as $2 \cdot S_6^+ \cong 2 \cdot S_6^-$, which holds for no other n. These similar but non-isomorphic groups are an example of *isoclinism* (see ref. [25], p. xxiii), and a complex faithful representation of one is obtained from a complex faithful representation of the other by multiplying the matrices representing elements in the outer half by i. Note that this will transform outer elements of order 2 into elements of order 4 which square to the central involution. It should be stressed that these covering groups can exhibit subtle behaviour. For instance, the group Aut $A_6 \cong A_6.2^2$, which, as we have seen, is isomorphic to the group $P\Gamma L_2(9)$, the projective general linear group $PGL_2(9)$ extended by the field automorphism of the field of order 9, possesses three subgroups of index 2, namely S_6, $PGL_2(9)$ and M_{10}, the point-stabilizer in the Mathieu group M_{11}. The covering group $2 \cdot A_6$ extends to either $2 \cdot S_6$ or $2 \cdot PGL_2(9)$; however, it cannot be extended to $2 \cdot M_{10}$. A brief description of the groups $2 \cdot S_n^\pm$ is given in the ATLAS (ref. [25], p. 236).

It is not surprising that the highly exceptional triple covers $3 \cdot A_6$ and $3 \cdot A_7$ frequently reveal themselves in the sporadic simple groups, and $3 \cdot A_7$ can be used to define a progenitor whose smallest 'true' image is the Held group; see Part III.

Construction of $3 \cdot$ Aut A_6

We constructed the group G defined in Equation (4.1) as permutations on 12 letters and, using Equation (4.3), we see that

$$t_\infty t_0 = (\infty \ 0)(1 \ 2 \ 4 \ 3)(\star \ \infty 0)(\infty 1 \ \infty 3 \ \infty 4 \ \infty 2).$$

Thus, the relation $(t_\infty t_0)^2 = (1 \ 4)(2 \ 3)$ holds in G. In other words, our G, which we now know to be isomorphic to Aut A_6, is a homomorphic image of

$$\hat{G} \cong \frac{2^{\star 6} : PGL_2(5)}{(t_\infty t_0)^2 = (1 \ 4)(2 \ 3)}. \tag{4.5}$$

Moreover, as with G, Lemma 4.1 implies that $\hat{G}/\hat{G}' \cong V_4$ and that $\hat{G}'' = \hat{G}'$. As before, to obtain the order of $\hat{G}$ we must now find all double cosets of the form NwN and calculate how many single cosets each of them contains. Certainly, $[\star] = N$ contains just one single coset, and we can now assert that $[\infty] = Nt_\infty N$ contains six single cosets. Our relation in Equation (4.5) tells us that $Nt_\infty t_0 t_\infty t_0 = N$, and so $Nt_\infty t_0 = Nt_0 t_\infty$. Thus, every single coset in $[\infty 0] = Nt_\infty t_0 N$ has at least two names, and so the double coset $[\infty 0]$ contains at most 15 single cosets. The triple transitivity of $\mathrm{PGL}_2(5)$ means that the only new double coset NwN with w of length 3 in the symmetric generators is $[\infty 01] = Nt_\infty t_0 t_1 N$. Denoting the permutation $(1\ 4)(2\ 3)$ by $\pi_{\infty 0}$, conjugation by elements of N allows us to write down π_{ij} for all $i, j \in \{\infty, 0, \dots, 4\}, i \neq j$. Thus,

$$Nt_\infty t_0 t_1 = Nt_\infty \pi_{01} t_1 t_0 = Nt_\infty (\infty\ 3)(2\ 4) t_1 t_0 = Nt_3 t_1 t_0 = Nt_\infty t_0 t_1 (\infty\ 3)(0\ 1),$$

and we see that $N^{(\infty 01)} \geq \langle (\infty\ 0)(2\ 3), (\infty\ 3)(0\ 1) \rangle \cong D_{10}$, the dihedral group of order 10. Thus, $[\infty 01]$ contains at most $120/10 = 12$ single cosets. Moreover, the group $N^{(\infty 01)}$ has orbits of length 5 and 1 on the six symmetric generators, and multiplication by any generator in the 5-orbit takes us back to the double coset $[\infty 0]$. So the only double coset NwN to consider with w of length 4 is $[\infty 014]$. As above, we have

$$\begin{aligned} Nt_\infty t_0 t_1 t_4 &= Nt_\infty t_0 \pi_{14} t_4 t_1 = Nt_\infty t_0 (\infty\ 0)(2\ 3) t_4 t_1 \\ &= Nt_0 t_\infty t_4 t_1 = Nt_\infty t_0 t_1 t_4 (\infty\ 0)(1\ 4), \end{aligned}$$

and so $N^{(\infty 014)} \geq \langle N^{(\infty 01)}, (\infty\ 0)(1\ 4) \rangle \cong \mathrm{PSL}_2(5)$. Thus $[\infty 014]$ contains at most $120/60 = 2$ single cosets. Since $N^{(\infty 014)}$ acts transitively on the six symmetric generators, multiplication by any of them takes us back to the double coset $[\infty 01]$, and we appear to have constructed the Cayley diagram shown in Figure 4.10. We know that $\hat{G}$ is a pre-image of Aut A_6, and our Cayley diagram appears to be showing that this pre-image possesses a normal subgroup of order 3 which, when factored out, yields Aut A_6. It remains for us to exhibit a group satisfying the presentation of Equation (4.5) in which this subgroup of order 3 survives. The most natural way to proceed is to exhibit $\hat{G}$ as a subgroup of S_{36} acting on the right cosets of N. Firstly, note that the action of $N \cong \mathrm{PGL}_2(5)$ on each of the sub-orbits of lengths 1, 6, 15, 12 and 2 is well defined. We must check that multiplication by t_∞ is well defined, commutes with the stabilizer in N of ∞, and satisfies the additional relation $(t_\infty t_0)^2 = (1\ 4)(2\ 3)$. This may be achieved by exhibiting our symmetric generators as permutations on the $(1+6+15+12+2 = 36)$ letters which we claim represent cosets of our control subgroup in the image group. In order to do this we need a more concise notation since, for instance, cosets in the 12-orbit have $5 \times 2 = 10$ labels as ordered triples of $\Lambda = \{\infty, 0, \dots, 4\}$. It turns out that all the information can be read directly from the synthematic total displayed in Table 4.2, in which the five synthemes must be regarded as possessing a cyclic ordering.

Figure 4.10. Cayley diagram of $3 \cdot$ Aut A_6 over $\mathrm{PGL}_2(9)$.

Table 4.4. Verification of the additional relation

	$(\star \, \infty0)$	$(\infty1$	01	$\infty \, 4$	$04)$	$(\infty$	$\|$	$0)$	$(1$	2^+	4	$3^+)$
$t_\infty t_0:$	$(\star^+ \;\; 23)$	$(34$	42	21	$13)$	$(\infty^-$	$\|$	$0^-)$	$(1^-$	2	4^-	$3)$
	$(\star^- \;\; 14)$	$(02$	$\infty3$	03	$\infty2)$	$(\infty^+$	$\|$	$0^+)$	$(1^+$	2^-	4^+	$3^-)$

Thus, straightforwardly the 1-orbit consists of $\{\star\}$; the 6-orbit consists of Λ; and the 15-orbit consists of the unordered pairs of elements of Λ. More subtly, the 12-orbit can be labelled by i^+ and i^- for $i \in \Lambda$, where, if $ij.kl.mn$ is a syntheme in Table 4.2, then $i^+ \sim mnj \sim nmj$ and $i^- \sim klj \sim lkj$; thus,

$$\infty^+ \sim 230, 320, 341, 431, 402, 042, 013, 103, 124, 214$$

and

$$\infty^- \sim 140, 410, 201, 021, 312, 132, 423, 243, 034, 304.$$

The 2-orbit can be labelled $\{\star^+, \star^-\}$, where a representative for $\star^+$ is $\infty014$ and that for $\star^-$ is $\infty023$. Using this we may write down the following action:

$$
\begin{aligned}
t_\infty : \;& (\star \; \infty)(0 \; \infty0)(1 \; \infty1)(2 \; \infty2)(3 \; \infty3)(4 \; \infty4) \\
& (\star^+ \; \infty^+)(0^+ \; 23)(1^+ \; 34)(2^+ \; 40)(3^+ \; 01)(4^+ \; 12) \\
& (\star^- \; \infty^-)(0^- \; 14)(1^- \; 20)(2^- \; 31)(3^- \; 42)(4^- \; 03),
\end{aligned}
$$

which visibly commutes with the subgroup $\langle(0\;1\;2\;3\;4), (1\;2\;4\;3)(+\;-)\rangle$. The notation here is supposed to convey that odd elements of $N \cong \mathrm{PGL}_2(5)$ interchange $+$ and $-$. This enables us to write down the action of $t_\infty t_0$ in Table 4.4, which confirms that $(t_\infty t_0)^2 = (1\;4)(2\;3)$; the columns of Table 4.4 demonstrate how these 36 symbols fall into the 12 blocks of imprimitivity permuted by our group $\mathrm{Aut}\,A_6$. A 'central' element of order 3 cycles these columns downwards and is realized by, for example, $t_\infty t_0 t_1 t_4 (\infty\;1\;2)(0\;4\;3)$. We have proved that the derived group of $\hat{G}$ is perfect, has order $3 \times |A_6|$, and has A_6 as a homomorphic image. Thus we have constructed the group $3^{\cdot}\,\mathrm{Aut}\,A_6$ acting as permutations on 36 letters as required. Note that odd permutations of $\mathrm{PGL}_2(5)$ invert the central elements of order 3, as does t_i.

An alternative to the above would be to exhibit $\hat{G}$ as a matrix group (extended by various automorphisms), which satisfies the symmetric presentation in Equation (4.5) and has a centre of order 3. Consider the group $\mathrm{SL}_3(4)$, which acts on the 21 points and 21 lines of the projective plane of order 4 (see ref. [25], p. 23).

DEFINITION 4.2 An *oval* in a projective plane is a maximal set of points, no three of which lie on a line.

Clearly an oval can contain at most $q+2$ points, where q is the order, as the lines from a given point of the oval to the other points must all be distinct.

Ovals in the plane of order 4 contain six points; an example[2] would be as follows:

$$\mathcal{O} := \left\{ \begin{array}{cccccc} (1\ 1\ 1), & (0\ 0\ 1), & (0\ \omega\ 1), & (1\ \omega\ 1), & (\omega\ 1\ 1), & (\omega\ 0\ 1) \\ (p_\infty) & (p_0) & (p_1) & (p_2) & (p_3) & (p_4) \end{array} \right\},$$

where in each case we give a vector p_i generating the 1-dimensional subspace P_i which is the point. Now there are $\binom{6}{2} = 15$ lines cutting this oval in two points, and so all five lines through a point of the oval are among them. This leaves six lines which do not intersect the oval; they are as follows:

$$\mathcal{L} := \left\{ \begin{array}{cccccc} (1\ 1\ 1)^t, & (0\ 0\ 1)^t, & (0\ \omega\ 1)^t, & (1\ \omega\ 1)^t, & (\omega\ 1\ 1)^t, & (\omega\ 0\ 1)^t \\ (l_\infty) & (l_0) & (l_1) & (l_2) & (l_3) & (l_4) \end{array} \right\},$$

where in each case we give a vector l_i defining the hyperspace which is the line L_i as described in Section 4.1. Note that α, the outer automorphism of $\mathrm{SL}_3(4)$ which simply transposes the vectors representing points and lines, interchanges these two sets, and if $A \in \mathrm{SL}_3(4)$ then

$$\alpha : A \mapsto (A^{-1})^t.$$

As stated in the ATLAS [25], an oval is stabilized by a subgroup of $\mathrm{L}_3(4)$ isomorphic to A_6, but we do not need to assume that here. Instead, observe that if we label the points of the oval $\mathcal{O}$ with the symbols $\{\infty, 0, 1, 2, 3, 4\}$, respectively, as displayed, then the following matrices:

$$A = \begin{pmatrix} 0 & \omega & \bar{\omega} \\ 1 & 0 & 0 \\ 0 & \bar{\omega} & \omega \end{pmatrix} ; B = \begin{pmatrix} \bar{\omega} & \omega & 0 \\ \omega & \bar{\omega} & 0 \\ 1 & 1 & 1 \end{pmatrix},$$

act on $\mathcal{O} \cup \mathcal{L}$ as follows:

$$A : (p_\infty)(p_0\ \omega p_1\ \bar{\omega} p_2\ \bar{\omega} p_3\ \omega p_4)(l_\infty)(l_0\ \omega l_1\ \bar{\omega} l_2\ \bar{\omega} l_3\ \omega l_4);$$
$$B : (p_\infty\ p_0)(p_1\ p_4)(p_2)(p_3)(l_\infty)(l_0)(l_1\ \bar{\omega} l_2)(l_3\ \omega l_4);$$

and satisfy $A^5 = B^2 = (AB)^3 = 1$; so $\langle A, B \rangle \cong \mathrm{A}_5 \cong \mathrm{L}_2(5)$ acting transitively on the six points of $\mathcal{O}$ and intransitively on the six lines of $\mathcal{L}$. (Of course, the 36 letters in the permutation action above are the scalar multiples of the vectors in $\mathcal{O} \cup \mathcal{L}$.) Moreover, the matrix given by

$$C = \begin{pmatrix} 0 & 1 & 0 \\ 1 & 0 & 0 \\ 0 & 0 & 1 \end{pmatrix} = B^{A^2 BA^{-1}}$$

fixes P_∞ and $A^t A = C^t C = 1$; so both A and C are fixed by α. We thus choose our symmetric generator $t_\infty = \alpha$ and may label our symmetric generators

---

[2] We have chosen this oval so that $\alpha : A \mapsto (A^t)^{-1}$ interchanges it with the six lines which do not intersect it. Other ovals are fixed by the field automorphism σ, but do not have this property.

so that their labels correspond to the labels of the points P_i. We then have the following:

$$t_\infty t_0 = t_\infty t_\infty^B = \alpha B \alpha B = B^\alpha B = \begin{pmatrix} \bar\omega & \omega & 1 \\ \omega & \bar\omega & 1 \\ 0 & 0 & 1 \end{pmatrix} \begin{pmatrix} \bar\omega & \omega & 0 \\ \omega & \bar\omega & 0 \\ 1 & 1 & 1 \end{pmatrix}$$

$$= \begin{pmatrix} 0 & 1 & 1 \\ 1 & 0 & 1 \\ 1 & 1 & 1 \end{pmatrix} \sim (\infty\ 0)(1\ 2\ 4\ 3).$$

Thus $(t_\infty t_0)^2 = (1\ 4)(2\ 3)$ as required. So far we have constructed an image of the group

$$\frac{2^{*6} : \mathrm{L}_2(5)}{(1\ 4)(2\ 3)(t_\infty t_0)^2},$$

so we need to extend the control subgroup to $\mathrm{PGL}_2(5)$. This we do by adjoining the element

$$D = \sigma \begin{pmatrix} \bar\omega & \omega & 0 \\ \omega & \bar\omega & 0 \\ 0 & 0 & 1 \end{pmatrix} \sim (1\ 2\ 4\ 3),$$

where σ represents the field automorphism which interchanges ω and $\bar\omega$. Then we have $A^D = A^2$ and $B^D = BC$; so $\langle A, B, D \rangle \cong \mathrm{PGL}_2(5)$ as required. Moreover, $\alpha = t_\infty$ commutes with σ and with the matrix in the definition of D, which is symmetric of order 2. So α commutes with $\langle A, D \rangle \cong 5 : 4$, a Frobenius group of order 20. Thus $t_\infty = \alpha$ has just six images under conjugation by the extended control subgroup, and we have constructed an image of

$$G = \frac{2^{*6} : \mathrm{PGL}_2(5)}{(1\ 4)(2\ 3)(t_\infty t_0)^2}.$$

We know that G' has index 4 in G and is perfect; it is an image of a group of shape $\mathrm{A}_6.2^2$ or $3\,{}^{\cdot}\mathrm{A}_6.2^2$ which contains elements of order 5, and so is isomorphic to one or other of these two groups. It remains to show that the normal subgroup of order 3 survives and so the latter case holds. But

$$((\infty\ 0)(1\ 4)t_\infty t_1)^4 = (B\alpha\alpha^{BA})^4 = (B\alpha A^{-1}B^{-1}\alpha BA)^4$$

$$= (BA^tB^tBA)^4 = \bar\omega I_3,$$

which commutes with all the matrices in G and is inverted by α and D. Thus

$$G \cong 3\,{}^{\cdot}\mathrm{A}_6.2^2$$

as asserted.

Construction of $3\,{}^{\cdot}\mathrm{S}_7$

We first prove the following theorem.

THEOREM **4.2**

$$H = \frac{2^{\star 7} : L_3(2)}{[(0)(1\ 3)(2\ 4\ 6\ 5)t_0t_1]^2} \cong S_7.$$

Proof   To see that S$_7$ is an image of H, we take $N \cong L_3(2)$ to be the set of all permutations of $\Omega = \{0, 1, \ldots, 6\}$ which preserve the seven *lines* $\{013, 124, 235, 346, 450, 561, 602\}$ in the standard way. We further define the symmetric generator t_i to be the unique involution of cycle shape 1.2^3 which fixes the three lines through i; thus, $t_0 = (0)(1\ 3)(2\ 6)(4\ 5)$. It is an easy matter to see that $\langle N, t_0 \rangle \cong S_7$ (for instance, $(2\ 6)(4\ 5) \in N$ and so $(1\ 3) \in \langle N, t_0 \rangle$), and N clearly permutes the seven t_i in the prescribed manner. But

$$(0)(1\ 3)(2\ 4\ 6\ 5)t_0t_1 = (0)(1\ 3)(2\ 4\ 6\ 5).(0)(1\ 3)(2\ 6)(4\ 5).(1)(2\ 4)(3\ 0)(5\ 6)$$
$$= (0\ 3)(2\ 6),$$

which has order 2 as required. Thus S$_7$ is an image of H. If we can show that $|H| \le 7!$, then we are done. As before, we find all double cosets of form $[w] = NwN$. Now, $[0] = Nt_0N$ contains at most the seven single cosets Nt_i. Since our relator can be rewritten as $(2\ 6)(4\ 5)t_0t_3t_0t_1$, we see that $Nt_0t_3 = Nt_1t_0$, and so $(0\ 1\ 3\ )(4\ 5\ 6) \in N^{(01)}$. Thus,

$$N^{(01)} \ge \langle (2\ 4)(5\ 6), (2\ 5)(4\ 6), (0\ 1\ 3\ )(4\ 5\ 6) \rangle \cong A_4,$$

and [01] contains at most $168/12 = 14$ single cosets. Under the action of $N^{(01)}$, we see that the only double coset $[w]$ with w of length 3 is [012]. But

$$Nt_0t_1t_2 = Nt_0 \cdot t_1t_2t_4t_2 \cdot t_2t_4 = Nt_0(3\ 5)(0\ 6)t_2t_4$$
$$= Nt_6t_2t_4 = Nt_0t_1t_2(1\ 2\ 4)(0\ 6\ 3),$$

and so $(1\ 2\ 4)(0\ 6\ 3) \in N^{(012)}$ and

$$N^{(012)} \ge \langle (0\ 1\ 3\ )(4\ 5\ 6), (1\ 2\ 4)(0\ 6\ 3) \rangle \cong 7 : 3,$$

a Frobenius group of order 21 acting transitively on Ω. Thus, [012] contains at most $168/21 = 8$ single cosets. This tells us that

$$|H| \le 168 \times (1 + 7 + 14 + 8) = 7!,$$

and so $|H| = 7!$ and $H \cong S_7$. In particular, we have confirmed that the Cayley diagram in Figure 4.11 does not collapse.   □

We can now prove Theorem 4.3.

Figure 4.11. Cayley diagram of S$_7$ over $L_3(2)$.

THEOREM **4.3**

$$K = \frac{2^{\cdot 7} : \mathrm{L}_3(2)}{[(0\ 1\ 2\ 3\ 4\ 5\ 6)t_0]^6} \cong 3^{\cdot}\mathrm{S}_7,$$

a group of order $3 \times 7!$ which possesses a homomorphic image isomorphic to S_7, and whose derived group is a perfect central extension of A_7 of shape $3^{\cdot}\mathrm{A}_7$. Outer elements invert the 'central' elements of order 3.

Proof We first show that S_7 is a homomorphic image of K. Take the same copy of $N \cong \mathrm{L}_3(2)$ as in Theorem 4.2, and again define $t_0 = (1\ 3)(2\ 6)(4\ 5)$. It remains to check the additional relation. But $(0\ 1\ 2\ 3\ 4\ 5\ 6)t_0 = (0\ 1\ 2\ 3\ 4\ 5\ 6).(1\ 3)(2\ 6)(4\ 5) = (0\ 3\ 5\ 2\ 1\ 6)$ has order 6 as required.

Next note that by Theorem 3.1 we have $|K : K'| = 2$ and K' is perfect.

We now proceed in the familiar manner to find all double cosets of form $[w] = NwN$. Let $\pi = (0\ 1\ 2\ 3\ 4\ 5\ 6)$. Then a permutation $\rho \in \mathrm{L}_3(2)$ is conjugate to π if, and only if, $\rho = (a_0\ a_1\ a_2\ a_3\ a_4\ a_5\ a_6)$, where $\{a_i, a_{i+1}, a_{i+3}\}$ is a line for $i = 0, 1, \ldots, 6$, all subscripts being read modulo 7. Our additional relation then becomes $(\pi t_0)^6$ and, expanding, we have the following:

$$(\pi t_0)^6 = 1 \Leftrightarrow \pi^6 t_0^{\pi^5} \cdots t_0^{\pi} t_0 = 1$$

$$\Leftrightarrow \pi = t_5 t_4 t_3 t_2 t_1 t_0 \tag{4.6}$$

$$\Rightarrow Nt_0 t_1 t_2 = Nt_5 t_4 t_3 \tag{4.7}$$

$$\Rightarrow (0\ 5\ 1\ 4\ 6\ 2\ 3) \in N^{(012)}. \tag{4.8}$$

Thus we conclude that the double coset [012] contains at most $168/7 = 24$ single cosets, but the 42 single cosets of form $Nt_i t_j$ contained in [01] appear to be distinct. We must now examine the double cosets [010] and [013], but we will find it easier to look first at [0103]. Using Equation (4.8), we see that

$$Nt_0 t_1 t_0 t_3 = Nt_0 t_1 t_2 \cdot t_2 t_0 t_3$$

$$= Nt_4 t_2 t_5 \cdot t_2 t_0 t_3$$

$$= Nt_4 t_2 \cdot t_5 t_2 t_0 t_3 t_6 t_1 \cdot t_1 t_6$$

$$= Nt_4 t_2 \cdot (1\ 6\ 3\ 0\ 2\ 5\ 4) \cdot t_1 t_6$$

$$= Nt_1 t_5 t_1 t_6, \tag{4.9}$$

in which calculation we have used a conjugate of Equation (4.6). But this shows us that $(0\ 1\ 5\ 3\ 6\ 2\ 4) \in N^{(0103)}$, and so

$$N^{(0103)} \geq \langle (0\ 1\ 5\ 3\ 6\ 2\ 4), (2\ 4)(6\ 5), (2\ 5)(4\ 6) \rangle \cong \mathrm{L}_3(2).$$

This shows that [0103] contains just one single coset, which we denote by $\star^+$; using the result proved in Equation (4.2), that $|[w]| = |[w^{-1}]|$, we deduce that [0131] also contains just one single coset, which we denote by $\star^-$. But since the cosets $Nt_0t_1t_0t_3$ and $Nt_0t_1t_3t_1$ are fixed by the whole of $N \cong L_3(2)$, we see that:

(1) right multiplication of either of these cosets by a symmetric generator t_i leads to a coset of form $Nt_jt_kt_j$ or $Nt_jt_it_k$, where $\{i, j, k\}$ is a line;

(2) the coset stabilizing subgroups $N^{(010)}$ and $N^{(013)}$ contain the stabilizers in N of 3 and 1, respectively, both of which are isomorphic to S_4;

(3) thus, the double cosets [010] and [013] contain at most $168/24 = 7$ single cosets.

Together these remarks lead to the Cayley diagram shown in Figure 4.12. Note that, provided there is no further collapse, it is sensible to label the coset $Nt_it_jt_i$ by k^- and the coset $Nt_it_kt_j$ by k^+, where as above $\{i, j, k\}$ is a line (since they are both fixed by the stabilizer of k). To prove that no collapsing of the Cayley diagram in Figure 4.12 occurs, we can exhibit t_0 as a permutation of the $3 \times (15 + 15) = 90$ symbols:

$$\{\star, \star^{\pm}, i, i^{\pm}, ij, ijk \mid i, j, k \in \{0, 1, \ldots, 6\}, \ \{i, j, k\} \text{ distinct and } not \text{ a line}\},$$

where it is understood that each coset ijk has seven such names which can be obtained using conjugates of Equations (4.7) and (4.8). This is straightforward, if slightly tedious. Finally, note from Theorem 4.2 that $Z(K')$, the centre of K', must be generated by $z = [(1\ 3)(2\ 4\ 6\ 5)t_0t_1]^2 = (2\ 6)(4\ 5)t_0t_3t_0t_1$, since factoring out this relator results in the symmetric group S_7. Of course, z may be exhibited cycling each of the 30 blocks of size 3 which K permutes; however, we prefer to show directly that z is inverted by conjugation by t_0. We have

$$Nz^{t_0}z = Nt_3t_0t_1t_0 \cdot t_0t_3t_0t_1 = Nt_1t_0t_3t_0 \cdot t_0t_3t_0t_1 = N,$$

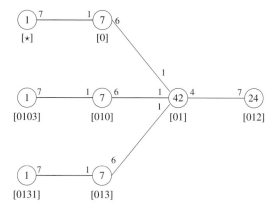

Figure 4.12. Cayley diagram of $3 \cdot S_7$ over $L_3(2)$.

where we have used the fact that $Nt_3t_0t_1t_0$ is fixed by the whole of $N \cong L_3(2)$. Thus $z^{t_0}z = (t_3t_0t_1)^2$ belongs to N and, moreover,

$$z^{t_0}z \in C_N(\langle (2\,6)(4\,5), (2\,4)(6\,5) \rangle) = \langle ((2\,6)(4\,5), (2\,4)(6\,5) \rangle.$$

In particular, $z^{t_0}z$ commutes with t_3. Then

$$\begin{aligned} z^{t_0}z &= (z^{t_0}z)^{t_3} = t_3 \cdot t_3t_0t_1t_3t_0t_1 \cdot t_3 \\ &= t_0t_1t_3t_0t_1t_3 = (t_3t_0t_1t_3t_0t_1)^{(3\,0\,1)(6\,4\,5)}, \end{aligned}$$

and so $z^{t_0}z \in C_N((3\,0\,1)(6\,4\,5)) = \langle (3\,0\,1)(6\,4\,5) \rangle$. Thus $z^{t_0}z = 1$ and so $z^{t_0} = z^{-1}$. $\square$

Rather than exhibit our group as permutations on 90 letters in order to show that the derived group has a non-trivial centre, it is easier and more revealing to give an alternative representation of the group defined in Theorem 4.3. We shall, in fact, produce a monomial semi-linear 21-dimensional representation of the group $3\dot{\,}S_7$. Let V be a complex vector space with basis $\mathcal{B} = \{ v_{ij} \mid i, j \in \mathbb{Z}_7, i \neq j, v_{ij} = v_{ji} \}$. In other words, the basis elements correspond to the 21 unordered pairs of integers modulo 7. Let our control subgroup $N \cong L_2(7)$ permute $\mathcal{B}$ in the natural way; that is to say, $v_{ij}^{\pi} = v_{\pi(i)\pi(j)}$ for $\pi \in N$. Our seven symmetric generators, which will of course be closely related to the elements of S_7 described in Theorem 4.2, are in fact semi-linear monomial permutations of the 1-dimensional subspaces spanned by the basic vectors. Thus we define the following:

$$t_i : (v_{jk})(\omega v_{jk}\ \bar{\omega}v_{jk})(v_{ij}\ \omega v_{ik})(\bar{\omega}v_{ij}\ \bar{\omega}v_{ik})(v_{jl}\ \bar{\omega}v_{km})(\omega v_{jl}\ \omega v_{km}),$$

where ijk and ilm are distinct lines in the projective plane and ω is a primitive complex cube root of unity. This element clearly only has seven images under conjugation by N and so it remains to verify that $(0\,1\,2\,3\,4\,5\,6)t_0$ has order 6 and, finally, that the derived group of $\langle N, t_0 \rangle$ does indeed have a centre of order 3. The first of these tasks is left as an exercise for the interested reader. In order to achieve the other, we simply show that the element $(1\,3)(2\,4\,6\,5)t_0t_1$ squares to the element which maps $v_{ij} \mapsto \omega v_{ij}$ for all unordered pairs ij.

4.3   The 11-point biplane and $PGL_2(11)$

The Fano plane, which occurred in our construction of the Mathieu group M_{24} in Chapter 1 and as our first example of symmetric generation of a group, consists of seven points and seven lines with the properties that any pair of points lie together on just one line, and any pair of lines intersect in a unique point. We took a 3-dimensional vector space over the field of order 2 and let its 1-dimensional subspaces be the points and its 2-dimensional subspaces be the lines. A point is said to lie on a line if the corresponding subspaces are contained in one another. Clearly two distinct 1-dimensional subspaces are contained in a unique 2-dimensional subspace, and a consideration of dimension soon reveals that any two distinct 2-dimensional

Table 4.5. The 7-point biplane

0365
1406
2510
3621
4032
5143
6254

subspaces of a 3-dimensional space must intersect in a 1-dimensional sub-space. Thus this construction produces a projective plane no matter what field we work over. If the field has order $q = p^m$, for p a prime number, then the corresponding *classical projective plane* contains $q^2 + q + 1$ points, and the same number of lines. These beautiful structures are visibly preserved by the projective general linear group $\mathrm{PGL}_3(q)$, by any automorphisms possessed by the underlying field, and by the inverse-transpose duality (see Section 4.4.1) which interchanges points and lines.

Biplanes are a fascinating variation of projective planes. Again we have a finite set of *points* and a collection of subsets of the points which we call *lines*; and we require that the structure shares a further property with projective planes, namely that the number of points equals the number of lines. However, instead of requiring that each pair of points has a unique line passing through it, we require that there are precisely *two* lines passing through every pair of points.

As it happens, we have already come close to seeing an example of a biplane. Recall that the 7-points of the Fano plane were labelled by $\mathbb{Z}_7$, the integers modulo 7, and the lines were taken to be the set of quadratic residues $\{1, 2, 4\}$ and its translates. To obtain a biplane we take as *lines* the complements of the lines of the Fano plane, so *lines* contain four points. Clearly, given any two points there is a unique line passing through both of them and $2 \times 2 = 4$ lines passing through one but not both. Thus $7 - 5 = 2$ lines of the Fano plane contain neither point, which is to say each pair of points is contained in two *lines*. We display the *lines* of the 7-point biplane in Table 4.5.

Just as this biplane was essentially obtained, complete with its group of automorphisms, directly from a symmetrically generated group, we obtain a further example by consideration of the following.

EXAMPLE **4.1**

Let $G = \dfrac{2^{\star 6} : \mathrm{L}_2(5)}{[(0\ 1\ 2\ 3\ 4)t_0]^4}.$

Using the fact that $(\pi t)^m = \pi^m t^{\pi^{m-1}} t^{\pi^{m-2}} \cdots t^\pi t$, we see that our additional relator is equivalent to the following relation:

$$(4\ 3\ 2\ 1\ 0)t_3 t_2 t_1 t_0 = 1.$$

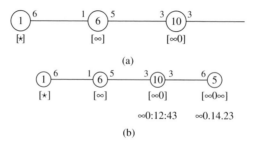

(a)

$\infty0{:}12{:}43 \qquad \infty0{.}14{.}23$

(b)

Figure 4.13. Cayley diagram for Example 4.1: (a) in part; (b) in full.

Table 4.6. The 10-orbit in Example 4.1

$\infty0 : 12 : 43$	$0\infty : 42 : 13$
$\infty1 : 23 : 04$	$1\infty : 03 : 24$
$\infty2 : 34 : 10$	$2\infty : 14 : 30$
$\infty3 : 40 : 21$	$3\infty : 20 : 41$
$\infty4 : 01 : 32$	$4\infty : 31 : 02$

This shows us that

$$Nt_3t_2t_1t_0 = N, \text{ and so } Nt_0t_1 = Nt_3t_2;$$

thus any permutation in N which maps 0 to 3 and 1 to 2 lies in the coset stabilizing subgroup $N^{(w)}$. In particular, we see that when $w = t_0t_1$

$$N^{(w)} \geq \langle x \mapsto 3 - x \equiv (0\ 3)(1\ 2), x \mapsto \frac{3x}{x+2} \equiv (\infty\ 3)(2\ 4) \rangle \cong S_3,$$

and so $Nt_0t_1 = Nt_3t_2 = Nt_\infty t_4$, which we write in abbreviated form as $01 \sim 32 \sim \infty4$. This enables us to deduce that the double coset Nt_0t_1N contains at most $60/6 = 10$ single cosets. At this stage we appear to have constructed part of a collapsed Cayley diagram of form indicated in Figure 4.13, in which the ten cosets in Nt_0t_1N may be labelled as shown in Table 4.6.

We must now investigate the double coset $Nt_\infty t_0 t_\infty N$, which at the moment seems only to be stabilized by the group $\langle (1\ 4)(2\ 3) \rangle \cong C_2$. However, in abbreviated notation our additional relation takes the following form:

$$0123 = (4\ 3\ 2\ 1\ 0),$$

and of course all conjugates of this under permutations in N also hold. We can now observe that

$$\infty0\infty = \infty.0\infty24.42 = \infty.(4\ 2\ \infty\ 0\ 3).42 \sim 042 \sim 232,$$

where the substitution corresponds to a conjugate of the additional relation and the permutation used is $x \mapsto -1/(x+3)$. The last step of the calculation comes from the second row of Table 4.6. Thus,

$$N^{(\infty 0 \infty)} \geq \langle x \mapsto -x \equiv (1\ 4)(2\ 3),\ x \mapsto \frac{2x+2}{x-1} \equiv (\infty\ 2\ 1)(0\ 3\ 4) \rangle \cong \mathrm{A}_4,$$

and we see that

$$\infty 0 \infty \sim 0 \infty 0 \sim 141 \sim 414 \sim 232 \sim 323.$$

So this coset can conveniently be labelled by the *syntheme* $\infty 0.14.23$. We note in particular that this coset stabilizing subgroup is transitive, and so multiplication by any of the symmetric generators returns us to the '10-orbit'. This shows that N has at most index $(1+6+10+5) = 22$ in G, and the Cayley graph appears to have the form shown in Figure 4.13(b), although we have not yet proved that further collapsing does not occur.

Since $|G| \leq 22 \times |N| = 22 \times 60 = 2 \times 660$, the reader will not be surprised to learn that, in fact, $G \equiv \mathrm{PGL}_2(11)$. To prove this we need only show that $\mathrm{PGL}_2(11)$ is an image of G, as we should then have that $2 \times 660 \geq |G| \geq 2 \times 660$, and so equality will hold. In order to see this, we define certain elements of $\mathrm{PGL}_2(11)$; thus,

$$\beta \equiv \quad x \mapsto 3x \quad \equiv (1\ 3\ 9\ 5\ 4)(2\ 6\ 7\ \mathrm{X}\ 8),$$

$$\gamma \equiv \quad x \mapsto -\frac{1}{x} \quad \equiv (\infty\ 0)(1\ \mathrm{X})(2\ 5)(3\ 7)(4\ 8)(6\ 9),$$

$$\delta \equiv x \mapsto \frac{x+5}{x-1} \quad \equiv (\infty\ 1)(3\ 4)(2\ 7)(5\ 8)(0\ 6)(9\ \mathrm{X}),$$

where elements are displayed as permutations of the projective line $\mathrm{P}_1(11)$ and the symbol X stands for '10'; see Definition 4.1. We readily check that $\beta\delta$ has order 3, and so $\beta^5 = \delta^2 = (\beta\delta)^3 = 1$ and $\langle \beta, \delta \rangle$ is a non-trivial homomorphic image of the simple group A_5. Thus $\langle \beta, \delta \rangle \cong \mathrm{A}_5$. Now, γ visibly inverts β, and so $\langle \beta, \gamma \rangle \cong \mathrm{D}_{10}$. Perhaps the easiest way to see that $\gamma \in \langle \beta, \delta \rangle$ is to note that each of the three permutations given is even and preserves the pairing $\infty 0.16.37.9\mathrm{X}.58.42$. But this pairing must have at least 11 images under the action of $\mathrm{PSL}_2(11)$, and since it is preserved by a subgroup isomorphic to A_5 it has exactly 11 images. Its stabilizing subgroup is thus $\langle \beta, \delta \rangle$, and so $\gamma \in \langle \beta, \delta \rangle$.

Now let

$$t_\infty \equiv x \mapsto -x \equiv (1\ \mathrm{X})(2\ 9)(3\ 8)(4\ 7)(5\ 6),$$

and observe that t_∞ commutes with both β and γ and so has just six images under conjugation by $\langle \beta, \delta \rangle$, namely.

$$
\begin{aligned}
t_\infty &\equiv (1\ \mathrm{X})(2\ 9)(3\ 8)(4\ 7)(5\ 6), \\
t_0 &\equiv (\infty\ 9)(7\ \mathrm{X})(4\ 5)(3\ 2)(8\ 0), \\
t_1 &\equiv (\infty\ 4)(8\ 2)(3\ 1)(5\ 7)(6\ 0), \\
t_2 &\equiv (\infty\ 3)(6\ 7)(5\ 9)(1\ 8)(\mathrm{X}\ 0), \\
t_3 &\equiv (\infty\ 5)(\mathrm{X}\ 8)(1\ 4)(9\ 6)(2\ 0), \\
t_4 &\equiv (\infty\ 1)(2\ 6)(9\ 3)(4\ \mathrm{X})(7\ 0).
\end{aligned}
$$

Table 4.7. Labelling of the points and lines of
the 11-point biplane

Points		Lines		
In graph	$\mathbb{Z}_{11}$	As subsets	$\mathbb{Z}_{11}$	In graph
⋆	0	267X8	0	∞
2∞:14:30	1	37809	x	∞4.03.12
∞0:43:12	2	4891X	9	3
0∞:42:13	3	59X20	8	∞2.31.40
4∞:31:02	4	6X031	7	∞1.34.20
1∞:03:24	5	70142	6	∞3.42.01
∞3:21:40	6	81253	5	1
∞1:04:23	7	92364	4	4
∞2:10:34	8	X3475	3	0
3∞:20:41	9	04586	2	∞0.14.23
∞4:32:01	X	15697	1	2

The $L_2(5)$-action of $N = \langle \beta, \delta \rangle$ on these six generators is given by $\beta \equiv (t_0\ t_3\ t_1\ t_4\ t_2)$, $\delta \equiv (t_\infty\ t_0)(t_2\ t_3)$. Since N is maximal in $\mathrm{PGL}_2(11)$ and $t_\infty \notin N$, we see that $\mathrm{PGL}_2(11)$ *is* an image of the progenitor $2^{\star 6} : L_2(5)$. That the element $(0\ 1\ 2\ 3\ 4)t_0 = \beta^2 t_0$ has order 4 is readily checked, and so $\mathrm{PGL}_2(11)$ is indeed an image of G and isomorphism is proved.

Now recall that in the transitive graph Γ_T we join the coset Nw to the cosets Nt_iw, so in this case Γ_T is bipartite of valence 6 between two subsets of 11 vertices, the *points* and the *lines*, say. Moreover, the group of automorphisms plainly acts doubly transitively on the points (and on the lines) as the stabilizer of a point has an orbit of length 10 on the remaining points. So if we take our two points to be ⋆ and $\infty 0 : 12 : 43$, we see that ⋆ is joined to $\{\infty, 0, 1, 2, 3, 4\}$ and $\infty 0 : 12 : 43$ is joined to $\{0, 2, 3, \infty 0.14.23, \infty 4.03.12, \infty 1.20.34\}$. So three lines are joined to both points, and six lines are joined to one but not the other. Thus precisely two lines are joined to neither, and this must hold for every pair of points. If we define a point to lie on a line if they are *not* joined in Γ_T, then we see that we have constructed an 11-point biplane which is preserved by the whole of $\mathrm{PGL}_2(11)$.

In order to describe the joins in our bipartite graph more fully, we introduce some notation. Let the syntheme *ij.kl.mn* be denoted by any one of $[ij], [kl]$ or $[mn]$. The ten vertices labelled *ij:kl:mn* correspond to three disjoint *ordered* pairs and, of course, every ordered pair occurs in just one of them; let $ij : kl : mn$ be denoted by any one of $\overline{ij}, \overline{kl}$ or $\overline{mn}$. Then

$$
\begin{array}{rl}
\star & \text{joins} \quad \{\infty, 0, 1, 2, 3, 4\} \\
ij{:}kl{:}mn & \text{joins} \ \{j, l, n, [ij], [kl], [mn]\} \\
i & \text{joins} \quad \{\star, \overline{ji} \text{ for } j \neq i\} \\
ij.kl.mn & \text{joins} \ \{\overline{ij}, \overline{ji}, \overline{kl}, \overline{lk}, \overline{mn}, \overline{nm}\}
\end{array}
$$

The 11 points and 11 lines can best be labelled with the elements of $\mathbb{Z}_{11}$, the integers modulo 11, and we exhibit in Table 4.7 the correspondence with the notation produced by our construction. Here points are denoted by the larger integers modulo 11, lines by the smaller integers, and we take as lines the set of quadratic non-residues, namely $\{2, 6, 7, X, 8\}$ and its translates. Right multiplication by a single symmetric generator gives rise to a duality interchanging points and lines, and the notation is chosen so that t_∞ corresponds to

$$(0\ 0)(1\ 1)(2\ 2)(3\ 3)(4\ 4)(5\ 5)(6\ 6)(7\ 7)(8\ 8)(9\ 9)(X\ X).$$

4.4 The group of the 28 bitangents

A tangent which touches a curve twice is referred to as a double tangent or a *bitangent*. As was mentioned in Chapter 2, the maximum number of bitangents a quartic curve can possess is 28; indeed, it is possible to realize 28 distinct real bitangents as follows. Let $S = 0$ and $T = 0$ be two ellipses centred on the origin, one with its major axis horizontal and the other with it vertical, chosen so that they intersect in four real points; see Figure 4.14(a). Then the curve $ST = 0$ is a degenerate quartic. Now perturb this quartic slightly by increasing the right hand side to obtain $ST = a$, where a is small. This can be done in such a way that the curve separates into four disjoint components taking the form of 'kidney beans'; see Figure 4.14(b). It is clear that each component has a unique bitangent touching it twice, and that for any pair of components there are four bitangents touching each of them once. This gives a total of $4 + \binom{4}{2} \times 4 = 28$ bitangents.

It turns out that certain sets of four bitangents are special in the sense that their eight points of contact with the quartic curve lie on a conic;

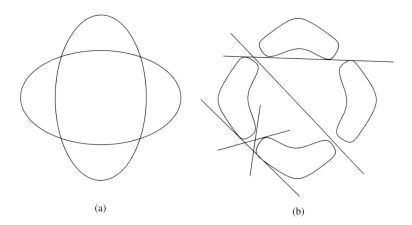

(a)   (b)

Figure 4.14. Quartic curve with 28 bitangents. (a) $ST = a$; $ST = 0$.

call these *special tetrads*. Cayley (see ref. [49], p. 334)[3] denoted these 28
bitangents by the set of unordered pairs of eight letters $\{a, b, c, d, e, f, g, h\}$,
say; in other words, to the edges of a complete graph on eight letters. In
this notation, the special tetrads correspond either to the four edges of a
quadrilateral such as $\{ab, bc, cd, da\}$, of which there are $\binom{8}{4} \times 3 = 210$, or
to four non-intersecting lines such as $\{ab, cd, ef, gh\}$, of which there are
$7 \times 5 \times 3 = 105$. The famous group of the 28 bitangents may be thought
of combinatorially as the set of all permutations of the 28 objects which
preserve the set of $210 + 105 = 315$ special tetrads. Clearly, permutations of
the original eight letters preserve this set, with the two orbits just described.
Moreover, it is readily checked that Hesse's so-called *bifid maps*, which
correspond to the 35 ways in which we can partition the eight letters into
two fours, also preserve it. A bifid map is defined to fix an unordered pair
which has a letter on each side of the partition, and to complement an
unordered pair which lies on one side. Thus,

$$\frac{abcd}{efgh} \equiv (ab\ cd)(ac\ bd)(ad\ bc)(ef\ gh)(eg\ fh)(eh\ fg),$$

fixing the other 16 pairs. Note, though, that

$$\frac{abcd}{efgh} : \{ab, be, ef, fa\} \mapsto \{cd, af, be, gh\},$$

which is shown diagrammatically in Figure 4.15, and so the bifid maps fuse
the two orbits of S_8 acting on the 315 tetrads.

As we shall see shortly, the bifid maps, together with the identity element, form a set of coset representatives for the symmetric group S_8 in the
group we want, which thus has order $(36 \times 8!)$.

The set of bifid maps furnishes us with a ready made set of symmetric
generators and, since they are defined as permutations, we are able to
calculate relations which hold between them. This leads naturally to the
following definition in which the symmetric generator corresponding to the
partition $abcd/efgh$ is, for ease of notation, labelled $abcd/efgh$, rather than
$t_{abcd/efgh}$.

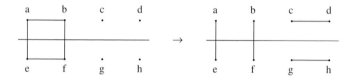

Figure 4.15. Fusion of the two orbits of special tetrads.

---

[3] Hilton states here that any two of the 28 bitangents lie in precisely five of these special
tetrads, which thus form a *2-design* of type 2-(28,4,5).

EXAMPLE **4.2** Let

$$G = \frac{2^{\star 35} : S_8}{[(a\ b)(c\ e\ g\ d\ f\ h)(abcd/efgh)]^3, [(a\ e)(abcd/efgh)]^3}.$$

We first expand the two additional relators to obtain

$$\frac{abgh}{cdef}\frac{abef}{ghcd}\frac{abcd}{efgh} = (a\ b)(c\ d)(e\ f)(g\ h);$$

$$\frac{abcd}{efgh}\frac{ebcd}{afgh}\frac{abcd}{efgh} = (a\ e). \tag{4.10}$$

We must now investigate all double cosets of the form NwN, where $N \cong S_8$ and w is a word in the 35 symmetric generators which correspond to the bifid maps. But the above relations which we have deduced from the defining relators show that

$$N\,\frac{abgh}{cdef}\frac{abef}{ghcd} = N\,\frac{abcd}{efgh}$$

and

$$N\,\frac{abcd}{efgh}\frac{ebcd}{afgh} = N\,\frac{abcd}{efgh}.$$

Thus the only double cosets of form NwN are the trivial double coset N and

$$N\,\frac{abcd}{efgh}\,N = \bigcup_{\text{B a bifid map}} NB,$$

which consists of the union of 35 apparently distinct single cosets. So the maximal order of G is $36 \times |S_8|$. In fact, we can see that these 35 cosets are distinct by embedding the group G in the symmetric group S_{28} by way of the action on the unordered pairs of eight letters defined above. We readily check that the two additional relators hold:

$$(a\ b)(c\ e\ g\ d\ f\ h)\frac{abcd}{efgh} \equiv (ab\ cd\ gh)(ef)(ac\ be\ ag)(ad\ bf\ ah)$$
$$(ae\ bg\ bc)(af\ bh\ bd)(ce\ fh\ ch)$$
$$(cf\ fg\ dh)(cg\ de\ eh)(df\ eg\ dg);$$

$$(a\ e)\frac{abcd}{efgh} \equiv (ab\ be\ cd)(ac\ ce\ bd)(ad\ de\ bc)(ae)$$
$$(af\ gh\ ef)(ag\ fh\ eg)(ah\ fg\ eh)$$
$$(bf)(bg)(bh)(cf)(cg)(ch)(df)(dg)(dh);$$

both have order 3, as required. But the two relations in Equation (4.10) show that when considered as permutations of 28 letters no product of two bifid maps is in N, and so the double coset

$$N\,\frac{abcd}{efgh}\,N$$

really does contain 35 distinct single cosets. We thus have that $|G| = 36 \times 8!$.

Of course, we have also constructed the action of the group G on 36 letters, namely the 36 cosets of N in G, where the action is given by Equation

Table 4.8. *Some primitive actions of* $S_6(2)$

Degree	28	36	63
Notation for the points	unordered pairs of eight letters	{bifid maps} $\cup \{\star\}$	{transpositions of S_8} $\cup$ {bifid maps}
Action of $\dfrac{abcd}{efgh}$	$ab \leftrightarrow cd$	$\star \leftrightarrow \dfrac{abcd}{efgh}$ $\dfrac{abef}{cdgh} \leftrightarrow \dfrac{abgh}{cdef}$	$(ae) \leftrightarrow \dfrac{ebcd}{afgh}$
Cycle shape of transpositions	$1^{16}.2^6$	$1^{16}.2^{10}$	$1^{31}.2^{16}$

(4.10). For convenience, we denote these 36 letters by $\star$, which stands for the coset N and the 35 bifid maps. Thus, permutations of $N \cong S_8$ fix $\star$ and act on the bifid maps in the natural way, whilst

$$\frac{abcd}{efgh} : \star \leftrightarrow \frac{abcd}{efgh}, \quad \frac{abef}{cdgh} \leftrightarrow \frac{abgh}{cdef}, \quad \frac{ebcd}{afgh} \leftrightarrow \frac{ebcd}{afgh}.$$

Thus, as a permutation of 36 letters, a bifid map has cycle shape $1^{16}.2^{10}$, as does a transposition of N. In fact, the second of the two relations in Equation (4.10) shows that bifid maps and tranpositions are conjugate and, since any element of G can be written *canonically* (and uniquely) as a permutation of S_8 or as a permutation of S_8 times a bifid map, we see that conjugating a bifid map by any element of G gives a bifid map or a transposition. So G possesses a conjugacy class of $28 + 35 = 63$ involutions. We may multiply canonically represented elements to obtain

$$\pi B \times \sigma C = \pi \sigma B^\sigma C,$$

where π and σ are permutations of S_8, and B and C are bifid maps, and we may then use Equation (4.10) if necessary to put the result in canonical form. Thus we can work with elements of G without writing them as permutations on 28, 36 or 63 letters, or indeed as matrices. However, for the convenience of the reader we include in Table 4.8 the conversion from canonically represented elements to the aforementioned three primitive permutation actions.

4.4.1   Identification with $S_6(2)$

The group G that we have constructed has order $36 \times 8! = 1\,451\,520$; it is, in fact, isomorphic to the symplectic group $S_6(2)$, and the 63 involutions forming a conjugacy class are the *Steiner transpositions*, an example of *symplectic transvections*. In order to see this, we take a vector space of dimension 8 over the field of order 2, fix a basis so that the vectors are 8-tuples of 0s and 1s, and restrict to the subspace consisting of all vectors with an even number of non-zero entries. Now factor out the 'all 1s' vector to obtain a 6-dimensional space V_6. The vectors of V_6 correspond to the various partitions of eight letters into two even subsets: $\{0/8 \text{ (1 such)}, 2/6 \text{ (28 such)}, 4/4 \text{ (35 such)}\}$. The transvection corresponding to the vector u is defined by

$$t_u : v \mapsto v + B(u, v)u,$$

where B is the non-degenerate alternate bilinear form inherited from the standard inner product of vectors in the original 8-dimensional space. Of course, the transpositions of S_8 correspond to u being a partition of type $2/6$, and the bifid maps correspond to $4/4$-type partitions. Since there is a bijection between transvections and vectors, we can if we wish think of the 63 involutions in this conjugacy class as the non-zero vectors of V_6 when addition is given by

$$(a\ b) + (b\ c) + (c\ a) = (a\ b) + (c\ d) + \frac{abcd}{efgh}$$

$$= (a\ e) + \frac{abcd}{efgh} + \frac{afgh}{ebcd} = \frac{abcd}{efgh} + \frac{abef}{ghcd} + \frac{abgh}{cdef} = 0.$$

The group, of course, acts on this space by conjugation, and so any element of the group will commute with $2^m - 1$ of these transvection/vectors for some m. For instance, the involutions $(a\ b), (a\ b)(c\ d), (a\ b)(c\ d)(e\ f)$ and $(a\ b)(c\ d)(e\ f)(g\ h)$ of S_8 are soon seen to commute with 31, 15, 7 and 15 transvections, respectively.

EXERCISE 4.4

(1) Express the following elements of $S_6(2)$ in canonical form (as π or π B for $\pi \in S_8$ and B a bifid map):

(a) $(a\ b\ e)\dfrac{abcd}{efgh}.(b\ c\ g\ h)\dfrac{afgh}{ebcd}$;

(b) $\left((a\ e)\dfrac{abcd}{efgh}\right)^3$;

(c) $(\pi \mathrm{B})^{-1}$.

(2) Find all the elements in $G \cong S_6(2)$ which are conjugate to the 3-cycle $(a\ b\ c)$, and show that this conjugacy class, T say, contains 336 elements.

(3) Show that in the 28-point action of G on unordered pairs of eight letters, elements of T above fix precisely ten points. Regarding these unordered

pairs as edges of a graph on eight vertices, say what subgraphs of the complete graph on eight vertices the elements of T fix. Why might we refer to T as the Kuratowski class?

(4) Show that the fixed-point-free involution $(a\ b)(c\ d)(e\ f)(g\ h)$ of S_8 has 315 conjugates in $G \cong S_6(2)$. Identify their fixed subgraphs as in question (3), and verify that these subgraphs are precisely Cayley's special tetrads of bitangents whose eight points of contact with the quartic curve lie on a conic.

5

Sporadic simple groups

Having flexed our muscles on a number of rather small and familiar classical groups, we now turn our attention to the exceptional structures. As the reader will be aware, the *classification of finite simple groups* or CFSG states that any finite simple group is isomorphic to a member of one of the known infinite families or to one of the 26 so-called *sporadic* simple groups. Five of these are the Mathieu groups $M_{12}, M_{11}, M_{24}, M_{23}$ and M_{22}, which were discovered in the second half of the nineteenth century by Emil Mathieu [70, 71]; the subscript in each case indicates the degree of the group's natural permutation action. No sporadic simple group was then found until Z. Janko discovered his smallest simple group J_1 in 1965 [58, 59]. This discovery triggered one of the most exciting periods in mathematics, as a further 20 sporadic groups were unearthed in the following decade.

As was explained in Part I, it was the behaviour of the largest Mathieu group M_{24} which motivated this approach to defining and constructing groups, but it turns out that it is perhaps M_{22} which lends itself most readily to hand construction via symmetric generation.

5.1 The Mathieu group M_{22}

As control subgroup, we take the alternating group A_7 in its action on 15 letters, and so we form the progenitor

$$2^{\star 15} : A_7.$$

As is always the case, in order to proceed we need to understand in detail the permutation action of the control subgroup. The well known fact that A_7 acts doubly transitively on 15 letters in two different (non-permutation identical) ways, with the two actions being interchanged by outer elements of S_7, was demonstrated in the proof of Theorem 4.3. The point-stabilizer is isomorphic to $L_2(7)$, and the 2-point stabilizer is isomorphic to A_4, fixes a further point and acts regularly on the remaining 12 letters. (In the notation in the proof, the triple fixed by $N^{(01)} \cong A_4$ is $\{\star, 01, 10\}$.) The implied Steiner triple system $S(2, 3, 15)$ is, of course, inherited from the isomorphism $L_4(2) \cong A_8$, and the triples correspond to the non-zero vectors in the 35 2-dimensional subspaces of a 4-dimensional vector space over $\mathbb{Z}_2$.

5.1.1   The action of A_7 on 15 points

The reader who is familiar with the Mathieu group M_{24} can see the actions of A_7 on 7 and 15 letters occurring simultaneously as follows. The stabilizer of an octad in the Steiner system $S(5, 8, 24)$ is isomorphic to $2^4 : A_8$. Fixing a point in the octad and a point outside the octad is a subgroup A_7 with orbits $(1 + 7) + (1 + 15)$. The triples of the Steiner triple system referred to above are those subsets of the 15-orbit which can be completed to an octad by adjoining the two fixed points of the A_7 and three points of the 7-orbit; see Figure 5.1(a), where the triple of black points in the 7-orbit together with the two fixed points (denoted by asterisks) is completed to an octad of the Steiner system by the special triple of white points in the 15-orbit. Thus the triples correspond to the 35 subsets of size 3 in the 7-orbit, and the A_4 fixing each point of a special triple has orbits of size $(3 + 4)$ on the 7-orbit. This shows us that the A_4 in question centralizes an element of order 3 in A_7.

Let $N \cong A_7$ be transitive of degree 15, then N has two orbits on unordered triples: the *special* triples of the Steiner system and the *non-special* triples. The group acts regularly on the non-special triples or, to put it another way, given two ordered non-special triples $[a_1, a_2, a_3]$ and $[b_1, b_2, b_3]$, there is a unique element of the group mapping $a_i \mapsto b_i$ for $i = 1, 2, 3$. In the calculations that follow, we need to be able to work out this element.

5.1.2   The additional relation

Recall that by Lemma 3.3 the only elements of N which can be written as words in just two of the symmetric generators, t_i and t_j say, without causing collapse are those that lie in the centralizer in N of the 2-point

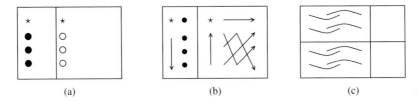

Figure 5.1. A_7 acting on 7 and 15 letters simultaneously.

stabilizer N_{ij}. In our case, $N_{ij} = N_{ijk} \cong A_4$ for $\{i, j, k\}$ a special triple, and $C_N(N_{ijk}) = \langle \pi_{ijk} \rangle \cong C_3$, generated by the element shown in Figure 5.1(b) for the special triple indicated by white dots in Figure 5.1(a), and so the only non-trivial elements of N which can be written as a word w in t_i, t_j and t_k are π_{ijk} and its inverse. Accordingly, we shall factor $P = 2^{\star 15} : A_7$ by a relation of the form $\pi_{ijk} = w(t_i, t_j, t_k)$, where w is chosen to have as short a length as possible. It is readily shown, and is left as an exercise for the reader, that $l(w) \leq 3$ causes collapse, and that the only possibility with $l(w) = 4$ is effectively given by

$$\pi_{ijk} = t_i t_j t_k t_i = (t_j t_k)^{t_i},$$

which is equivalent to requiring that $\pi_{ijk} t_i$ should have order 4. It also shows that $t_j t_k$ is conjugate to π_{ijk}, and so the product of any two symmetric generators has order 3. Since t_i has order 2 and π_{ijk} has order 3, this tells us that $\langle t_i, \pi_{ijk} \rangle = \langle t_i, t_j, t_k \rangle$ satisfies the following presentation:

$$\langle x, y \mid x^2 = y^3 = (xy)^4 = 1 \rangle \cong S_4.$$

This implies that $\langle t_i, t_j, t_k \rangle$ is a homomorphic image of S_4. But any proper image is readily seen to lead to collapse, and so we assume isomorphism. We shall now prove the following theorem.

THEOREM 5.1

$$G = \frac{2^{\star 15} : A_7}{(\pi_{ijk} t_i)^4},$$

is isomorphic to $2 \cdot M_{22} : 2$, the double cover of the automorphism group of M_{22}.

Proof that G maps onto a transitive subgroup of S_{22}. We first note that, by Theorem 3.1, $|G : G'| = 2$ and $G' = G''$; so G contains a perfect subgroup to index 2, but at this stage we cannot say whether G' is trivial or indeed infinite.

The first possibility is easily eliminated by attempting to embed the progenitor in the symmetric group S_{22}. As noted above, A_7 acts on 15 letters in two non-permutation identical ways and, in fact, the stabilizer of a point in one action has orbits of lengths 7 and 8 on the other. If we take A_7 acting on $7 + 15$ letters, restrict to a subgroup $L_3(2)$ of index 15, having orbits $7 + (7 + 8)$ on this set of 22 points, then we find that the two actions of $L_3(2)$ on seven points *are* permutation identical. Thus there is an involution of cycle shape $2^7.1^8$ interchanging these two 7-orbits, fixing each point in the 8-orbit, and commuting with this copy of $L_3(2)$. The adjunction of this element will generate $M_{22} : 2$ together with our original A_7, although without further work all we can observe is that we have obtained a transitive subgroup of S_{22}. Note, though, that this has been achieved without consideration of the additional relation obtained above. In order to show that this relation holds, we must carry out the process explicitly.

Now, the subgroup of M_{24} isomorphic to A_7 which fixes an octad, a point in that octad and a point outside it, also acts on the 15 octads disjoint from the fixed octad and not containing the fixed point outside it. Three such octads will form a special triple if they sum to the empty set (that is to say, if the symmetric difference of two of them is the third). The symmetric generators are odd permutations of S_{22} and so must interchange the two points we have fixed. In Figure 5.1(c) we exhibit a suitable symmetric generator t_i, and in Figure 5.1(b) we show a corresponding permutation π_{ijk} in A_7. It is readily checked that $\pi_{ijk} t_i$ has order 4 as required, and so this transitive subgroup of S_{22} satisfies our symmetric presentation.

5.1.3   Mechanical enumeration of double cosets

We label the 15 points $\{1, 2, \ldots, 15\}$ and the 7 letters $\{a, b, \ldots, g\}$ as shown in the MOG diagram in Figure 5.2. As generators for A_7 we take

$$x = (1\ 2\ 3\ 4\ 5\ 6\ 7)(8\ 9\ 10\ 11\ 12\ 13\ 14)(a\ b\ c\ d\ e\ f\ g)$$

and

$$y = (1\ 11\ 8\ 12)(2\ 6\ 13\ 9)(3\ 4\ 10\ 5)(7\ 15)(a\ g\ b\ d)(f\ e),$$

which are chosen so that $\langle x, y^2 \rangle = L_3(2)$. Note that the action on letters is irrelevant for our purposes but is included to assist those familiar with the MOG arrangement to check that these elements do indeed lie in M_{24} and so generate A_7. Now $\{1, 2, 4\}$ is a special triple (reverting now to the action on the 15 octads rather than the 15 points), and we have

$$\pi_{124} = (1\ 2\ 4)(3\ 13\ 11)(5\ 10\ 9)(6\ 12\ 8)(7\ 14\ 15)(a\ d\ b),$$

which of course generates the centralizer in N of the stabilizer of $1, 2$ and 4. Note that these elements π_{ijk} act fixed-point-free on the 15 points, but as 3-cycles on the seven letters. We are now in a position to use the double coset enumerator, thus we input the following to MAGMA:

```
> s15:=Sym(15);
> x:=s15!(1,2,3,4,5,6,7)(8,9,10,11,12,13,14);
```

o	g	0	7	15	14
a	c	1	3	8	10
b	f	2	6	9	13
d	e	4	5	11	12

Figure 5.2. The alternating group A_7 acting simultaneously on 7 and 15 letters as it appears in the Miracle Octad Generator (MOG).

```
> y:=s15!(1,11,8,12)(2,6,13,9)(3,4,10,5)(7,15);
> N:=sub<s15|x,y>;
> RR:=[<[1,2,4,1],s15!(1,2,4)(3,13,11)(5,10,9)(6,12,8)(7,14,15)>];
> HH:=[N];
> CT:=DCEnum(N,RR,HH:Print:=5,Grain:=100);
```

We obtain the following output:

```
Index: 704 === Rank: 10 === Edges: 22 === Status: Early
closed

> CT[4];
[
    [],
    [ 1 ],
    [ 1, 2 ],
    [ 1, 2, 1 ],
    [ 1, 2, 3 ],
    [ 1, 2, 1, 3 ],
    [ 1, 2, 3, 1 ],
    [ 1, 2, 1, 3, 1 ],
    [ 1, 2, 3, 1, 11 ],
    [ 1, 2, 3, 1, 11, 3 ]
]
> CT[7];
[ 1, 15, 105, 35, 252, 140, 105, 35, 15, 1 ]
```

where `CT[4]` gives the canonical double coset representatives and `CT[7]` gives the corresponding orbit lengths. Thus the fifth entry tells us that the double coset $Nt_1t_2t_3N$ contains 252 single cosets of N. In addition, the enumerator gives all the joins in the Cayley graph of G over N and so we are able to draw the collapsed Cayley diagram given in Figure 5.3. This enumeration shows that the double coset $Nt_1t_2t_3t_1t_{11}t_3N$ contains just one single coset, Nz say, and so the stabilizer of a coset fixes another coset at maximal distance from the first; thus the 704 cosets fall into 352 blocks of imprimitivity of size 2. A central element of G interchanges the cosets in each of these blocks. In fact, this group is the double cover of the automorphism group of the Mathieu group M_{22}, which we denote by $2^{\cdot}M_{22}:2$.

5.1.4   Factoring out the centre

From the enumeration and the shape of the Cayley graph we see that if we fix a coset of N in G a second coset at maximal distance from the first coset is also fixed; and so the group contains a central element which interchanges the two cosets in each of these pairs. Such a central element must have form $z = \pi t_1t_2t_3t_1t_{11}t_3 = \pi w$, say, for some $\pi \in N$. In order to factor out this

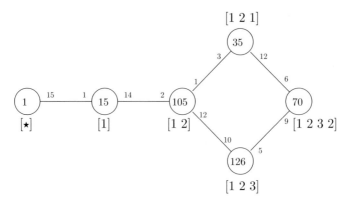

Figure 5.3. Cayley diagram of $M_{22} : 2$ over A_7.

centre we must set the word w equal to $\pi^{-1} = \tau$, say, and by consideration of the action of w on the cosets we can work out what τ must be. For[1]

$$2 \ 1\tau \sim 2^\tau \ 1^\tau \sim 2 \ 1 \ 1 \ 2 \ 3 \ 1 \ 11 \ 3 \sim 3 \ 1 \ 11 \ 3 \sim 3 \ 1 \ 11 \ 3 \ 14 \ 11.11 \ 14 \sim 11 \ 14,$$

where the element $(2, 1, 3, 11, 14, 6, 12)(4, 7, 13, 5, 9, 8, 10) \in N$ shows that $[3, 1, 11, 3, 14, 11]$ is a conjugate of $[1, 2, 3, 1, 11, 3]$. Thus we may choose $1^\tau = 14$ and $2^\tau = 11$. Furthermore,

$$3 \ 2 \ 1 \ \tau = 3 \ 2 \ 1 \ 1 \ 2 \ 3 \ 1 \ 11 \ 3 = 1 \ 11 \ 3 \sim 3^\tau \ 11 \ 14 \sim 6 \ 11 \ 14,$$

the triple $[6, 11, 14]$ being an alternative name for the coset $Nt_1 t_{11} t_3$. So $3^\tau = 6$ and, since $\{1, 2, 3\}$ is a non-special triple, τ is uniquely defined; we have

$$\tau = (1 \ 14 \ 2 \ 11 \ 12 \ 3 \ 6)(4 \ 5 \ 10 \ 13 \ 8 \ 7 \ 9) = [1, 2, 3, 1, 11, 3].$$

At this stage, it is natural to ask whether this relation implies our initial relation. Pleasingly, if we conjugate by the element

$$(1 \ 3)(2 \ 11 \ 4 \ 13)(5 \ 9 \ 6 \ 12)(8 \ 15 \ 10 \ 14) \in N$$

we obtain

$$(3 \ 8 \ 11 \ 4 \ 5 \ 1 \ 12)(13 \ 9 \ 14 \ 2 \ 15 \ 7 \ 6) = [3, 11, 1, 3, 4, 1],$$

and, multiplying, we have

$$(1 \ 2 \ 4)(3 \ 13 \ 11)(5 \ 10 \ 9)(6 \ 12 \ 8)(7 \ 14 \ 15) = \pi_{124}$$
$$= [1, 2, 3, 1, 11, 3, 3, 11, 1, 3, 4, 1] = [1, 2, 4, 1],$$

which is our original relation. The element τ of order 7 obtained here simply emerged from the construction of the Cayley diagram. Before using the double coset enumerator to verify what we have now proved by hand, we replace τ by our original element x and note that it is x^{-1} which is conjugate

---

[1] We are using the abbreviated notation for cosets here, so '2 1' stands for the coset $Nt_2 t_1$.

to τ and that the cycle of x which contains the relevant symmetric generators is $(8, 9, 10, 11, 12, 13, 14)$. The required additional relator now has the delightfully simple form $(t_{13}t_{14}x)^3$, which when multiplied out becomes $x^3 = [12, 11, 13, 12, 14, 13]$.

Note Since $13^x = 14$, this relator takes the form $([t, x]x)^3$, where t is any symmetric generator in the relevant 7-cycle. Explicitly, a 7-element in A_7 acting on 15 points has cycle shape 1.7^2. The 7-cycle we want is the one whose stabilizer (as a set) preserves the fixed point, so it does not consist of the non-zero vectors of a 3-dimensional subspace of the 4-dimensional space on which $A_7 \leq L_4(2)$ acts.

```
> s15:=Sym(15);
> x:=s15!(1,2,3,4,5,6,7)(8,9,10,11,12,13,14);
> y:=s15!(1,11,8,12)(2,6,13,9)(3,4,10,5)(7,15);
> N:=sub<s15|x,y>;
> RR:=[<[12,11,13,12,14,13],x^3>];
> CT:=DCEnum(N,RR,[N]:Print:=5,Grain:=100);
Index: 352 === Rank: 6 === Edges: 12 === Time: 0.07
> CT[4];
[
    [],
    [ 1 ],
    [ 1, 2 ],
    [ 1, 2, 3 ],
    [ 1, 2, 3, 2 ],
    [ 1, 2, 1 ]
]
> CT[7];
[ 1, 15, 105, 126, 70, 35 ]
```

5.1.5   Enumeration of double cosets by hand

It is perhaps surprising given the somewhat complicated appearance of Figure 5.3 that we are able to carry out these same calculations quite quickly by hand. This of course is due to the extent to which we are exploiting the symmetry of the definition: although in each case we have factored the progenitor by a single additional relator, all conjugates of that relator by elements of the control subgroup also hold and can be readily written down. We shall spare the reader the hand calculation for the double cover $2^{\cdot}M_{22}:2$, although there is little extra work involved, but we shall demonstrate the process by enumerating the double cosets NwN for

$$\frac{2^{\star 15} : A_7}{(1\ 14\ 2\ 11\ 12\ 3\ 6)(4\ 5\ 10\ 13\ 8\ 7\ 9) = [1, 2, 3, 1, 11, 3]}.$$

Table 5.1. Conjugates of the additional defining relation

Label	Word	Permutation in N
a_0	[1,2,4,1]	(1 2 4)(3 13 11)(5 10 9)(6 12 8)(7 14 15)(a d b)
b_0	[1,12,13,1]	(1 12 13)(2 15 9)(3 14 8)(4 5 7)(6 11 10)(b c f)
c_0	[1,8,15,1]	(1 8 15)(2 5 3)(4 13 10)(6 14 9)(7 11 12)(d g f)
d_0	[1,9,11,1]	(1 9 11)(2 6 7)(3 15 10)(4 14 12)(5 13 8)(c e d)
e_0	[1,10,14,1]	(1 10 14)(2 11 8)(3 4 6)(5 15 12)(7 13 9)(b g e)

We have already proved that the relations of the form $t_i t_j t_k t_i = \pi_{ijk}$ are a consequence of this relation, for $\{i, j, k\}$ a special triple; we shall assume, moreover, that, given two non-special triples $A = [a_1, a_2, a_3]$ and $B = [b_1, b_2, b_3]$, we are able to find the unique element referred to in Section 5.1.1 which maps A to B. This last operation can be performed manually using Figure 5.2 (which is why we have appended the action on the seven letters) or, of course, mechanically. For each double coset NwN, where w is a word in the 15 symmetric generators, we must obtain the *coset stabilizing subgroup* $N^{(w)}$ defined as usual by $N^{(w)} = \{\pi \in N \mid Nw\pi = Nw\}$. Recall that the index of $N^{(w)}$ in N gives the number of single cosets in NwN. Suppose that $N^{(w)}$ has r orbits on the 15 points and that $i_1, i_2, \ldots, i_r$ are representatives of those orbits. In order to prove closure, we must identify to which double coset Nwt_{i_k} belongs for $k = 1, 2, \ldots, r$.

In Table 5.1 we display a dictionary of the elements $\pi_{ijk} = t_i t_j t_k t_i$ for the various special triples $\{i, j, k\}$, but, instead of writing out all 35 of them, we shall simply give a representative for each orbit under conjugation by the element $x = (1\ 2\ 3\ 4\ 5\ 6\ 7)(8\ 9\ 10\ 11\ 12\ 13\ 14)(a\ b\ c\ d\ e\ f\ g)$ of order 7. We label these five orbit representatives $a_0, \ldots, e_0$, and denote $x^{-i} u_0 x^i$ by u_i for $u \in \{a, b, c, d, e\}$.

Conjugation by π_{ijk} and inversion show that

$$\pi_{ijk} = \pi_{jki} = \pi_{kij} = \pi_{ikj}^{-1} = \pi_{kji}^{-1} = \pi_{jik}^{-1}.$$

Proceeding, we note that the double coset $Nt_1 N$, whose canonical double coset representative (CDCR) t_1 is denoted by [1], appears to contain just the 15 single cosets Nt_i. Since N acts doubly transitively on the 15 symmetric generators, the double coset $Nt_1 t_2 N$ consists of the union of all single cosets of the form $Nt_i t_j$. But, since $t_1 t_2 t_4 t_1 \in N$, $\{1, 2, 4\}$ being a special triple, we have that $Nt_1 t_2 = Nt_1 t_4$, and so there are at most $(15 \times 14)/2 = 105$ such single cosets. The previous sentence also shows us that $Nt_1 t_2 t_4 N = Nt_1 N$, and so the only double cosets NwN with $l(w) = 3$ we need consider are $Nt_1 t_2 t_1 N$ and $Nt_1 t_2 t_3 N$, $\{1, 2, 3\}$ being a non-special triple. The CDCRs for these two double cosets are thus $[1, 2, 1]$ and $[1, 2, 3]$, respectively.

We must now do some calculations using the additional relation. Using the abbreviated notation whereby $Nt_i t_j$ is denoted by ij, for instance, we see that $1\ 2 \sim 1\ 4$ and so $1\ 2\ 1 \sim 1\ 4\ 1$. Moreover,

$$1\ 2\ 1 \sim 1.2\ 1\ 4\ 2.2\ 4 \sim 1^{\pi_{214}}\ 2\ 4 \sim 4\ 2\ 4,$$

and so

$$1\ 2\ 1 \sim 1\ 4\ 1 \sim 2\ 4\ 2 \sim 2\ 1\ 2 \sim 4\ 1\ 4 \sim 4\ 2\ 4.$$

Thus there are at most $105/3 = 35$ distinct single cosets in the double coset $Nt_1 t_2 t_1 N$, corresponding to the 35 special triples. Again, since $\{2, 3, 5\}$ is a special triple,

$$1\ 2\ 3 \sim 1.2\ 3\ 5\ 2.2\ 5 \sim 1^{\pi_{235}}\ 2\ 5 = 8\ 2\ 5,$$

where $\pi_{235} = a_1$. Of course, we also have that $1\ 2\ 3 \sim 1\ 4\ 3$ and, from our additional relation, $1\ 2\ 3 \sim 3\ 11\ 1$. In order to obtain generators for the coset stabilizing subgroup $N^{(w)}$ for $w = t_1 t_2 t_3$, we need elements of $N \cong A_7$ which map the ordered triple $[1, 2, 3]$ in turn to $[1, 4, 3], [8, 2, 5]$ and $[3\ 11\ 1]$. This is achieved as follows:

$$\rho_1 = (2\ 4)(6\ 5)(9\ 12)(11\ 13)(15\ 14)(8\ 10)(b\ d)(f\ e),$$

$$\rho_2 = (1\ 8)(3\ 5)(4\ 11)(6\ 14)(7\ 13)(10\ 12)(b\ c)(d\ f)$$

and

$$\rho_3 = (1\ 3)(2\ 11\ 4\ 13)(5\ 9\ 6\ 12)(8\ 15\ 10\ 14)(a\ g)(b\ e\ d\ f),$$

respectively. We soon see that when $w = t_1 t_2 t_3$,

$$N^{(w)} \geq \langle \rho_1, \rho_2, \rho_3 \rangle \cong F_{20},$$

a Frobenius group of order 20 with two orbits: $\{2, 11, 7, 13, 4\}$ and the remaining ten points. So, the double coset $Nt_1 t_2 t_3 N$ contains at most $2520/20 = 126$ single cosets. So ten of the symmetric generators return us to the double coset $Nt_1 t_2 N$, and we must consider the double coset $Nt_1 t_2 t_3 t_2 N$. Now, the coset $Nt_1 t_2 t_3 t_2$ is certainly fixed by

$$\rho_4 = \rho_3^{(\rho_2 \rho_1)^2} = (9\ 15)(1\ 6\ 8\ 14)(3\ 10\ 5\ 12)(4\ 7\ 11\ 13) \in F_{20},$$

since it fixes the coset $Nt_1 t_2 t_3$ and the symmetric generator t_2. Moreover,

$$1\ 2\ 3\ 2 \sim 1\ 2.3\ 2\ 5\ 3.3\ 5 \sim 1\ 2\pi_{325}3\ 5 \sim 15\ 5\ 3\ 5,$$

since we may write down $\pi_{325} = a_1^{-1}$ from Table 5.1. We thus seek the element which maps $[1, 2, 3]$ to $[15, 5, 3]$, and we obtain

$$\rho_5 = (1\ 15)(2\ 5)(4\ 12)(6\ 9)(7\ 10)(11\ 13)(b\ c)(f\ g)$$

when $\langle \rho_4, \rho_5 \rangle \cong (S_3 \wr C_2)^e$, the even permutations fixing e and preserving the partition abc/dfg. This group has order $6 \times 6 \times 2/2 = 36$, and so the double coset $Nt_1 t_2 t_3 t_2 N$ contains at most $2520/36 = 70$ single cosets. Moreover, it has (at most) two orbits of lengths 6 and 9 on the 15 points: $\{1, 6, 8, 9, 14, 15\}$ and the remainder. We are close to verifying that the Cayley diagram shown in Figure 5.3 has the maximum possible number of cosets of N in G. It remains to show that the double cosets $Nt_1t_2t_1N$ and $Nt_1t_2t_3t_2N$ are joined in the Cayley diagram. Note that if they are joined then there must be six joins from each coset of the latter double coset to the former, as indicated in Figure 5.3 Now,

$$1\ 2\ 1\ 3 \sim 1\ 2.1\ 3\ 7\ 1.1\ 7 \sim 3\ 12\ 1\ 7,$$

since

$$\pi_{137} = a_6 = (1\ 3\ 7)(2\ 12\ 10)(4\ 9\ 8)(5\ 11\ 14)(6\ 13\ 15)(a\ g\ c).$$

The element

$$\rho_6 = (1\ 3)(2\ 12)(4\ 9)(5\ 13)(6\ 11)(14\ 15)(a\ g)(e\ f)$$

maps $[1, 2, 3] \mapsto [3, 12, 1]$, and so the 5-orbit of the stabilizer of the coset $Nt_3t_{12}t_1$ is $\{2, 11, 7, 13, 4\}^{\rho_6} = \{12, 6, 7, 5, 9\}$. So 12 and 7 are in the same orbit, and we have

$$3\ 12\ 1\ 7 \sim u\ 7\ v\ 7$$

for some u and v. This implies $Nt_1t_2t_1t_3N = Nt_1t_2t_3t_2N$, as required.

5.1.6   Group presentations which describe our symmetric presentations

Our permutations x and y of degree 15 which generate A_7, see Section 5.1.3, satisfy the following presentation:

$$\langle x, y \mid x^7 = y^4 = (xy^2)^3 = (x^3y)^3 = (xyx^2y^{-1})^2 = 1 \rangle \cong A_7,$$

in which we have $\langle x, y^2 \rangle \cong L_3(2)$. Thus, if we adjoin an involution t, corresponding of course to t_{15}, which commutes with x and y^2, we obtain a presentation for the progenitor,

$$P = 2^{\star 15} : A_7,$$

in which x and y act by conjugation on the 15 symmetric generators in the manner given above. The element x^2yx^4y is a 3-cycle of A_7 in its 7-point action which has 15 in its 'special' 3-cycle when it acts fixed-point-free on 15 letters. So, our additional relator may be taken to be $(x^2yx^4yt)^4$, and we have proved that

$$\langle x, y, t \mid x^7 = y^4 = (xy^2)^3 = (x^3y)^3 = (xyx^2y^{-1})^2$$
$$= t^2 = [t, x] = [t, y^2] = (x^2yx^4yt)^4 = 1 \rangle \cong 2 \dot{} M_{22} : 2.$$

A diagram of the Cayley graph of $2 \dot{} M_{22} : 2$ acting on the cosets of A_7, together with the labelling of the double cosets, is shown in Figure 5.4. Now, $t_{15}^{yxy} = t_{11}$, and so an equivalent relation to that found above which factors out the centre (and implies the final relation in the preceding presentation) is $([t^{yxy}, x]x)^3$, so we have also shown that

$$\langle x, y, t \mid x^7 = y^4 = (xy^2)^3 = (x^3y)^3 = (xyx^2y^{-1})^2$$
$$= t^2 = [t, x] = [t, y^2] = ([t^{yxy}, x]x)^3 = 1 \rangle \cong M_{22} : 2.$$

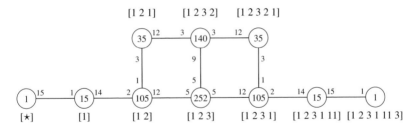

Figure 5.4. The Cayley graph of $2'M_{22}: 2$ over A_7.

The Mathieu group M_{22} is unique among the sporadic simple groups in that it has a Frobenius–Schur multiplier of order 12. By replacing the control subgroup A_7 by its triple cover $3'A_7$ in the above presentation and making the appropriate modifications, we obtain the following:

$$\langle x, y, t \mid x^7 = y^4 = (xy^2)^3 = (x^3 y)^3 = (yx)^5 (xyx^2 y^{-1})^2$$
$$= t^2 = [t, x] = [t, y^2] = (t(yx)^5)^2 = (x^2 yx^4 yt)^4 = 1\rangle$$
$$\cong 6'M_{22} : 2.$$

5.2   The Janko group J_1

Our definition and construction of the smallest Janko group J_1 takes as control subgroup the simple group $L_2(11)$ of order 660 acting on 11 letters. In fact, we shall study the progenitor

$$P := 2^{*11} : L_2(11),$$

and so we need a good understanding of the way in which the group $N \cong L_2(11)$ acts on the 11 letters $\Lambda = \{0, 1, 2, \ldots, X\}$, where X stands for 10. We recall a few well known facts about this action.

5.2.1   The group $L_2(11)$ acting on 11 letters

The following hold.

(i) N acts doubly transitively with point-stabilizer $N^0 \cong A_5$, the alternating group of order 60, and 2-point stabilizer $N^{01} \cong S_3$, the symmetric group of order 6.

(ii) N has two orbits on triples, one of which consists of the fixed point sets of the 55 involutions. We call these the 55 *special triples*.

(iii) N has an orbit of 11 *special pentads*, the lines of the famous 11-point *biplane*, such that every pair of points belongs to precisely two special pentads. For the copy of $L_2(11)$ that we use below, these special pentads are the set of quadratic non-residues modulo 11, namely

$\{2, 6, 7, X, 8\}$ and its translates. Indeed, $L_2(11)$ consists precisely of those permutations of the 11 letters which preserve this set of 11 pentads.

(iv) This copy of the group is generated by the permutations

$$(0, 1, 2, 3, 4, 5, 6, 7, 8, 9, X) \text{ and } (3, 4)(2, X)(5, 9)(6, 7),$$

and its elements of order 6 have cycle shape 2.3.6. An example of such an element in this group is $(3,4)(0,1,8)(2,5,6,X,9,7)$. Since the point-stabilizer of a special triple has order 2, there is precisely one element of order 6 containing a given 3-cycle. We thus denote the above element by σ_{810}, where $\sigma_{ijk} = \sigma_{jki} = \sigma_{kij} = \sigma_{kji}^{-1}$ for any special triple $\{i, j, k\}$.

5.2.2 An image of the progenitor $2^{\star 11} : L_2(11)$

We factor this progenitor by a relation which says that an element of order 6 in N times a symmetric generator in its 3-cycle has order 5, and we obtain Theorem 5.2,[2] which we shall prove both mechanically and manually in the following two sections.

THEOREM **5.2**

$$G = \frac{2^{\star 11} : L_2(11)}{(\sigma_{018} t_0)^5} \cong J_1,$$

the smallest Janko group.

Proof that the index of $L_2(11)$ in G is less than or equal to 266. Writing σ for σ_{018}, we see that

$$(\sigma t_0)^5 = \sigma t_0 \sigma t_0 \sigma t_0 \sigma t_0 \sigma t_0 = \sigma^5 t_0^{\sigma^4} t_0^{\sigma^3} t_0^{\sigma^2} t_0^{\sigma} t_0 = \sigma^{-1} t_8 t_0 t_1 t_8 t_0 = e,$$

where e is the identity. Thus, $\sigma_{018} = t_8 t_0 t_1 t_8 t_0$, and, analogously,

$$\sigma_{ijk} = t_i t_j t_k t_i t_j \text{ for any special triple } \{i, j, k\}.$$

As usual, we denote the symmetric generator t_i by i when this relation becomes

$$\sigma_{ijk} = ijkij \text{ for any special triple } \{i, j, k\}.$$

We also let i stand for the coset Nt_i when there is no fear of ambiguity; for words u and v in the symmetric generators, we write $u \sim v$ to mean $Nu = Nv$. Thus $ij \sim k$ would mean $Nt_i t_j = Nt_k$. The single coset N is denoted by $\star$, and so $ijk \sim \star$ means that $t_i t_j t_k \in N$.

Table 5.2. Involutions and 6-elements of $N = L_2(11)$
written as words in the symmetric generators

$01801 =$	$(0, 8, 1)(2, 7, 9, X, 6, 5)(3, 4)$	$= \sigma_{018}$
$03203 =$	$(0, 2, 3)(6, X, 5, 8, 7, 4)(9, 1)$	$= \sigma_{032}$
$09609 =$	$(0, 6, 9)(7, 8, 4, 2, X, 1)(5, 3)$	$= \sigma_{096}$
$05705 =$	$(0, 7, 5)(X, 2, 1, 6, 8, 3)(4, 9)$	$= \sigma_{057}$
$04X04 =$	$(0, X, 4)(8, 6, 3, 7, 2, 9)(1, 5)$	$= \sigma_{04X}$
$01010 =$	$(0, 1)(2, 6)(7, X)(3, 4)(8)(5)(9) =$	π_{01}
$03030 =$	$(0, 3)(6, 7)(X, 8)(9, 1)(2)(4)(5) =$	π_{03}
$09090 =$	$(0, 9)(7, X)(8, 2)(5, 3)(6)(1)(4) =$	π_{09}
$05050 =$	$(0, 5)(X, 8)(2, 6)(4, 9)(7)(3)(1) =$	π_{05}
$04040 =$	$(0, 4)(8, 2)(6, 7)(1, 5)(X)(9)(3) =$	π_{04}

Following ref. [34], we now deduce another family of relations:

$$85858 = 10.01801.10510.01801.10510.01801.10$$
$$= 10.\sigma_{018}.\sigma_{105}\sigma_{018}\sigma_{105}\sigma_{018}.10$$
$$= 10.(5, 8)(0, 1)(2, 4)(3, 6).10$$
$$= 1001.(5, 8)(0, 1)(2, 4)(3, 6)$$
$$= (5, 8)(0, 1)(2, 4)(3, 6) = \pi_{58},$$

say.

Of course, this calculation was possible because both $\{0, 1, 5\}$ and $\{0, 1, 8\}$ are both special triples. The involution π_{ij} is determined by i and j since $C_N(N^{ij}) = \langle \pi_{ij} \rangle$, where N^{ij} denotes the stabilizer in N of i and j. Thus we have the generic relation

$$\pi_{ij} = ijiji.$$

Representatives for the orbits of elements σ_{ijk} and π_{ij} under the action of $\alpha = (0, 1, 2, 3, 4, 5, 6, 7, 8, 9, X)$ are given in ref. [34], Table 1, and are reproduced here in Table 5.2.

As usual, we now seek the index of $N \cong L_2(11)$ in G by manually enumerating double cosets of the form $NwN = [w]$, where w is a word in the symmetric generators. We immediately have the double cosets N, Nt_0N and Nt_0t_1N, which appear to contain 1, 11 and 110 single cosets, respectively. Moreover,

$$Nt_0t_1t_0t_1t_0 = N \Rightarrow Nt_0t_1t_0 = Nt_0t_1 \Rightarrow [010] = [01];$$
$$Nt_0t_1t_8t_0t_1 = N \Rightarrow Nt_0t_1t_8 = Nt_1t_0 \Rightarrow [018] = [10] = [01];$$

and so the next double coset to investigate is $Nt_0t_1t_2N = [012]$, since $\{0, 1, 2\}$ is a non-special triple. Now,

$$0120 \sim 018.820 \sim 10\sigma_{820}28 \sim 32.28 \sim 38 \Rightarrow 012 \sim 380,$$

where σ_{820} has been obtained by adding 2 to the third row of Table 5.2. This says that the coset $Nt_0t_1t_2$ is fixed by any permutation which maps 0, 1 and 2 to 3, 8 and 0, respectively. Since N acts transitively and regularly on the ordered non-special triples, there is a unique such permutation. In this case, we have

$$\rho_1 = (0, 3, 5, 6, 2)(1, 8, 4, 9, X)(7) \in N^{(012)},$$

and so

$$012 \sim 380 \sim 543 \sim 695 \sim 2X6.$$

Thus there are at most $11 \times 10 \times 6/5 = 132$ single cosets in $[012]$. We must now investigate double cosets of the form $[012i]$; under the action of ρ_1 above, we may assume that $i = 0, 1$ or 7. But $012.0 \sim 380.0 \sim 38$, and so $[0120] = [01]$; also $012.1 \sim 0.12121.12 \sim 0\pi_{12}12 \sim 812$, a non-special triple, and so $[0121] = [012]$. This leaves the double coset $[0127]$. The single coset 0127 is fixed by ρ_1, and

$$0127 \sim 019.927 \sim 10.\sigma_{927}.29 \sim 4629.$$

So $\rho_2 = (0, 4, 3, 8, X)(1, 6, 5, 7, 9)$ also fixes coset 0127. But $\langle \rho_1, \rho_2 \rangle \cong 11 : 5$, a Frobenius group of order 55 acting transitively on the 11 letters. Thus the double coset $[0127]$ contains at most $132/11 = 12$ single cosets, and we *appear* to have constructed a group of order

$$660 \times (1 + 11 + 110 + 132 + 12) = 660 \times 266 = 175\ 560,$$

whose Cayley graph[3] over N with respect to the 11 generators has the form given in Figure 5.5.

5.2.3   Mechanical enumeration

Considering what is being attempted, this hand enumeration is quite concise. However, it is of interest to record just how rapidly the double coset enumerator deals with this example. We must, of course, feed in our two generators for the control subgroup $L_2(11)$ acting on 11 letters. The single additional relation then takes a particularly simple form as the permutation σ has order 6; thus $1 = (\sigma t_0)^5$ reduces to $\sigma = t_1 t_0 t_8 t_1 t_0$ or, in the sequence-permutation form in which we input it: $< [1, 11, 8, 1, 11], \mathrm{sig} >$ (note that 0 has been replaced by 11):

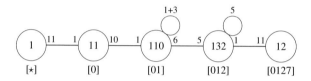

Figure 5.5. Cayley diagram of J_1 over $L_2(11)$.

---

[3] The corresponding transitive graph in which Nw is joined to Nt_iw is called the Livingstone graph.

```
> lll:=sub<Sym(11)|(1,2,3,4,5,6,7,8,9,10,11),(3,4)(2,10)(6,7)(5,9)>;
> #lll;
660
> sig:=lll!(3,4)(11,1,8)(2,5,6,10,9,7);
> RR:=[<[1,11,8,1,11],sig>];
> CT:=DCEnum(lll,RR,[lll]:Print:=5,Grain:=100);
Dealing with subgroup.
Pushing relations at subgroup.
Main part of enumeration.
PK          5 5 5    266 782 914    0 11    0.000    0
Index: 266 === Rank: 5 === Edges: 11 === Time: 0.015
> ^C
> CT[4];
[
    [],
    [ 1 ],
    [ 1, 2 ],
    [ 1, 2, 3 ],
    [ 1, 2, 3, 8 ]
]
> CT[7];
[ 1, 11, 110, 132, 12 ]
```

5.2.4 Existence of J_1

The smallest Janko group, J_1, is well known to be a simple permutation group of degree 266, and can easily be shown to satisfy the assumptions of our construction. In order to *prove* the existence of such a group from our construction, however, we need a little more.

THEOREM 5.3 The presentation of Theorem 5.2 either defines a simple group of order 175 560 acting as a permutation group on 266 letters in the manner described, or it defines the trivial group.

Proof Any relation which holds in the group defined by Theorem 5.2 has the form $\pi w = 1$, where $\pi \in N$ and w is a word in the 11 symmetric generators. Using the reduction process of the construction, we may assume that w has length at most 4. In particular, we may assume that w is trivial, or that one of t_0, $t_0 t_1$, $t_0 t_1 t_2$ and $t_0 t_1 t_2 t_7$ is in N. In the first case, the relation reduces to $\pi = 1$, so either the relation is trivial or some non-trivial element of N is set equal to the identity. This causes the simple group N to collapse to the trivial group and, since $(\sigma_{018} t_0)^5 = 1 = t_0^2$, t_0 also collapses to the identity and $G = 1$.

Now, $t_0 \in N$ implies that $t_0 \in C_N(N^0)$, which is trivial. Thus all the symmetric generators are trivial, and so $\pi_{01} = t_0 t_1 t_0 t_1 t_0 = 1$ as well, and we have $G = 1$.

Again, $t_0 t_1 \in N$ implies that $t_0 = \pi_{01}(t_0 t_1)^2 \in N$ and so $G = 1$, as above.

Now, changing to the abbreviated notation, we have that if $012 \sim \star$, then $01 \sim 2 \sim 2^{\pi_{34}} = X$, since the element $\pi_{34} = (3\ 4)(2\ X)(6\ 7)(5\ 9)$ fixes 0 and 1. But then $2X \sim \star$, and so $01 \sim \star$, by the double transitivity of N. This takes us back to the previous case, and so $G = 1$ as above.

Finally, if $0127 \sim \star$, then $01 \sim 72 \sim (72)^\pi \sim 24$, where $\pi = (5\ 8\ 9)(3\ X\ 6)(4\ 7\ 2)$ is chosen to fix 0 and 1. Thus $2427 \sim 247 \sim \star$, and so $012 \sim \star$, by the transitivity of N on ordered non-special triples. Again we have reduced our calculation to a previous case, and we may conclude that $G = 1$.

Thus, if any relation of which we are not aware holds in G, or if we factor G by any new relator, we get total collapse to the trivial group. Thus either G is simple with the degree and order given, or it is trivial.   □

In order to complete the existence proof, we must exhibit a non-trivial representation of the progenitor in which the additional relation holds. There are two approaches, which we present in Sections 5.2.5 and 5.2.6.

5.2.5   The permutation representation of G of degree 266

Each of the 266 nodes, which we claim correspond to cosets of $L_2(11)$ in a larger group, is labelled using the 11 *points* $\Lambda = \{0, 1, \ldots, X\}$ on which $L_2(11)$ acts doubly transitively. The fixed node is labelled $\star$, and the nodes in the 11-orbit and those in the 110-orbit are labelled uniquely by points and ordered pairs of points of Λ, respectively. Nodes in the 132-orbit correspond to elements of order 5 in a complete conjugacy class and are labelled by ordered non-special triples of points of Λ, each in five different ways which can be read off from the corresponding 5-element. For the convenience of the reader, representatives of the orbits on these triples under the action of the element $\alpha = (0\ 1\ 2\ 3\ 4\ 5\ 6\ 7\ 8\ 9\ 10)$ are displayed in Table 5.3, which is taken from ref. [34]. As we have seen, an element of order 5 of cycle shape 1.5^2 permutes the five possible *names* of a letter in this orbit, and the second and third points in a name fall in different 5-cycles. Thus, 120 of the letters in this orbit have unique names of the form $ij0$ or $i0k$. In the remaining $132/11 = 12$ cases, 0 is the fixed point of the aforementioned permutation of order 5, and so if ijk is any name of such a node then $ijk0$ is in the 12-orbit. For convenience, we choose the top name in each of the 12 blocks of Table 5.2 for these 12 points, and this name followed by 0 for the points of the 12-orbit. Of course, each node in the 12-orbit, which corresponds to the set of Sylow 11-subgroups, has 55 names of which five end in 0. The action of $L_2(11)$ on the 266 nodes is well defined, and the symmetric generator 0 is claimed to act on the nodes by right multiplication. Indeed we see that

$$t_0 : (\star\ 0)(i\ i0)(0i)(ij\ ji)(kl\ kl0)(k0l\ k^{\pi_{0l}}0l)(klm\ klm0),$$

Table 5.3. *The 132-orbit showing the correspondence with pairs of points of the projective line*

(X)361		(X)136		(8)397		(2)95X		(6)548		(7)412	
(8)973		(7)241		(X)613		(8)739		(2)X95		(6)854	
(2)5X9	0∞	(5)9X2	X9	(4)586	85	(1)427	24	(3)16X	61	(9)378	73
(6)485		(3)879		(9)2X5		(5)684		(4)721		(1)X63	
(7)124		(4)658		(1)742		(3)X16		(9)837		(5)29X	
(6)1X3		(3)6X1		(9)783		(5)X29		(4)865		(1)274	
(2)471		(5)846		(4)217		(1)63X		(3)798		(9)X52	
(8)564	∞0	(7)938	9X	(X)592	58	(8)456	42	(2)147	16	(6)31X	37
(X)925		(X)259		(8)645		(2)714		(6)X31		(7)893	
(7)389		(4)172		(1)3X6		(3)987		(9)52X		(5)468	

The corresponding 5-elements are read vertically.

where $\{0, i, j\}$ is special, $\{0, k, l\}$ is non-special and 0 is fixed by the 5-element corresponding to the ordered non-special triple $\{k, l, m\}$. The brackets in the expression for t_0 above indicate the action on 2, 20, 10, 30, 120, 60 and 24 letters, respectively, and we observe that t_0 fixes just ten letters. This *is* a well defined permutation on 266 letters; we must show that it commutes with N^0, the subgroup of $L_2(11)$ isomorphic to A_5 fixing 0. Let $\nu \in N^0$. Then ν clearly preserves the fixed points of t_0 and the transpositions of each of the six types, except possibly $(k0l\ k^{\pi_{0l}}0l)$. We must verify that

$$(k0l\ k^{\pi_{0l}}0l)^\nu = (k^\nu 0l^\nu\ k^{\nu\pi_{0l^\nu}}0l^\nu),$$

i.e. that $k^{\pi_{0l}\nu} = k^{\nu\pi_{0l^\nu}}$. But this is plainly true, because $(\pi_{0l})^\nu = \pi_{0l^\nu}$ and so $\nu\pi_{0l^\nu} = \pi_{0l}\nu$. This shows that the permutation group $\langle N, t_0 \rangle$ on 266 letters *is* an image of $2^{\star 11} : L_2(11)$. It remains to check that the additional relation holds, namely that $(3\ 4)(0\ 1\ 8)(2\ 5\ 6\ X\ 9\ 7)t_0$ does indeed have order 5. This we may do long-hand. Thus, $\sigma_{081}t_0$:

($\star$ 0 10 18 8)(34)(309)(20X) (7830 1X3 907 X05 709)(2740 397 90X 102 406)
(1 80 01 81 08)(43)(405)(X02)(95X0 361 506 209 605)(4120 X29 502 10X 307)

(2 50 610 603 78) (25 560 6X10 783 804)(X6 9X0 903 X290 136)
(X 90 710 704 68)(X9 970 5480 95X 803)(27 520 504 3970 865)

(3 40 310 806 48)(03 140 X2 950 84)(32 450 3610 6X1 74)(3X 94 730 1360 274)
(4 30 410 807 38)(04 130 2X 590 83)(4X 390 1X30 548 63)(42 53 640 8650 412)

(5 60 X10 17 28)(6 X0 19 870 58)(31 480 401 106 408)(21 580 201 604 708)
(9 70 210 16 X8)(7 20 15 860 98)(41 380 301 107 308)(X1 980 X01 703 608)

(13 840 85 06 1X0)(12 850 59 76 2X0)(37 420 X06 72 250)(39 470 9X 790 306)
(14 830 89 07 120)(1X 890 95 67 X20)(46 3X0 207 6X X90)(45 360 52 650 407)

(02 51 680 701 87)(05 160 908 71 82)(35 460 204 205 24)(36 X4 930 60X 103)
(0X 91 780 601 86)(09 170 508 61 8X)(49 370 X03 X09 X3)(47 23 540 702 104)

(29 75 62 X50 509)(92 57 26 5X0 30X)(73 240 805 79 270)(93 740 97 720 50X)
(X5 69 7X 290 905)(5X 96 X7 920 402)(64 X30 809 65 X60)(54 630 56 6X0 902),

which does indeed have order 5 as required. Note that this element must commute with $\sigma_{081}^3 = (3\ 4)(2\ \text{X})(5\ 9)(6\ 7)$. Using the fact that $(\sigma t_0)^{-1} = t_0\sigma^{-1} = \sigma^{-1}t_{\sigma^{-1}(0)} = \sigma^{-1}t_8$, we may also see that this element is inverted by $(0\ 8)(5\ 7)(6\ 9)(3\ 4)$. The calculation is straightforward, if somewhat tedious, and is facilitated by the lists of corresponding names as given in ref. [34] and reproduced in Table 5.3.

5.2.6   The matrix representation of G of degree 7

In the remainder of this section we describe an alternative representation of the group G defined by Theorem 5.2 and, at the same time, prove by hand some important results about J_1 that have previously been proved by machine. We shall, in fact, build a matrix representation of G of degree 7 over the field $\mathbb{Z}_{11}$ (as was exhibited in Janko's original papers [58, 59]). Explicitly, we take the irreducible 7-dimensional representation of $L_2(11)$ over $\mathbb{Z}_{11}$ and restrict to a subgroup isomorphic to the alternating group A_5. The underlying module then decomposes as the direct sum of submodules of degrees 3 and 4. If we take a basis for the module consisting of the union of bases of the submodules, the only involution in $SL_2(11)$ commuting with this A_5 is clearly of the form $\text{diag}(1^3, -1^4)$. This constructs a representation of the progenitor, and it remains to show that the additional relation holds. The Coxeter diagram (see Section 3.8) in Figure 5.6 represents a presentation for J_1 obtained computationally by Soicher [81] and reproduced in the ATLAS [25]. Note that the order of de is placed in square brackets to indicate that it is redundant, in the sense that it is implied by the other relations. In what follows we shall show that that this presentation defines our progenitor factored by the additional relation. We shall then exhibit elements of $SL_2(11)$, not all trivial, which satisfy the presentation and which therefore, by Theorem 5.3, generate a simple group of order 175 560.

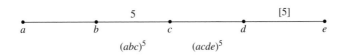

Figure 5.6. Coxeter diagram for $G \cong J_1$.

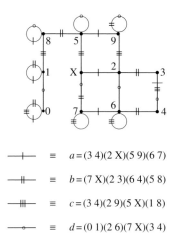

$$\underline{\quad\vdash\quad} \;\equiv\; a = (3\ 4)(2\ X)(5\ 9)(6\ 7)$$

$$\underline{\quad\Vdash\quad} \;\equiv\; b = (7\ X)(2\ 3)(6\ 4)(5\ 8)$$

$$\underline{\quad\VVdash\quad} \;\equiv\; c = (3\ 4)(2\ 9)(5\ X)(1\ 8)$$

$$\underline{\quad\multimap\quad} \;\equiv\; d = (0\ 1)(2\ 6)(7\ X)(3\ 4)$$

Figure 5.7. Cayley diagram of $G_2 = \langle a, b, c, d \rangle$ over $G_1 = \langle a, b, c \rangle$, showing the action on 11 points.

In Figure 5.6, $G_1 = \langle ab, c \rangle$ is a (2,3,5)-group, and thus is an image of the alternating group A_5. But $(bc)^5 = 1 = b.c.c^b.c.c^b.c$ and $c^b = c^{ab}$, since $[a, c] = 1$. Thus $b = (c.c^{ab})^2.c \in G_1$, and hence $G_1 = \langle ab, c \rangle = \langle a, b, c \rangle$.

In order to identify the subgroup $G_2 = \langle a, b, c, d \rangle$, we construct a Cayley graph of the action of G_2 on the cosets of its subgroup G_1, to obtain the graph shown in Figure 5.7.

This is probably best achieved by extending a Cayley graph for $\langle a, b, c \rangle$ over $\langle a, b \rangle$, the details being left to the reader. Indeed, since d commutes with $\langle a, b \rangle \cong S_3$, we see that

$$G_2 \cong \frac{2^{\star 10} : A_5}{[(0\ 3)(1\ 2)t_{01}]^3},$$

where the ten symmetric generators correspond to the unordered pairs of $\Lambda = \{0, 1, 2, 3, 4\}$. The techniques of symmetric generation can then be used to give the same result. By one or other of these methods, we see that G_2 is a homomorphic image of a group of permutations of 11 letters of order $11 \times 60 = 660$. Since the permutations a, b, c, d given above preserve the special pentads described Section 5.2.1(iii) the group defined by the subgraph on $\{a, b, c, d\}$ is plainly $L_2(11)$.

Now, e commutes with $\langle a, b, c \rangle \cong A_5$, and so $G_3 = \langle a, b, c, d, e \rangle$ is an image of $2^{\star 11} : L_2(11)$, in which e corresponds to the symmetric generator t_0. But $acd = (3\ 4)(0\ 1\ 8)(2\ 5\ 6\ X\ 9\ 7\ ) = \sigma_{018}^{-1}$, and so

$$G_3 \cong \frac{2^{\star 11} : L_2(11)}{(\sigma_{018}t_0)^5},$$

where we are ignoring the 'redundant' relation $(de)^5$. But we have shown that this group is trivial, or simple of order $175\,560$.[4] It remains to produce non-trivial elements of $SL_7(11)$ which satisfy the presentation of Figure 5.6. As explained at the beginning of this section, we represent our element e by the matrix $\mathrm{diag}(1^3, -1^4)$. That the matrices in Section 5.2.7 satisfy the presentation can readily be checked by hand. They thus generate a simple group of order $175\,560$, isomorphic to the permutation group on 266 letters obtained earlier.

5.2.7 The matrices

It is readily checked that the following matrices satisfy the presentation given by the Coxeter diagram of Figure 5.5 and thus that they generate a simple group of order $175\,560$ isomorphic to the permutation group of degree 266 defined above.

$$
a = \begin{pmatrix}
-1 & . & . & . & . & . & . \\
. & . & 1 & . & . & . & . \\
. & 1 & . & . & . & . & . \\
. & . & . & 1 & . & . & . \\
. & . & . & . & -1 & . & . \\
. & . & . & . & . & 1 & . \\
. & . & . & . & . & . & 1
\end{pmatrix}, \quad
b = \begin{pmatrix}
-1 & . & . & . & . & . & . \\
. & -1 & -1 & . & . & . & . \\
. & . & 1 & . & . & . & . \\
. & . & . & 1 & . & . & . \\
. & . & . & . & -1 & . & . \\
. & . & . & . & . & -1 & -1 \\
. & . & . & . & . & . & 1
\end{pmatrix},
$$

$$
c = \begin{pmatrix}
-5 & 1 & -1 & . & . & . & . \\
-1 & 2 & -3 & . & . & . & . \\
1 & -3 & 2 & . & . & . & . \\
. & . & . & 3 & . & -2 & -2 \\
. & . & . & . & -4 & 4 & \\
. & . & . & 2 & 4 & 4 & 4 \\
. & . & . & 2 & -4 & 4 & 4
\end{pmatrix}, \quad
d = \begin{pmatrix}
-2 & . & . & . & 1 & . & . \\
. & -4 & . & . & . & 4 & . \\
. & . & -4 & . & . & . & 4 \\
. & . & . & -1 & . & . & . \\
-3 & . & . & . & 2 & . & . \\
. & -1 & . & . & . & 4 & . \\
. & . & -1 & . & . & . & 4
\end{pmatrix},
$$

$$
e = \begin{pmatrix}
1 & . & . & . & . & . & . \\
. & 1 & . & . & . & . & . \\
. & . & 1 & . & . & . & . \\
. & . & . & -1 & . & . & . \\
. & . & . & . & -1 & . & . \\
. & . & . & . & . & -1 & . \\
. & . & . & . & . & . & -1
\end{pmatrix}.
$$

---

[4] Note that our relation $(acde)^5 = 1$ is equivalent to the relation $(cde)^5 = a$ given in the ATLAS [23] (since a commutes with $\langle c, d, e \rangle$).

5.3 The Higman–Sims group

5.3.1 Introduction

The Higman–Sims group, HS, was discovered [47] at the height of the most exciting period in modern group theory. Following the announcement of the Hall–Janko group J_2 as a rank 3 permutation group on 100 letters with subdegrees {1, 36, 63}, D. G. Higman and C. C. Sims considered the possibility of subdegrees {1, 22, 77}, with point-stabilizer the Mathieu group M_{22}. The investigation proved succcessful and, thus, two new rank 3 permutation groups on 100 letters were discovered within months of one another! Soon after, G. Higman [48] constructed a geometry, consisting of 176 points and 176 *quadrics*, whose automorphism group had the same order as HS. Various people showed the two groups to be isomorphic [79, 80], and when HS appeared as a subgroup of the Conway group [20] stabilizing a 2-dimensional sublattice of the Leech lattice, the identification became delightfully clear. The automorphism group HS:2 could be seen, simultaneously permuting sets of 100 and (176 + 176) lattice points in the prescribed manner.

With its rich geometric pedigree, it is not surprising that HS lends itself to the techniques of symmetric generation developed in refs. [33] and [34]. Generally, construction of a new group takes the following form: start with a well known group H, and use this to build some combinatorial structure Λ which possesses additional symmetries to those we have built in. Thus,

$$H \longrightarrow \Lambda \longrightarrow G$$

Here Λ may be, for example, a lattice, a geometry, a graph, a code or a block design. In what follows we attempt to pass *directly* from H to G, but, since groups are invariably best studied through their action on some structure, we aim to produce the most important possible Λs as a by-product.

We shall in fact obtain HS:2 as an image of the progenitor

$$P := 2^{\bullet 50} : (U_3(5) : 2),$$

and so a detailed knowledge of the action of this unitary group on the 50 vertices of the Hoffman–Singleton graph [50] will be needed to facilitate a manual construction of HS:2. This approach will yield generating permutations on both 100 and (176 + 176) letters.

5.3.2 The Hoffman–Singleton graph

Following Sylvester [62], we call the 15 partitions of six points into three pairs *synthemes*. Certain sets of five synthemes, termed *synthematic totals* or simply *totals*, are preserved by a subgroup of the symmetric group S_6 acting as $PGL_2(5)$ on the six points; see Section 4.2. Thus, such a total, see the following, has just $6!/120 = 6$ images under S_6; these six totals and the six points are interchanged by the outer automorphism of S_6. Thus, we have the following:

$$\left.\begin{array}{l} ab.cf.de \\ ac.db.ef \\ ad.ec.fb \\ ae.fd.bc \\ af.be.cd \end{array}\right\} \equiv a/bcdef.$$

This synthematic total, which may conveniently be denoted by $a/bcdef$, is fixed by the group $\langle(b, c, d, e, f), (a, c, b, f)\rangle \cong \mathrm{PGL}_2(5)$. The six totals are obtained from this one by permuting $\{d, e, f\}$, say. We may now define a graph Γ on 50 vertices as in ref. [50]: the vertex set consists of a single vertex labelled ∞, 7 vertices labelled from $J_\infty = \{0, 1, \ldots, 6\}$, and 42 vertices labelled (i, T), where $i \in J_\infty$ and T is a total on $J_\infty \setminus \{i\}$. The joins are as follows: ∞ joins i for all $i \in J_\infty$, i joins (i, T) and (i, T) joins $(j, T^{(i,j)})$ for $i \neq j$. The resulting graph, which visibly has diagram given by Figure 5.8 possesses a transitive automorphism group isomorphic to $U_3(5):2$ with point-stabilizer S_7. One can easily make the following observations about Γ.

(i) Γ has diameter 2.

(ii) Γ possesses no triangles or quadrangles.

(iii) An edge and a disjoint vertex of Γ complete to a unique pentagon.

(iv) The edge stabilizer has orbits as shown in Figure 5.9.

(v) The pentagon stabilizer has orbits as shown in Figure 5.10.

(vi) Figure 5.10 defines a partition of the 50 vertices into (5+5) disjoint pentagons (see Figure 5.11).
These form the cycles of an element of conjugacy class 5A, and the stabilizer of such a configuration is a maximal subgroup of $U_3(5):2$ of shape $5^{1+2}:8:2$; see ref. [25], p. 34. An example of such a partition is given by

$$[\{\infty, 1, (1, 0/23456), (0, 1/23456), 0\},$$
$$\{2, (2, 0/13456), (0, 2/13456), (1, 2/03456), (2, 1/03456)\},$$
$$\{(2, 0/13465), (3, 0/14526), (4, 0/15632), (5, 0/16243), (6, 0/12354)\}],$$

together with images under the subgroup $\langle(2, 3, 4, 5, 6), (3, 4, 6, 5)\rangle$.

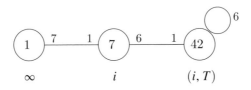

Figure 5.8. Diagram of the Hoffman–Singleton graph.

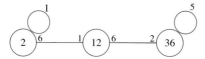

Figure 5.9. Orbits of an edge stabilizer in the Hoffman–Singleton graph.

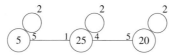

Figure 5.10. Orbits of a pentagon stabilizer in the Hoffman–Singleton graph.

Figure 5.11. Partition of the Hoffman–Singleton graph into ten disjoint pentagons.

(vii) The stabilizer of a point in Γ acts faithfully on the seven points joined to it, and, thus, a permutation which fixes a point is uniquely defined by its action on the joins of that point. In particular, if α is joined to β is joined to γ, the involution (α, γ) which interchanges α and γ and fixes the other five points joined to β is well defined.

Note that Γ may also be defined as follows. The stabilizer of an octad in the Mathieu group M_{24} is isomorphic to $2^4 : A_8$; see, for example, Todd [83] or Curtis [29]. If we fix a point in the octad and a point outside it, we are left with a group A_7 acting simultaneously on 7 *letters* and 15 *points*. Now define a graph whose vertices are the 35 unordered triples of the 7 letters and the 15 points. Join 2 triples if they are disjoint, and join a triple to the 3 points which complete it, together with the 2 fixed points, to an octad. The 3 points joined to a particular triple can easily be read off using the Miracle Octad Generator [29] (see Figure 5.12(a)). The resulting graph,

Figure 5.12. The alternating group A_7 acting on the Hoffman–Singleton graph.

which visibly has a diagram as shown in Figure 5.12(b) can be shown to be isomorphic to our Γ as defined above.

5.3.3 The progenitor and relation

As stated above, we intend to take as our control subgroup $N \cong U_3(5):2$ and our progenitor of shape $2^{\star 50}:(U_3(5):2)$; thus our 50 symmetric generators will form the vertices of a Hoffman–Singleton graph. We now seek finite homomorphic images of such a progenitor, and must consider what additional relation or relations to factor by. As stated above, this is equivalent to asking how to write permutations of N in terms of the symmetric generators $\mathcal{T}$. Now, Lemma 3.3 tells us that in order to identify which permutations of N can be written in terms of two symmetric generators without causing collapse, we must consider the centralizers in N of its 2-point stabilizers. In our case there are two 2-point stabilizers, depending on whether i and j are joined or not joined. If i is joined to j, then $N^{ij} \cong S_6$, which has a trivial centralizer in N, and so the only element of N which could be written as a word in t_i and t_j for i joined to j is the identity. However, if i is distance 2 from k, with j the unique point joined to both, then $N^{ik} = N^{ijk} \cong S_5$. This subgroup is centralized by just the involution we labelled (i,k) in the previous section, which interchanges i and k and fixes the other five points joined to j.

LEMMA 5.1 The involution (i,k) has action $(j)(i,k)(l,m)$ on any pentagon $(ijklm)$.

Proof Without loss of generality, we may take $i = 0$, $j = \infty$, $k = 1$, in which case l, m are $(1, T^{(0,1)})$, $(0, T)$, respectively, where T is any total on $\{1, 2, 3, 4, 5, 6\}$. Clearly our involution is the transposition $(0\ 1)$ which has the effect claimed. $\square$

The extension to Lemma 3.3 states that (i,k) is the only permutation of N which can be written in terms of t_i, t_j and t_k. We now make the assumption that

$$(i,k) = w(t_i, t_j, t_k),$$

a word in the three symmetric generators t_i, t_j, t_k, and ask what is the shortest length, $l(w)$, which does not lead to collapse. Now, Lemmas 3.4 and 3.5 show that $l(w) < 4$ leads to nothing interesting except possibly the case $(i,k) = t_i t_k t_i$, which is left as an exercise. So suppose $l(w) = 4$. All three generators t_i, t_j, t_k must be involved in w since $\langle t_i, t_j \rangle \cap N = \langle 1 \rangle$, and $(i,k) = ikik$ would imply that $\langle t_i, t_k \rangle \cong D_8$ with (i,k) in its centre – a contradiction. Thus, without loss of generality, one of t_i and t_j must be repeated. The possibilities (allowing reversals in ordering and conjugation by (i,k)) are as follows:

$$(i,k) = ijik,\ ijki,\ ikij,\ jikj \text{ or } jijk.$$

Now,

$$(i, k) = ijik \implies i(i, k) = jik \implies (i, k)k = jik$$
$$\implies (i, k) = ji, \text{ which is a contradiction to Lemma 3.4;}$$
$$(i, k) = jikj \implies j(i, k)j = ik \implies (i, k) = ik, \text{ which is a contradiction to Lemma 3.4;}$$

$$(i, k) = jijk \implies (i, k)k = jij \implies i(i, k) = jij$$
$$\implies (i, k) = ijij, \text{ which is a contradiction to Lemma 3.3;}$$
$$(i, k) = ijki \implies (i, k) = jkik = kikj = ikij.$$

Thus the only non-collapsing possibility with $l(w) \leq 4$ is given by

$$(i, k) = ikij.$$

5.3.4 Manual double coset enumeration

We now claim that the following theorem results.

THEOREM 5.4

$$G = \frac{2^{*50} : (U_3(5) : 2)}{(i, k) = ikij} \cong \text{HS} : 2,$$

the automorphism group of the Higman–Sims group, where j is joined to i and k in the Hoffman-Singleton graph Γ and (i, k) denotes the unique element of $U_3(5) : 2$ which interchanges i and k and fixes the other five points joined to j.

Proof Firstly note that

$$(i, k) = ikij = kikj = jkik$$

and

$$i(i, k) = kij = (i, k)k \implies (i, k) = kijk$$
$$\implies kjik = jkik$$
$$\implies kj = jk.$$

So any two symmetric generators which are joined in Γ commute. Moreover,

$$ikij.kikj = 1 = (ik)^3.$$

Thus any two symmetric generators which are not joined in Γ have product of order 3, and so $\mathcal{T}$ forms a Fischer system for HS in the sense of ref. [39]. Note further that

$$jk \sim ki_r \sim kj \sim jl_s \text{ for } r, s \in \{1, 2, \ldots, 6\},$$

where j is joined to k, $\{i_1, i_2, \ldots, i_6\}$ are the six points joined to j but not k and $\{l_1, l_2, \ldots, l_6\}$ are the six points joined to k but not j. Thus the

Table 5.4. Double coset enumeration of G over $U_3(5):2$

Label $[w]$	Coset stabilizing subgroup $N^{(w)}$	Cosets
$[\star]$	N	1
$[0]$	$N^{(0)} = N^0 \cong S_7$, with orbits $1+7+42$ on the 50 points.	50
$[01]$	$N^{01} \cong S_6$, $N^{(01)} \cong S_6 : 2$, with orbits $2+12+36$ on the 50 points	175
$[02] \equiv [01]$	Since $0201 \sim \star$	
$[012] \equiv [0]$	Since $012 = 102 = 1020.0 \sim 0$	
$[013]$	Such a triple defines the pentagon $(0,1,2,3,4)$.[a]	126

[a] Now, $013 \sim 0.1312.21 \sim 0^{(1,3)}21 \sim 421 \sim 4243.341 \sim 341$. Thus this coset may be named by any edge of the pentagon, followed by the opposite vertex. As in Section 5.3.2(v), there are 25 points joined to just one point of this pentagon and 20 joined to none of its points. Suppose that 5 is a new vertex joined to 1 only and that $(5,1,2,3,6)$ is a pentagon. Then we have $013 \sim 053 \sim 0.5356.65 \sim 0^{(3,5)}65 \sim 865$, say, $\sim 8687.785 \sim 785$, where 8 is a new vertex joined to 2 and $(5,6,7,8,9)$ is the pentagon defined by $\{5,6,8\}$. Note that any coincidence among these vertices would lead to triangles or quadrangles. Similarly, our coset may be named by any edge of this last pentagon, followed by the opposite vertex. This can be any one of the five pentagons in a 25-orbit of Section 5.3.2(v). Applying the process again, the coset may be labelled by an edge and an opposite vertex of a pentagon from either five. Thus our coset has $10 \times 5 \times 2$ such names and is fixed by the transitive, maximal subgroup of shape $5^{1+2} : 8 : 2$. Thus, this double coset contains 126 cosets.

coset $jk \equiv Nt_j t_k$ has $(2+12)$ 'names' as words of length 2 in the symmetric generators.

We are now in a position to perform a manual double coset enumeration of G over N. That is to say, we shall find all double cosets of the form NwN and say how many single cosets of N each contains. Firstly let $(0, 1, 2, 3, 4)$, in that order, be the vertices of a pentagon in Γ. We have the double cosets and coset stabilizing subgroups as given in Table 5.4.

The double coset enumeration yields the Cayley diagram of G over N (Figure 5.13). It should be stressed, as in ref. [34], that Figure 5.13 is a diagram of the Cayley graph of G acting on the cosets of N. Here a coset labelled w is joined to the 50 cosets labelled wi, possibly reduced. The regular bipartite graph whose automorphism group is G is obtained from the above by joining w to the 50 cosets iw. Note that the action of the symmetric generators on these $(176+176)$ letters can be read off from our coset enumeration, and so the relation can be verified. Moreover, G visibly has order

$$252\,000 \times 176 \times 2 = 44\,352\,000 \times 2,$$

and it has a subgroup of index 2 which acts doubly transitively on the 176 points. It is, of course, the group found by D. G. Higman and C. C. Sims acting on the geometry of G. Higman. $\qquad \square$

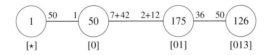

Figure 5.13. Cayley diagram of HS : 2 acting on the cosets of $U_3(5) : 2$.

5.3.5   Mechanical enumeration

As in the $2\,^{\cdot}M_{22} : 2$ and J_1 cases, we supplement our hand calculations with a mechanical enumeration. This time we need our control subgroup $U_3(5) : 2$ acting on 50 letters, and so we feed in the three permutations a, b and c. We can now appeal directly to Lemma 3.3 and ask which of the 50 letters are fixed by the stabilizer of letters 1 and 2. We obtain the answer $\{1, 2, 50\}$, which incidentally guarantees that 1 and 2 are *not* joined in the Hoffman–Singleton graph as the stabilizer of an edge would fix no further point. We conclude that the unique point joined to both 1 and 2 is 50. The required additional relation is thus $\pi = t_1 t_2 t_1 t_{50}$, where π generates the centralizer of the stabilizer in N of points 1 and 2. The MAGMA code then takes the following form:

```
s50:=SymmetricGroup(50);

a:=s50!(1, 2, 4, 8, 14, 22, 34, 49)(3, 6, 10, 5, 9, 16, 26, 40)
(7, 12)(11, 18, 30,21)(13, 20, 33, 48, 50, 25, 38, 29)
(15, 24, 36, 44, 39, 41, 31, 45)
(17,28, 43, 27, 42, 19, 32, 46)(23, 35, 37, 47);


b:=s50!(1, 3, 7, 13, 21)(2, 5, 6, 11, 10)(4, 9, 17, 29, 40)
(8, 15, 25, 39, 28)(12,19, 18, 31, 20)(14, 23, 22, 35, 48)
(16, 27, 33, 32, 47)(24, 37, 43, 34,50)
(26, 41, 38, 42, 46)(30, 44, 45, 49, 36);

c:=s50!(1, 2)(3, 37)(4, 42)(6, 17)(7, 35)(8, 34)(9, 25)
(10, 40)(11, 20)(13, 47)(14,18)(16, 31)(19, 26)(21, 29)
(22, 30)(23, 44)(24, 39)(27, 49)(33, 38)(36,46)(41, 43)(45, 48);


> u35a:=sub<s50|a,b,c>;
> #u35a;
252000
> Fix(Stabilizer(u35a,[1,2]));
1, 2, 50

> RR:=[<[1,2,1,50],Centralizer(u35a,Stabilizer(u35a,[1,2])).1>];
> CT:=DCEnum(u35a,RR,[u35a]:Print:=5,Grain:=100);
```

```
Index: 352 === Rank: 4 === Edges: 8 === Time: 0.015
> CT[4];
[
    [],
    [ 1 ],
    [ 1, 2 ],
    [ 1, 2, 3 ]
]
> CT[7];
[ 1, 50, 175, 126 ]
```

5.3.6 The centralizer in G of a symmetric generator

Let the point ∞ be joined in Γ to the points $\{0,1,2,3,4,5,6\}$, and so t_∞ commutes with the subgroup $\langle t_0, t_1, t_2, \ldots, t_7 \rangle$. Thus t_∞ commutes with $\langle s_0, s_1, \ldots, s_6 \rangle$, where $s_i = t_i t_\infty$. But

$$s_i s_j s_i = t_i t_\infty t_j t_\infty t_i t_\infty = t_i t_j t_i t_\infty = (i, j).$$

Thus $\langle s_0, s_1, \ldots, s_6 \rangle$ is a homomorphic image of

$$\frac{2^{*7} : S_7}{(0, 1) = s_0 s_1 s_0} = K, \tag{5.1}$$

say. But double coset enumeration of K over $L \cong S_7$, as above, immediately gives the Cayley diagram in Figure 5.14, and, as shown in ref. [33], $K \cong S_8$. Thus, $C_G(t_0)$ contains a subgroup isomorphic to $C_2 \times S_8$, which in fact, is the complete centralizer [25]. In particular, from their action on the eight cosets in Figure 5.14, we see that $t_1 t_2$ has order 3, $t_1 t_2 t_3$ has order 4, and so on.

5.3.7 The 100-point action

Recall our alternative definition of Γ (see Section 5.3.2), in which the alternating group A_7 acts with orbits 15 and 35 on the vertices. It is natural to ask what subgroup of HS:2 is generated by such a set of 15 symmetric generators. It is certainly a homomorphic image of the progenitor $2^{*15} : A_7$.

Figure 5.14. Cayley diagram of S_8 over S_7.

Now let i, j, k be three generators in such a set which are joined to the same point, l say, in the 35-orbit. Then we have

$$ijil = (i, j) \text{ and } ikil = (i, k) \Rightarrow ijki = (i, j, k)$$

$$\Rightarrow [(i, j, k)i]^4 = 1,$$

where the order of the last element follows from Section 5.3.6. Now, applying Lemma 3.3 to the progenitor $2^{\star 15} : A_7$, we see that

$$\langle t_i, t_j, t_k \rangle \cap M \leq C_M(M^{ijk}) = \langle (i, j, k) \cdots \rangle \cong C_3,$$

where the control subgroup $M \cong A_7$. If we denote this element of order 3, centralizing the A_4 and fixing i, j, and k by

$$(i, j, k)(\cdots) \cdots (\cdots) = \pi_{(ijk)},$$

then we see that the subgroup we wish to identify is a quotient of

$$\frac{2^{*15} : A_7}{(\pi_{(ijk)}i)^4}. \tag{5.2}$$

This group is precisely the group that was investigated in Section 5.1 and found to be isomorphic to $2^{\cdot}M_{22} : 2$, the double cover of the automorphism group of the Mathieu group M_{22}. A subgroup of this shape would have index 50 in G, which would contradict the maximality of $U_3(5) : 2$ in S_{50}. The only other possibility is that such a set of 15 symmetric generators generates a subgroup of G isomorphic to $M_{22} : 2$ with index 100.

5.3.8 The symmetric generators as permutations of 100 letters

In Part I, which was based on refs. [33] and [34], we were able to describe in a simple manner our symmetric generators for the Mathieu groups M_{12} and M_{24} acting on the 12 faces of a dodecahedron and the 24 faces of the Klein map, respectively. In order to perform a similar task for HS:2, let us first observe (from ref. [25], p. 34) that $U_3(5) : 2$ has two conjugacy classes of subgroups isomorphic to A_7: the 50 of *type I* extend to copies of S_7, while the 100 of *type II* are self-normalizing. Consider $N \cong U_3(5) : 2$ acting on the 100 cosets of an A_7 of type II. Then N', the derived group of N, has two orbits of length 50 and, if $K \cong A_7$ is of type I, K acts on these two 50s with orbits $(15+35)+(15+35)$. Now, A_7 has two distinct permutation actions on 15 letters, both of which are realized here, but only one action on 35 letters (see ref. [25], p. 10) Thus there is a correspondence, preserved by K, between the points of the two 35-orbits. Our generator, which has cycle shape $1^{30}.2^{35}$, fixes the 15-orbits and interchanges the pairs. Of course there are 50 such involutions corresponding to the 50 choices of K.

5.3.9 A presentation for HS:2 in terms of the symmetric generators

In order to give a presentation for G which (implicitly) involves the symmetric generating set $\mathcal{T}$, we must first produce a presentation for the control subgroup $N \cong U_3(5) : 2$. In the spirit of the current work, we choose to give a symmetric presentation which, although less efficient than could be achieved, reveals the subgroup structure of N in an informative manner. The presentation hinges on the fact that $U_3(5)$ is generated by six elements of order 5 whose associated cyclic subgroups are normalized by $2^{\cdot}L_2(5) \cong SL_2(5)$, the special linear group of 2×2 matrices over the field of order 5. In order to achieve the full symmetry of the situation, and to include the outer automorphism, we extend this control subgroup to $2^{\cdot}L_2(5) : 2$, acting on two sets of six cyclic subgroups of order 5, and then adjoin an element of order 4 which squares to the centre of this group, commutes with its derived group and is inverted by its outer elements. Thus we take

$$M \cong \langle x, y, a; x^5 = y^4 = (xy^2)^3 = a^2, (x^3y)^2 = 1 = [a, x] = ay^{-1}ay\rangle.$$

If $\mathcal{S} = \{r_\infty, r_0, \cdots, r_4\} \cup \{s_\infty, s_0, \cdots, s_4\}$, then a conjugates r_i into its cube and s_i into its square for $i \in \{\infty, 0, 1, 2, 3, 4\}$, while

$$x^2 : (r_\infty)(r_0, r_1, r_2, r_3, r_4)(s_\infty)(s_0, s_1, s_2, s_3, s_4),$$

$$y : (r_\infty, s_2, r_4, s_1^{-1}, \dots)(r_0, s_0^2, r_0^3, s_0, \dots)(r_1, s_\infty^{-2}, r_2, s_4^3, \dots)(r_3, s_3, r_3^2, s_3^2, \dots).$$

Of course, $y^4 = a^2$ inverts all the symmetric generators. This action of the control subgroup M on the symmetric generators corresponds to a faithful 5-modular monomial representation, obtained by inducing up a linear representation of a subgoup of index 12. We must now build this action into a presentation of our progenitor:

$$\langle x, y, a, r; x^5 = y^4 = (xy^2)^3 = a^2, (x^3y)^2 = 1 = [a, x] = ay^{-1}ay = r^5$$

$$= r^x r = r^a r^2, r^{y^2 x^2 y^2} = (r^{y^2 x^2})^3\rangle \cong 5^{\star(6+6)} : ((2^{\cdot}L_2(5) : 2) : 2).$$

Finally we must factor the progenitor by $(x^3yr)^3$ to obtain the $3^{\cdot}U_3(5) : 2$, and a further $(y^3r)^7$ to obtain $U_3(5) : 2$. In this presentation r is mapped onto the symmetric generator r_∞; thus, other symmetric generators can be written in terms of x^2, y and r using the monomial representation given above. Explicitly,

$$r_\infty = r, r_i = r^{y^2 x^{2(i+1)}}, s_\infty = r^{yx^4 y^{-2}}, s_i = r^{yx^{2(i-2)}}, \text{ for } i = 0, 1, \cdots, 4.$$

From here we may identify the following subgroups:

$$\langle r_i, s_i \rangle \cong 5^{1+2}, \langle r_0, y \rangle \cong 5^{1+2} : 8, \langle r_0, y, a \rangle \cong 5^{1+2} : 8 : 2;$$

$$\langle r_\infty, r_4 \rangle = \langle r_\infty, r_4, s_0, s_3 \rangle \cong A_6, \langle r_\infty, y^2 \rangle \cong M_{10}, \langle r_\infty, y^2, a \rangle \cong \text{Aut}A_6;$$

$$\langle r_\infty, r_0, r_2 \rangle = \langle r_\infty, r_0, r_2, s_1, s_3, s_4 \rangle \cong A_7, \langle r_\infty, r_0, r_2, a \rangle \cong S_7;$$

$$\langle x, y \rangle \cong 2^{\cdot}S_5, \langle x, y, a \rangle = N \cong 2^{\cdot}S_5 \cdot 2.$$

The automorphism of $U_3(5)$ of order 3 may be taken to commute with x and y, be inverted by a, and conjugate r_∞ to $s_\infty r_\infty s_\infty$. The three sets of symmetric generators then become

$$\{r_i, s_i\}, \{s_i r_i s_i, r_i s_i^{-2} r_i\}, \{s_i r_i^{-1} s_i, r_i s_i^2 r_i\}.$$

We may thus obtain representatives of the other classes of maximal subgroups:

$$\langle r_\infty s_\infty^{-2} r_\infty, xy^2 \rangle \cong \langle r_\infty s_\infty^2 r_\infty, xy^2 \rangle \cong A_7;$$

$$\langle r_\infty s_\infty^{-2} r_\infty, y^2 \rangle \cong \langle r_\infty s_\infty^2 r_\infty, y^2 \rangle \cong M_{10}.$$

To obtain a presentation for $G \cong HS:2$, we must adjoin a further generator which commutes with a subgroup isomorphic to S_7 and satisfies the additional relation. Thus,

$$\langle x, y, r, a, t : x^5 = y^4 = (xy^2)^3 = a^2, r^5 = (x^3 y)^2 = 1 = r^x r = (y^3 r)^7$$

$$= [a, x] = ay^{-1}ay = r^a r^2, r^{y^2 x^2 y^2} = (r^{y^2 x^2})^3, (x^3 yr)^3, t^2 = 1 = [r, t]$$

$$= [t, r^{yxy}] = [t, r^{yxyx}] = [t, a], ay^2 = (t^{y^2 x} t^x)^{t^{ty^2}} \rangle \cong HS:2.$$

5.3.10 Other maximal subgroups of HS

A list of maximal subgroups of HS, obtained by Magliveras [69], is given in the ATLAS (ref. [25], p. 80).

The two classes of subgroups isomorphic to $U_3(5) : 2$ are, of course, represented by N and N^{t_0}.

Representatives for the classes of subgroups isomorphic to M_{22} and S_8 are described in Sections 5.3.6 and 5.3.7.

Now, any element of our control subgroup $N \cong U_3(5) : 2$ must commute with the subgroup of HS:2 generated by the symmetric generators which it fixes. The involution y^4 in the centre of our control subgroup fixes ten symmetric generators which, together with its centralizer in N, generate its full centralizer in HS:2; i.e.

$$\langle x, y, a, t \rangle \cong 4{\cdot}2^4 : S_5.$$

An involution π such as ay^2 of N, which belongs to class $2B$ of HS, fixes six symmetric generators which form a 5-hook in Γ, and is of the class we have labelled (i, k) in our symmetric presentation. In addition to interchanging i and k, this element interchanges 6 pairs of $\mathcal{T}$ which commute with one another and 15 pairs which do not. Then

$$\langle \mathrm{Fix}(\pi), t_r t_s \rangle \cong (2 \times A_6{\cdot}2^2).2,$$

where π interchanges r and s, and r is joined to s in Γ.

Let $A_7 \cong L \leq N$ be of type II (see Section 5.3.8) and $A_5 \cong K \leq L$ with $N_L(K) \cong S_5$. Then K has orbits $(5+10+10+15)$ on the elements of $\mathcal{T}$. The 5-orbit generates a subgroup of G of shape $2^4 : S_6$, which completes to the maximal subgroup of HS:2 of shape $2^5.S_6$ if we adjoin $N_L(K)$. Two of the 10-orbits generate subgroups isomorphic to $M_{22} : 2$, while the third generates the whole group; see below. The 15-orbit generates a subgroup isomorphic to $M_{22} : 2$ as described in Section 5.3.7.

Again, let $L_2(7) \cong M \leq L$ with $N_N(M) \cong L_2(7) : 2$ (see ref. [25], p. 34). Such a normalizer is maximal in N, although M is not maximal in N', and was termed a *novelty* in ref. [87]. Such a subgroup has orbits $(7+7+8+28)$ on $\mathcal{T}$, and the 8-orbit generates a subgroup of G isomorphic to $M_{21} : 2$, which completes to the maximal subgroup of HS:2 of shape $M_{21} : 2^2$ if we adjoin $N_N(M)$.

Now recall from Section 5.3.2 that the 50 vertices of $\mathcal{T}$ may be labelled ∞, 7 vertices labelled from $J_\infty = \{0, 1, \ldots, 6\}$, and 42 vertices labelled (i, T) where $i \in J_\infty$ and T is a total on $J_\infty \setminus \{i\}$. Any permutation which fixes ∞ is uniquely specified by its action on J_∞ and, similarly, any permutation which fixes a point is uniquely specified by its action on the seven joins of that point. Now, our double coset enumeration shows that any element of HS:2 may be written, not necessarily uniquely, as the product of a permutation on 50 letters belonging to $N \cong U_3(5) : 2$ followed by a word of length at most 3 in the symmetric generators. Then the following form the non-trivial elements of an elementary abelian group of order 8:

$$(1, 3)(2, 6)(4, 5)t_\infty t_0,$$
$$(2, 4)(3, 0)(5, 6)t_\infty t_1,$$
$$(3, 5)(4, 1)(6, 0)t_\infty t_2,$$
$$(4, 6)(5, 2)(0, 1)t_\infty t_3,$$
$$(5, 0)(6, 3)(1, 2)t_\infty t_4,$$
$$(6, 1)(0, 4)(2, 3)t_\infty t_5,$$
$$(0, 2)(1, 5)(3, 4)t_\infty t_6.$$

To see this, let $\pi_i = (1, 3)(2, 6)(4, 5)^{(0,1,2,3,4,5,6)^i}$ and $w_i = t_\infty t_i$ for $i \in \{0, 1, 2, \ldots, 6\}$. Then

$$(\pi_i w_i)^2 = \pi_i^2 w_i^{\pi_i} w_i = t_\infty t_i t_\infty t_i = 1,$$

since t_∞ commutes with t_i. Again,

$$\pi_0 w_0 \times \pi_1 w_1 = (1, 3)(2, 6)(4, 5).(2, 4)(3, 0)(5, 6)t_\infty t_3 t_\infty t_1$$

$$= (0, 3, 1)(2, 5)(4, 6)t_3 t_1 = (0, 3, 1)(2, 5)(4, 6)t_3 t_1 t_3 t_\infty . t_\infty t_3$$

$$= (0, 3, 1)(2, 5)(4, 6).(1, 3).t_\infty t_3 = (0, 1)(2, 5)(4, 6)t_\infty t_3 = \pi_3 w_3.$$

This elementary abelian 2^3 is visibly normalized by

$$\langle (0, 1, 2, 3, 4, 5, 6), (1, 2, 4)(3, 6, 5), (2, 4)(6, 5) \rangle \cong L_2(7),$$

and commutes with t_∞ to give a subgroup isomorphic to $2 \times (2^3 : L_2(7))$. The above seven involutions belong to class $2A$ in HS:2 and so have cyclic subgroups of order 4 associated with them, whose normalizers are the centralizers of the involutions. We must identify these elements of order 4 which square to our elements $\pi_i w_i$. The seven joins of 0 are ∞ and $(0, T)$, where T is a synthematic total on $\{1, 2, \ldots, 6\}$. Just two of these six totals contain the syntheme 13.26.45, namely 1/23645 and 1/23654. If we let $\alpha = (0, 1/23645)$ and $\beta = (0, 1/23654)$, then we claim that the required elements are $(\infty, \beta, \alpha)t_0 t_\alpha$ and $(\infty, \alpha, \beta)t_0 t_\beta$, because

$$(\infty, \beta, \alpha)t_0 t_\alpha \times (\infty, \beta, \alpha)t_0 t_\alpha = (\infty, \alpha, \beta)t_0 t_\infty t_0 t_\alpha$$

$$= (\infty, \alpha, \beta)t_\infty t_\alpha$$

$$= (\infty, \alpha, \beta) \cdot t_\infty t_\alpha t_\infty t_0 \cdot t_0 t_\infty$$

$$= (\infty, \alpha, \beta) \cdot (\infty, \alpha) \cdot t_\infty t_0$$

$$= (\alpha, \beta)t_\infty t_0 = (1, 3)(2, 6)(4, 5)t_\infty t_0.$$

Also note that

$$[(\infty, \beta, \alpha)t_0 t_\alpha]^{t_\infty} = t_\infty(\infty, \beta, \alpha)t_0 t_\alpha t_\infty$$

$$= (\infty, \beta, \alpha)t_\beta t_0 t_\alpha t_\infty$$

$$= (\infty, \beta, \alpha)t_\beta t_\alpha t_\beta t_0 \cdot t_\beta t_\infty t_\beta t_0 \cdot t_0 t_\beta$$

$$= (\infty, \beta, \alpha) \cdot (\alpha, \beta) \cdot (\infty, \beta) \cdot t_0 t_\beta$$

$$= (\infty, \alpha, \beta)t_0 t_\beta.$$

Adjoining these 'square roots', we obtain the following:

$$\langle (\infty, \alpha, \beta)t_0 t_\beta, t_\infty, (0, 1, 2, 3, 4, 5, 6), (2, 4)(6, 5) \rangle \cong 4^3 : (L_2(7) \times 2).$$

An element of class $5B$ in N (see ref. [25], p. 35) fixes five symmetric generators and thus commutes with the subgroup of G which they generate. These, in fact, correspond to a pentagon in Γ and generate a copy of S_5. An element of order 5 in a copy of A_7 clearly belongs to this class and is normalized within the A_7 by a subgroup of order 20. Together these give a maximal subgroup of shape $5 : 4 \times S_5$.

We have now seen a representative of each of the classes of maximal subgroups of HS:2 except those isomorphic to M_{11}. Perhaps the easiest way to obtain one of these is to take $N_L(K) \cong S_5$ with orbits (5+10+20+15) on $\mathcal{T}$. Such a group has orbit lengths (1+2+5+5+10+12+15+20+30) on the 100 points. The stabilizer of the union of the two orbits with lengths 2 and 10 is isomorphic to M_{11} with orbit lengths (12+22+66). The other class is, of course, obtained by conjugating by a symmetric generator.

The presentation given above has been chosen to reveal the subgroup structure of HS:2, and is inevitably rather cumbersome. However, using the ten symmetric generators mentioned above which generate G and are permuted by a subgroup isomorphic to S_5, we obtain a far simpler presentation. Indeed, it can be shown mechanically that

$$\frac{2^{*10} : S_5}{[(0, 1, 2, 3, 4)t_{01}]^{10}, (1, 2)(3, 4) = t_{01}^{t_{02}} . t_{03}^{t_{04}}, [(0, 1)t_{02} t_{03}]^5} \cong \text{HS:2}, \qquad (5.3)$$

which is realized by the following presentation:

$$\langle x, y, t; x^5 = y^2 = (xy)^4 = [x^2, y]^2 = 1 = t^2 = [t, y] = [t, y^{x^2}] = [t, y^{x^{-2}}]$$

$$= (xt)^{10} = (yt^x t^{xyx^{-1}})^5 = (xyt^{(xy)^2 t})^2 \rangle.$$

Here $x \sim (0, 1, 2, 3, 4)$, $y \sim (0, 1)$ and $t = t_{01}$. The subgroup generated by a pentagon in the associated Petersen graph is isomorphic to $M_{22} : 2$, as is the subgroup generated by the six symmetric generators which do not involve 0, say. Thus,

$$\langle x, t \rangle = \langle t_{01}, t_{12}, t_{23}, t_{34}, t_{40} \rangle \cong M_{22} : 2$$

and

$$\langle xy, y^x, t^x \rangle = \langle t_{12}, t_{13}, t_{14}, t_{23}, t_{24}, t_{34} \rangle \cong M_{22} : 2,$$

and so coset enumeration over either of these gives the required action on 100 letters.

Alternatively, we may use the double coset enumerator. We write the control subgroup S_5 as permutations on ten letters, the unordered pairs of five objects; and similarly write the three relations in the above presentation as a sequence of integers from 1 to 10, representing the symmetric generators followed by a permutation on ten letters. The coset enumeration is to be performed over a subgroup M, say, isomorphic to $M_{22} : 2$, and so the enumerator finds all double cosets of form MwN. Note that the group M, which is labelled HH in the input, is generated by the first generator a together with t_1:

```
> N:=PermutationGroup<10|(1,2,3,4,5)(6,7,8,9,10),
>   (2,6)(5,10)(7,9)>;
> RR:=[<[1,2,3,4,5,1,2,3,4,5], Id(N)>,
>   <[10,5,10,4,8,4], N!(2,9)(4,8)(5,10)(6,7)>,
> <[10,8,1,6,10,8,1,6,10,8], N!(1,10)(4,9)(6,8)>];
> HH:=[*N.1,[1]*];
> CT:=DCEnum(N,RR,HH:Print:=5,Grain:=300);
```

This produces the following output, which tells us that the index of M in G is 100, the rank of the action of N on the cosets of M is 9, and the number of orbits on edges in the Cayley graph is 22. Coset representatives for each of the double cosets are given, as are generators of coset stabilizing subgroups for each of them. Finally, the 22 types of join are recorded but not reproduced here, so that we can tell to which double coset each symmetric generator takes each canonical double coset representative (CDCR).

```
Index: 100 === Rank: 9 === Edges: 22 === Time: 0.859

> CT[4]
[
```

```
    [ ],
    [ 6 ],
    [ 6, 1 ],
    [ 6, 9, 2, 7 ],
    [ 6, 9, 2, 7, 6, 9 ],
    [ 6, 7, 1 ],
    [ 6, 7, 1, 5, 8, 3, 7, 8 ],
    [ 6, 9, 2, 7, 4, 1, 5, 7 ],
    [ 6, 7, 4, 2, 8, 7 ]
  ],

> CT[7]
[ 12, 15, 30, 20, 10, 5, 5, 1, 2 ]
>
```

So, for instance, the double coset Mt_6N contains 15 single cosets of M; but, as usual, the CDCRs are not necessarily the shortest possible. Indeed, in this case, we see there are CDCRs of length 8 but none of length 7, so they cannot possibly be the shortest possible.

5.4   The Hall–Janko group and the Suzuki chain

Section 5.3 on the Higman–Sims group demonstrates the effectiveness of Lemma 3.3, but perhaps an even more dramatic application is the manner in which it yields J_2, the Hall–Janko group, and the other groups in the Suzuki chain. It was mentioned in Section 3.6 that relations of the form $t_i t_j t_i = \pi \in N$ are often very powerful. We shall concentrate on such relations here and so we shall study a group defined by

$$\frac{2^{\star n} : N}{\pi = t_i t_j t_i},$$

where, as usual, N is a transitive permutation group of degree n acting on the n involutory symmetric generators by conjugation. Lemma 3.3 tells us that π must lie in the centralizer in N of the 2-point stabilizer N_{ij}.

We first recall some elementary permutation group theory. If N is a permutation group acting on the set Λ, so here $\Lambda = \{1, 2, \ldots, n\}$, then N also acts on the Cartesian product $\Lambda \times \Lambda$ in the natural manner:

$$(a, b)^\pi = (a^\pi, b^\pi) \text{ for } a, b \in \Lambda.$$

The orbits of N in this induced action are called *orbitals*, and the number of orbitals is called the *rank* of N. If N acts transitively on Λ, the rank of the permutation action on Λ is equal to the number of orbits on Λ of the stabilizer in N of a point of Λ. Thus a doubly transitive group has rank 2. It is readily shown using elementary character theory that the norm of the corresponding permutation character is equal to the rank and that the multiplicity of the trivial character is equal to the number of orbits. So, for instance, the permutation character of a doubly transitive action is the trivial character plus an irreducible.

Now, if X is an orbital of N then the orbit *paired* with X is defined by

$$X^* = \{(b, a) \mid (a, b) \in X\}.$$

If $X = X^*$, we say that X is *self-paired.* In this case, if we draw a graph with edges the points of Λ and (directed) edges the pairs $(a, b) \in X$, then this *orbital graph* is undirected. All the graphs we consider in this section are of this type; indeed, we shall usually be concerned with rank 3 groups in which the two non-trivial orbits of the point-stabilizer have distinct lengths.

So, let us assume that $(i, j) \in X$, a self-paired orbital. Then if $t_i t_j t_i = \pi \in N$, $t_j t_i t_j \in N$ as well, and

$$Nt_i \pi = Nt_{i^\pi} = Nt_i \cdot t_i t_j t_i = Nt_j t_i = Nt_j t_i t_j \cdot t_j = Nt_j.$$

Thus $i^\pi = j$ and, since $\pi = t_i t_j t_i = t_j^{t_i}$ is clearly an involution, we also have $j^\pi = i$. In particular, we have the following:

$$\pi = t_i t_j t_i = \pi^\pi = t_j t_i t_j \text{ and so } (t_i t_j)^3 = 1,$$

and π is an involution interchanging i and j. We shall label it π_{ij}.

In Section 4.1 we found that

$$\frac{2^{*4} : S_4}{(3\ 4)t_1 t_2 t_1 t_2} \cong L_3(2) : 2 \cong \mathrm{PGL}_2(7),$$

and we obtained the group as permutations of the seven points and seven lines of the Fano plane. We now take this group as our control subgroup acting on $(7+7)$ symmetric generators and so consider the progenitor

$$P = 2^{*(7+7)} : \mathrm{PGL}_2(7).$$

We display the standard notation for the points and lines in Figure 5.15; thus both points and lines are labelled $\mathbb{Z}_7$, the integers modulo 7, with the points being denoted by large numerals and the lines by small numerals. The lines are chosen to be $\{1, 2, 4\}$, the quadratic residues modulo 7, and its translates; the labelling of the lines is then given by 124 - 0, 013 - 1, 602 - 2, 561 - 3, 450 - 4, 346 - 5, 235 - 6. Note that this labelling means that the element $\alpha = (0\ 1\ 2\ 3\ 4\ 5\ 6)(6\ 5\ 4\ 3\ 2\ 1\ 0)$ preserves the lines, and it is readily checked that

$$\gamma = (0\ 0)(1\ 1)(2\ 2)(3\ 3)(4\ 4)(5\ 5)(6\ 6)$$

interchanges points and lines and preserves incidence. We shall need to be able to write down conjugates of this element γ, which is more properly denoted by π_{00}, and so if O is a point and $l = PQR$ is a line not passing through O, we demonstrate in Figure 5.16 how to read off the element

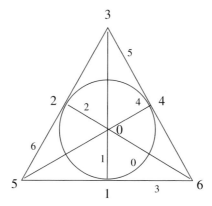

Figure 5.15. The Fano plane.

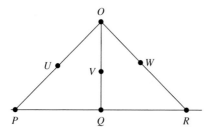

Figure 5.16. Duality in the Fano plane.

π_{Ol}. Let us denote the line through the points X and Y by $\overline{XY}$. Then the permutation which we denote by π_{Ol} is given by

$$\pi_{Ol} = (O\ l)(P\ \overline{OP})(Q\ \overline{OQ})(R\ \overline{OR})(U\ \overline{VW})(V\ \overline{WU})(W\ \overline{UV}).$$

We may now apply Lemma 3.3. The stabilizer of a point, 0 say, is isomorphic to S_4, and the stabilizer of a point and a line *not* passing through this point, 0 say, is isomorphic to S_3. In this case, we have

$$N_{00} = \langle (1\ 2\ 4)(3\ 6\ 5)(1\ 2\ 4)(3\ 6\ 5), (2\ 4)(6\ 5)(2\ 4)(6\ 5) \rangle \cong S_3.$$

The centralizer in N of this subgroup is cyclic of order 2 (in fact $\mathrm{PGL}_2(p)$ has maximal subgroups isomorphic to $D_{2p\pm2}$) and, in this case, is clearly generated by our element $\gamma = \pi_{00}$, so Lemma 3.3 says that this is the only non-trivial element that can be written in terms of t_0 and t_0 without causing collapse. We choose the shortest possible length for the word and so we are led to considering the following:

$$G = \frac{2^{\bullet(7+7)} : (L_3(2) : 2)}{(0\ 0)(1\ 1)(2\ 2)(3\ 3)(4\ 4)(5\ 5)(6\ 6) = t_0 t_0 t_0}.$$

Using our abbreviated notation in which i stands for the symmetric generator t_i, we deduce, using conjugates of the additional relation, that

$$1\,2 = 1\,4\,4 \quad 2 \sim 1\,4 \quad 2\,4\,4 \sim 5\,4 \text{ and } 1\,2 = 1\,5\,5 \quad 2 \sim 1\,5 \quad 2\,5\,5 \sim 4\,5 \sim 2\,1,$$

where the $\sim$ stands for 'represents the same coset' and $14 \sim 1$. So we have shown that

$$Nt_1 t_2 = Nt_2 t_1 = Nt_4 t_5 = Nt_5 t_4.$$

It can be shown by the usual manual double coset enumeration (with double cosets of lengths $1 + 14 + 21 + 42 + 14 + 14 + 2$) that

$$G \cong (3 \times U_3(3)) : 2,$$

but, since we are not particulary interested in this group, we shall give the MAGMA printout for the mechanical enumeration and immediately adjoin a further relation which factors out the normal subgroup of order 3. The points and lines have been labelled with the numbers 1–14, with the point i being labelled i (except that 0 is labelled 7) and the line i is labelled $i+7$ (except that 0 is labelled 14). Thus:

```
> s14:=Sym(14);
> a:=s14!(1,2)(3,6)(8,9)(10,13);
> b:=s14!(2,4)(5,6)(9,11)(12,13);
> c:=s14!(1,6)(2,3)(14,12)(8,9);
> d:=s14!(1,8)(2,9)(3,10)(4,11)(5,12)(6,13)(7,14);
> nn:=sub<s14|a,b,c,d>;
> #nn;
336

> RR:=[<[7,14,7],d>];

> CT:=DCEnum(nn,RR,[nn]:Print:=5,Grain:=100);

Index: 108 === Rank: 7 === Edges: 24 === Time: 0.032
> CT[4];
[
    [],
    [ 1 ],
    [ 1, 2 ],
    [ 1, 8 ],
    [ 1, 2, 4 ],
    [ 1, 8, 1 ],
    [ 1, 2, 4, 14 ]
]
> CT[7];
[ 1, 14, 21, 42, 14, 14, 2 ]
>
```

In order to find a suitable second relation, we note that the image we seek has just three double cosets of lengths $1 + 14 + 21$, so the double cosets of lengths 21 and 42 in G above must fuse. That is to say, $Nt_1t_2 = Nt_Pt_l$ for some point P and line l with $P \in l$. But the only flag fixed by the stabilizer in N of 1 and 2 is $\{4, 0\}$, so our relation will have form

$$t_1t_2t_4t_0 = \sigma \in N,$$

or, using our abbreviated notation, $1\ 2\ 4\ 0 = \sigma$. In order to find σ, we note that the pointwise stabilizer in N of $\{1, 2, 4,\ 0\}$ is given by

$$N_{124\ 0} = \langle (0\ 3)(6\ 5)(2\ 6)(4\ 5), (0\ 6)(5\ 3)(4\ 5)(1\ 3) \rangle \cong V_4,$$

which is self-centralizing. It must thus contain the element σ that we seek. A further calculation shows that

$$0\sigma \sim 0^{\sigma} \sim 01240 \sim 6\ 5240 \sim 6\ .525.540 \sim 3540 \sim 6440 \sim 60 \sim 6.$$

So, the only non-trivial possibility is $124\ 0 = (0\ 6)(5\ 3)(4\ 5)(1\ 3)$. Accordingly, we now examine

$$G = \frac{2^{\star(7+7)} : (L_3(2) : 2)}{(0\ 0)(1\ 1)(2\ 2)(3\ 3)(4\ 4)(5\ 5)(6\ 6) = t_0t_0t_0}$$
$$(0\ 6)(5\ 3)(4\ 5)(1\ 3) = t_1t_2t_4t_0.$$

Given these two relations, it is an easy matter to show that there are just two non-trivial double cosets; after all,

$$0\ _0 \sim 0 \text{ and } 1\ 2 \sim 2\ 1 \sim 5\ 6 \sim 6\ 5 \sim\ 40,$$

thus

$$Nt_1t_2N = Nt_5t_6N = Nt_4t_0N.$$

We claim that $G = N \cup Nt_0N \cup Nt_1t_1N$, which appears to consist of the coset N, 14 cosets corresponding to the points and lines, and 21 cosets corresponding to the *flags* (P, l), where l is a line and P is a point on l. To show that there are no further double cosets, we need simply show that $Nt_0t_3t_i$ is in this set of cosets for all choices of i. But if $N^{(t_0t_3)}$ denotes the coset stabilizing subgroup of Nt_0t_3, then

$$N^{(t_0t_3)} \geq \langle \gamma, (0\ 3)(5\ 6)(2\ 6)(4\ 5), (2\ 6)(4\ 5)(0\ 3)(5\ 6) \rangle,$$

with orbits $\{1, 1\}$, $\{0, 3, 0, 3\}$ and $\{2, 6, 5, 4, 2, 6, 5, 4\}$. We have

$$Nt_0t_3t_1 = Nt_1t_1t_1 = Nt_1,$$
$$Nt_0t_3t_0 = Nt_3t_0t_0 = Nt_3$$

and finally

$$Nt_0 t_3 t_2 = Nt_0 . t_3 t_2 t_3 . t_3 = Nt_1 t_3,$$

since $t_3 t_2 t_3 \in N$ maps 0 to 1. This yields the Cayley diagram shown in Figure 5.17. Recall that in the Cayley graph Γ_C of G on the cosets of N we join Nw to Nwt_i, but that in the transitive graph Γ_T which is preserved by G we join Nw to $Nt_i w$. We can thus readily work out that in Γ_T the joins are as follows:

(i) the vertex $\star$ joins the seven points and the seven lines;

(ii) a line joins the four points *not* on it;

(iii) a flag (P, l) joins all points on l and all lines through P;

(iv) two flags (P, l) and (Q, m) are joined if, and only if, $\overline{PQ} = l$ or m.

Of course, the elements of our control subgroup $N \cong L_3(2) : 2$ act on points, lines and flags in the standard manner, and preserve incidence. Moreover, using our two additional relations and their conjugates we can readily write down the action of right multiplication by a symmetric generator, t_0 say, on the 36 right cosets; thus

$$t_0 = (\star\ 0)(1\ 31)(2\ 62)(4\ 54)(3\ 11)(6\ 22)(5\ 44)(0)(3)(6)(5)$$

$$36\ 7\ 1\ 31\ 2\ 27\ 4\ 19\ 3\ 15\ 6\ 30\ 5\ 25\ 14\ 10\ 12\ 13$$

$$(1\ 01)(2\ 02)(4\ 04)(10\ 13)(20\ 26)(40\ 45)(35\ 36)(63\ 65)(56\ 53)$$

$$8\ 28\ 9\ 21\ 11\ 35\ 22\ 29\ 16\ 23\ 32\ 18\ 24\ 17\ 20\ 34\ 33\ 26,$$

where we also display a convenient labelling of the points, lines, flags and the symbol $\star$ with the integers from 1 to 36. We can observe that this permutation commutes with

$$N_0 = \langle (1\ 2\ 4)(3\ 6\ 5)(1\ 2\ 4)(3\ 6\ 5),\ (1\ 5)(3\ 4)(1\ 4)(0\ 6) \rangle \cong S_4,$$

the stabilizer in N of 0. Thus t_0 has just 14 images under conjugation by N. Conjugating this element by γ above, we see that

$$t_0 = (\star\ 0)(1\ 13)(2\ 26)(4\ 45)(3\ 11)(6\ 22)(5\ 44)(0)(3)(6)(5)(1\ 10)$$

$$(2\ 20)(4\ 40)(01\ 31)(02\ 62)(04\ 54)(53\ 63)(36\ 56)(65\ 35),$$

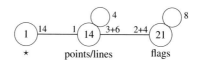

Figure 5.17. Cayley diagram of G acting on the cosets of $L_3(2)$: 2.

and we can readily check that $t_0 t_0 t_0 = (0\ 0)(1\ 1)(2\ 2)(3\ 3)(4\ 4)(5\ 5)(6\ 6)$, as required. We leave it as an exercise for the reader to verify that the second relation also holds within S_{36}. Thus, there is no collapse and $|G| = 36 \times |PGL_2(7)| = 2 \times 6048$. We know that the derived group of G is perfect, and so it is not surprising that $G \cong U_3(3) : 2$, the automorphism group of the special unitary group in 3-dimensions over the field of order 9. The fact that our relation $\pi = t_i t_j t_i$ was insufficient to define the group we desired is due to the fact that the control subgroup was not primitive, but acted with two blocks of size 7. Other groups in this *chain* are better behaved and the single relation is sufficient. Nonetheless, for the sake of completion, we give without proof a single relation which suffices in this case:

$$\frac{2^{\bullet(7+7)} : (L_3(2) : 2)}{t_0 t_1 t_2 t_4 t_6 = (0\ 0\ 6\ 6)(1\ 3\ 5\ 4\ 4\ 5\ 3\ 1)(2\ 2)} \cong U_3(3) : 2.$$

Theorem 5.5 makes explicit the 'better behaviour' of the larger groups in the chain.

THEOREM 5.5 Suppose the control subgroup N has degree n and that its derived subgroup N' has index 2 in N, is perfect, and acts transitively on the n letters. Then, if G is defined by

$$G = \frac{2^{\star n} : N}{\sigma = t_1 t_2 t_1},$$

where $\sigma \in N \setminus N'$, then $|G : G'| = 2$ and G' is perfect.

Proof Let $K = \{\pi w \mid \pi \in N', l(w) \text{ even}\} \leq G$; then

$$G = K \cup K\sigma \cup Kt_1 \cup K\sigma t_1,$$

where $\sigma \in N \setminus N'$. Now, for $i, j \in \{1, 2, \ldots, n\}$, since N' is transitive there exists a $\pi \in N'$ with $i^\pi = j$. Then $G' \geq \langle N', [t_i, \pi] = t_i t_j \ \forall i, j \rangle = K$. Moreover, $N'' = N' \leq G'' \lhd G$, and so $[t_i, \pi] = t_i t_j = (\pi^{-1})^{t_i} \pi \in G''$, for $\pi \in N'$. Thus $K' = K$, and K is perfect. But the additional relation implies that $K\sigma t_1 = Kt_2 t_1 = K$, and so $G = K \cup Kt_1$ as required. $\qquad \square$

This chain of subgroups of $\cdot O$ lends itself to the Y-diagram presentations described in Section 3.8; see ref. [25], pp. 42, 97 and 131. Indeed, if we take

$$a = (1\ 2)(3\ 6)_{(1\ 2)(3\ 6)},$$
$$b = (2\ 4)(5\ 6)_{(2\ 4)(5\ 6)},$$
$$c = (1\ 6)(2\ 3)_{(0\ 5)(1\ 2)},$$
$$d = (0\ 0)(1\ 1)(2\ 2)(3\ 3)(4\ 4)(5\ 5)(6\ 6),$$

then we see that the elements satisfy the following Coxeter diagram:

$$a = (cd)^4.$$

Now, $\langle a, b, c, d \rangle \cong L_3(2) : 2$, and so with $e = t_0$ we see that this is a presentation for

$$\frac{2^{\star(7+7)} : (L_3(2) : 2)}{t_0 t_0 t_0 = d},$$

which was stated above to define $(3 \times U_3(3)) : 2$.

EXERCISE 5.1

Proceed as follows to show that the Coxeter diagram

$$
\begin{array}{ccccc}
\bullet & \text{---} & \bullet & \overset{8}{\text{---}} & \bullet \qquad\qquad a = (cd)^4 \\
a & & b & c & d
\end{array}
$$

defines the group $L_3(2) : 2$. Note that $N = \langle a, b, c \rangle \cong S_4$ (or a homomorphic image thereof); taking $t_1 = d$, note that $|t_1^N| = 4$ (or a divisor of 4). Now show that this diagram is equivalent to the symmetric presentation

$$\frac{2^{\star 4} : S_4}{(1\ 2) = t_1 t_2 t_1, (3\ 4) = (t_1 t_2)^2},$$

which was shown in Section 4.1 to define $L_3(2) : 2$. (In fact, we have shown that the first of these relations is redundant and we should check that it does indeed hold.)

We now proceed precisely as before, but this time we take as control subgroup $N \cong U_3(3) : 2$ with its action on 36 points, which we have just constructed. Thus we consider the progenitor

$$P = 2^{\star 36} : (U_3(3) : 2),$$

and we may label the 36 symmetric generators $s_\star$ and s_i for i a point, a line or a flag. Since this action has rank 3, we have two 2-point stabilizers to consider: fixing $\star$ and the point 0, we have a $N_{\star 0} \cong S_4$, whose centralizer in N is $\langle t_0 \rangle \cong C_2$; and fixing $\star$ and the flag 11, we have a subgroup isomorphic to D_{16}, the dihedral group of order 16, whose centralizer is its centre. Clearly it is the first case which is relevant here as $t_0 = \pi_{\star 0}$ interchanges $\star$ and 0; thus we consider

$$G = \frac{2^{\star 36} : (U_3(3) : 2)}{t_0 = s_\star s_0 s_\star}. \qquad (5.4)$$

We now introduce a lemma which is useful in this situation. Let the permutation group N act transitively on the set Λ, where $|\Lambda| = n$, and let $a \in \Lambda$. Suppose that $\Delta(a)$ is an orbit of N_a, the stabilizer in N of a, whose corresponding orbital is self-paired. Draw a graph Γ with Λ as vertices and edges $\{(a, b) \mid b \in \Delta(a)\}$. By remarks earlier in this section, Γ is undirected.

Table 5.5. *The edges of the 36-point graph in Figure 5.17 and their correspondence with conjugates of* t_0

Conjugate of t_0	Simplified form	Involution	Number
t_0	t_0	$\pi_{\star\,0}$	14
$t_0^{t_0}$	$(0\ 0)(1\ 1)(2\ 2)(3\ 3)(4\ 4)(5\ 5)(6\ 6)$	$\pi_{0\,0}$	28
$t_0^{t_1}$	$(2\ 5)(4\ 6)(0\ 3)(2\ 4)t_0$	$\pi_{1\ 31}$	14×6
$t_0^{t_1}$	$(2\ 6)(4\ 5)(0\ 3)(5\ 6)t_0$	$\pi_{1\ 01}$	14×3
$t_0^{t_0 t_1}$	$(0\ 0)(1\ 1)(2\ 2)(3\ 3)(4\ 4)(5\ 5)(6\ 6)t_1 t_1$	$\pi_{31\ 10}$	21×4

LEMMA 5.2 *Suppose in the above situation that for a and b joined in Γ*

$$C_N(N_{ab}) = \langle \pi_{ab} \rangle \cong C_2,$$

where π_{ab} interchanges a and b. Form the progenitor $2^{\star n} : N$ as usual and factor by the additional relation $\pi_{ab} = t_a t_b t_a$, where the symmetric generators are $\{t_i \mid i \in \Lambda\}$ and b is joined to both a and c but a is not joined to c. Then the coset $Nt_a t_c$ is fixed by π_{bc}; i.e., $\pi_{bc} \in N^{(t_a t_c)}$, the coset stabilizing subgroup of $Nt_a t_c$.

Proof We have

$$Nt_a t_c = Nt_a t_b t_b t_c = Nt_a t_b t_a \cdot t_a \cdot t_b t_c t_b \cdot t_b = Nt_a \pi_{bc} t_b = Nt_a t_c \pi_{bc},$$

since $t_a t_b t_a = \pi_{ab} \in N$ and π_{bc} interchanges b and c. □

In order to use Lemma 5.2 to investigate the group defined by Equation (5.4), we need to be able to write down the element π_{ab} for a given edge (a, b) in the 36-point graph. There are clearly $36 \times 14/2$ such elements as they are in one-to-one correspondence with the edges of Γ. Starting with $t_0 = \pi_{\star 0}$, we may conjugate to obtain the involution corresponding to any edge of Γ. We display in Table 5.5 a representative of each orbit under the action of the vertex stabilizer $N_\star$.

It is now our aim to show that the group G defined by Equation (5.4) decomposes into just three double cosets of the form NwN, where $N \cong U_3(3) : 2$, namely

$$G = N \cup Ns_\star N \cup Ns_\star s_i N,$$

where i is not joined to $\star$ in the graph Γ. Building the Cayley graph of G over N, in which Nw is joined to Nws_i, we see that so far we have the partial diagram shown in Figure 5.18.

Now $\star$ is not joined to the 21 flags and so we consider the double coset $Ns_\star s_{11} N$, and use Lemma 5.2 to obtain the coset stabilizing subgroup $N^{(s_\star s_{11})}$. Certainly the subgroup of $L_3(2)$ which fixes the flag 11, namely a copy of D_{16} generated by our element γ and $(0\ 3)(2\ 4)(2\ 5)(4\ 6)$, is contained in this group. But by Lemma 5.2 we see that also contained are

$$\pi_{1\ 11} = (2\ 4)(6\ 5)(2\ 4)(5\ 6)t_1 \text{ and } \pi_{0\ 11} = (2\ 4)(6\ 5)(2\ 4)(5\ 6)t_3,$$

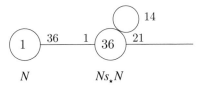

Figure 5.18. Partial Cayley diagram of G acting on the cosets of $U_3(3) : 2$.

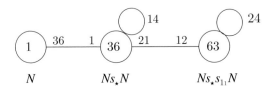

Figure 5.19. Cayley diagram of G acting on the cosets of $U_3(3): 2$.

since 1 and 0 are joined to both $\star$ and 11. These permutations were obtained by conjugating $\pi_{1\,0_1}$ from Table 5.5 by $(0\ 1)(4\ 6)(0\ 2)(4\ 3)$, and $\pi_{1\,3_1}$ by $(3\ 1\ 0)(6\ 5\ 4)(0\ 2\ 6)(3\ 4\ 5)$. Since the permutation part visibly fixes 1_1 in each case, we conclude that t_1 and t_3 lie in our coset stabilizing subgroup. Thus

$$N^{(s_\star s_{11})} \geq \langle \gamma, (0\ 3)(2\ 4)(2\ 5)(4\ 6), t_1, t_3 \rangle.$$

We work out the orbits of this subgroup on the 36 points and find that there are just two:

$$\{\star, 11, 1, 1, 0, 3, 0, 3, 01, 31, 10, 13\}$$

and the remainder. Each of the 12 symmetric generators in this orbit returns our coset $Ns_\star s_{11}$ to the double coset $Ns_\star N$, so the double coset $Ns_\star s_{11} N$ appears to contain $21 \times 21/12 = 63$ single cosets. It remains to show that $Ns_\star s_{11} s_2 \in Ns_\star s_{11} N$. But

$$Ns_\star s_{11} s_2 = Ns_1 s_1 s_2 = Ns_1 s_1 s_2 s_1 s_1 = Ns_1 \pi_{12} s_1 = Ns_0 s_1,$$

as required, where we have used the facts that $Ns_\star s_{11} = Ns_\star s_{11} t_1 = Ns_1 s_1$ and that $\pi_{12} = (0\ 2)(1\ 0)(2\ 1)(3\ 6)(4\ 3)(5\ 5)(6\ 4)$. Thus the Cayley graph of G over N closes and appears to take the (collapsed) form shown in Figure 5.19.

This manual enumeration enables us to write down explicitly permutations on 100 letters and to verify that they do satisfy the additional relation; they will of course preserve the associated transitive graph Γ_T. The group they generate is the automorphism group of the the Hall–Janko group, thus we have the following theorem.

THEOREM 5.6

$$G = \frac{2^{\star 36} : (U_3(3) : 2)}{\pi = t_i t_j t_i} \cong HJ : 2.$$

This process continues for two further steps and can be verified manually in the same manner; however, since the working is very similar to the above, and since the permutations are now more cumbersome, we prefer to give the significant steps of a mechanical enumeration.

Firstly, we need our control subgroup HJ : 2 acting on 100 letters; that is to say acting on the cosets of $\langle a, b, c, d, e \rangle \cong U_3(3) : 2$. If the homomorphism from the Coxeter presentation (see ref. [25], p. 42)

$$a = (cd)^4, (bcde)^8$$

to this action is given by f1, then we obtain

```
> nn:=sub<Sym(100)|f1(a*b*c*d*e),f1(f)>;
> #nn;
1209600
> 1^f1(f);
2
> RR:=[<[1,2,1],f1(f)>];
> CT:=DCEnum(nn,RR,[nn]:Print:=5,Grain:=100);

Index: 416 === Rank: 3 === Edges: 6  === Time: 0.031
> CT[4];
[
    [],
    [ 1 ],
    [ 1, 14 ]
]
> CT[7];
[ 1, 100, 315 ]
```

giving a rank 3 permutation group on 416 letters. This is, in fact, the automorphism group of the Chevalley group $G_2(4)$, and we have the following theorem.

THEOREM 5.7

$$G = \frac{2^{\star 100} : (HJ : 2)}{\pi = t_i t_j t_i} \cong G_2(4) : 2$$

with Cayley diagram as in Figure 5.20.

This in turn enables us to obtain $G_2(4) : 2$ as a permutation group on 416 letters and to use this as our control subgroup. Let f2 be the homomorphism from the following Coxeter diagram onto the action on the cosets of $\langle a, b, c, d, e, f \rangle \cong HJ : 2$ (see ref. [25], p. 97):

$$a = (cd)^4, (bcde)^8.$$

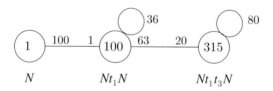

N Nt_1N Nt_1t_3N

Figure 5.20. Cayley diagram of $G_2(4) : 2$ over $HJ : 2$.

Then once again Lemma 3.3 allows an additional relation of the same form, and we obtain

```
> nn:=sub<g2|f2(a*b*c*d*e*f),f2(g)>;
> #nn;
503193600
> 1^f2(g);
2
> RR:=[<[1,2,1],f2(g)>];
> CT:=DCEnum(nn,RR,[nn]:Print:=5,Grain:=100);

Index: 5346 === Rank: 5 === Edges: 11 === Time: 0.328
> CT[4];
[
    [],
    [ 1 ],
    [ 1, 21 ],
    [ 1, 21, 249 ],
    [ 1, 21, 249, 274 ]
]
> CT[7];
[ 1, 416, 4095, 832, 2 ]
```

This time we see that there are blocks of imprimitivity of size 3. Theorem 3.5 will ensure that the group we have obtained contains a perfect subgroup to index 2, and so we have obtained the automorphism group of the triple cover of the Suzuki group.

THEOREM 5.8

$$G = \frac{2^{\star 416} : (G_2(4) : 2)}{\pi = t_i t_j t_i} \cong 3 \cdot Suz : 2.$$

This is in many ways the group we want as it will enable us to obtain the Conway group using a slightly different approach (see Section 7.5). However, in order to identify the centre of (the derived subgroup of) this group, we observe from the MAGMA output that $w = t_1 t_{21} t_{249} t_{274}$ must lie in it. By Lemma 3.3, w must be set equal to an element in the centralizer of the stabilizer in N of the sequence [1,21,249,274]. This centralizer is, in fact,

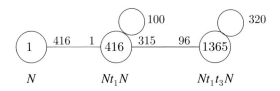

Figure 5.21. Cayley diagram of G acting on the cosets of $G_2(4) : 2$.

a copy of A_5 of order 60. Factoring by the correct such additional relation, we obtain the Cayley diagram shown in Figure 5.21.

Finally note that to obtain Suz : 2 from the Coxeter presentation

$$\bullet\!\!-\!\!\bullet\!\!-\!\!\overset{8}{\bullet}\!\!-\!\!\bullet\!\!-\!\!\bullet\!\!-\!\!\bullet\!\!-\!\!\bullet\!\!-\!\!\bullet \qquad a = (cd)^4, (bcde)^8$$
$$a\quad b\quad c\quad d\quad e\quad f\quad g\quad h$$

for the triple cover, Soicher factors out the relator $(bcdcdefgh)^{13}$; see ref. [25], p. 131.

5.5 The Mathieu groups M_{12} and M_{24}

At this stage, having seen how a number of sporadic groups can be defined by symmetric presentations, we return briefly to the two groups which motivated our investigations. Recall that M_{12} is generated by a set of five elements of order 3 whose set normalizer within M_{12} is the alternating group A_5. It is readily verified that the permutations given in Table 1.2 satisfy the relation $(s_2^{-1}s_1)^2 = (3\ 4\ 5)$, a 3-cycle in this copy of A_5. That is to say the permutation $(s_2^{-1}s_1)^2$ commutes with s_1 and s_2 and cycles the other three symmetric generators by conjugation. It turns out that this is almost enough to define M_{12}, and we have the following theorem.

THEOREM 5.9

$$\frac{3^{\star 5} : A_5}{(3\ 4\ 5) = (t_2^{-1}t_1)^2} \cong 3 \times M_{12}.$$

The corresponding definition of the large Mathieu group proved remarkably useful in writing down generators for the group simply from a knowledge of the linear group $L_3(2)$ (or, indeed, the Klein map). However, it turns out that we need rather more relations to define it in a relatively elegant manner, and we find that Theorem 5.10 holds.

THEOREM 5.10

$$\frac{2^{\star 7} : L_3(2)}{[t_0^{t_1}, t_3], (2\ 4)(5\ 6) = (t_0 t_3)^3, [(1\ 2\ 4)(3\ 6\ 5)t_3]^{11}} \cong M_{24}.$$

The last relation, which asserts that a certain element has order 11, is sufficient to ensure that a set of four of the seven symmetric generators which does not contain a line generates a copy of $L_2(23)$. We may then perform a coset enumeration over that group to verify the presentation.

5.6   The Janko group J_3

The third Janko group J_3 is surprisingly difficult to work with on account of the fact that, although it has a relatively modest order of around 50 million, its lowest permutation degree is over 6000. In comparison, the Higman–Sims group has order 44 million and lowest degree 100, and the McLaughlin group has order 900 million and lowest degree 275, not to mention the extraordinary Mathieu group M_{24} with order 200 million and degree 24. The group J_3 was predicted by Janko in 1968 [60], who considered simple groups whose single class of involutions had a centralizer of shape $2^{1+4} : A_5$. Thompson showed that this group would contain subgroups isomorphic to $L_2(16) : 2$, and Higman and McKay [46] constructed it on a computer. Weiss [86] constructed it theoretically as automorphisms of a graph on 17 442 vertices. Its embedding in $E_6(4)$ was shown computationally by Kleidman and Wilson [63] and was confirmed theoretically by Aschbacher [4].

Although J_3 has neither a low degree permutation representation nor a low dimensional complex matrix representation (as can be seen from the character table in the ATLAS (ref. [25], p. 83), its lowest degree irreducible representation is in 85 dimensions, and $J_3 : 2$ is twice this), it possesses a triple cover with a remarkable 9-dimensional matrix representation over $\mathbb{F}_4$ – found by Richard Parker when he was demonstrating his MeatAxe to Robert Wilson. This triple cover is constructed by Baumeister in ref. [9], and was used by John Bradley and the author [13] to verify that the following symmetric presentation is valid.

Our approach is based on the fact that if $Q < L < J$, where $Q \cong C_{17}$, $L \cong$ Aut($L_2(16)$) and $J \cong J_3 : 2$, then $N_L(Q)$ has shape $17 : 8$, whereas $N_J(Q) \cong 17 : 8 \times 2$. Thus the involution t centralizing $N_L(Q)$ has $|\text{Aut}(L_2(16))|/17 \times 8 = 120$ images under conjugation by L. Let this set of 120 involutions be denoted by $\mathcal{T}$, then $\langle \mathcal{T} \rangle = J$ since, as usual, $\langle \mathcal{T} \rangle$ is certainly normal in $\langle L, t \rangle = J$, by the maximality of L. But t is not contained in $J' \cong J_3$ and the only normal subgroup of J containing involutions outside J' is J itself.

This means that the group $J_3 : 2$ is a homomorphic image of the progenitor $2^{\star 120} : N$, where $N \cong \text{Aut}(L_2(16))$, a transitive permutation group of degree 120 acting on the Sylow 17-subgroups of N. It turns out that factoring this progenitor by a single additional relator will yield our group, but before describing this relator we need to say a little more about this 120-point permutation action. The point-stabilizer is, of course, $17 : 8$, and such a subgroup has orbits of lengths $1 + 17 + 34 + 68$ on the 120 points; we say that points in the 17-, 34- and 68-orbits are α-, β- and γ-joined to the fixed point, respectively. In particular, two points i and j are α-joined if, and only if, their stabilizer N_{ij} has order 8. Now, the field automorphism σ of order 4, which squares every element in the field, has cycle shape 4^{30} on the 120 Sylow 17-subgroups (since the only elements in the outer half of the Frobenius group of shape 17:8 have order 8), and σ commutes with a subgroup of L$_2$(16) isomorphic to L$_2$(2) $\cong$ S$_3$. Elements of order 3 clearly normalize no Sylow 17-subgroup, and so σ times an element of order 3 in this copy of S$_3$ has order[5] 12 and cycle shape 12^{10}. It turns out that there is a unique cycle in such an element with the property that adjacent terms are α-joined whilst terms two apart are γ-joined. Bray found that if we factor by a relation which says that such an element of order 12 times a symmetric generator in this 12-cycle has order 5, then we obtain J$_3$: 2. Bradley confirmed this computation by hand in his Ph.D. thesis [12], and further details of the argument may be found in Bradley and Curtis [13]. In this section we content ourselves by demonstrating how the double coset enumerator may be used to confirm this claim.

So we are asserting the following.

THEOREM **5.11**

$$G = \frac{2^{\star 120} : (\mathrm{L}_2(16) : 4)}{(\pi t_1)^5} \cong \mathrm{J}_3 : 2,$$

where the permutation action of $N \cong$ L$_2$(16) : 4 is on its Sylow 17-subgroups, π is an element of order 12 in N and t_1 is such that 1 is α-joined to 1^π and γ-joined to 1^{π^2}.

Verification We must first obtain the 120-point action of N. Thus,

```
> s17:=Sym(17);
> a:=s17!(1,10,11,15,17,14,5,12,9,6,13,4,16,3,7,8,2);
> b:=s17!(1,2,4,8)(3,6,12,9)(5,10)(7,14,13,11);
> autl216:=sub<s17|a,b>;
> #autl216;
16320

> na:=Normalizer(autl216,sub<autl216|a>);
> f,nn,k:=CosetAction(autl216,na);
> Degree(nn);
120
```

We have input two generators of the group Aut(L$_2$(16) $\cong$ L$_2$(16) : 4 as permutations on 17 letters: a of order 17 and b of order 4 (giving rise to the outer automorphism of order 4). To obtain our control

---

[5] A little caution is needed here as there are two classes of elements of order 12, but it does not matter which we choose.

subgroup as permutations on 120 letters, we take the action on the cosets of the normalizer of a Sylow 17-subgroup, which has shape $17:8$. We must now find an element of order 12 and identify the particular cycle we require. There is a single class of cyclic subgroups of order 12 in $\mathrm{Aut}(\mathrm{L}_2(16))$, a representative of which is generated by the field automorphism $\sigma : \lambda \mapsto \lambda^2$ times an element of order 3 in the simple group. Now,

$$C = \mathrm{C}_N(\sigma) \cong \langle \sigma \rangle \times \mathrm{L}_2(2) \cong 4 \times \mathrm{S}_3,$$

and so the derived group of C is isomorphic to C_3. We thus readily obtain the desired element of order 12 as the product of σ and a generator for this derived group:

```
> cb:=Centralizer(nn,f(b));
> #cb;
24
> e12:=f(b)*DerivedGroup(cb).1;
> Order(e12);
12
```

The element $\pi = e12$ has cycle shape 12^{10}; we must identify its (unique) cycle whose neighbouring terms are α-joined and terms two apart are γ-joined. So, if i and i^π are α-joined and i and i^{π^2} are γ-joined, then we define $rr := i$. The additional relation is then $(\pi t_{rr})^5$, which we may input as $\langle [rr], e12, 5 \rangle$:

```
> for i  in [1..120] do
for> if Order(Stabilizer(nn,[i,i^e12])) eq 8
for| if> and Order(Stabilizer(nn,[i,i^(e12^2)])) eq 2 then
for| if> rr := i;
for| if> break;
for| if> end if;
for> end for;
> RR:=[*<[rr],e12,5>*];
> CT:=DCEnum(nn,RR,[nn]:Print:=5,Grain:=100);
Index: 6156 === Rank: 7 === Edges: 62 === Time: 0.235
> CT[4]
[
    [],
    [ 1 ],
    [ 1, 18 ],
    [ 1, 4 ],
    [ 1, 4, 15 ],
    [ 1, 2 ],
    [ 1, 2, 65 ]
]
```

```
> CT[7];
[ 1, 120, 510, 2040, 680, 2720, 85 ]
```

The canonical double coset representatives (CDCRs) given by this output are indeed the shortest possible, and the Cayley graph of G over N does have diameter 3 as shown. But I repeat the *warning* that the enumerator does not necessarily produce the shortest possible CDCRs as this would slow things up considerably. In fact, I performed this enumeration several times before obtaining an output with minimal length CDCRs. □

5.7 The Mathieu group M_{24} as control subgroup

In order to make this chapter as self-contained as possible, we include here, albeit in a concise form, the information about the Mathieu groups that is required in order to follow the constructions. For more detailed information about these remarkable groups, the reader is referred, for example, to refs. [29] or [24].

The Mathieu group M_{24} and its primitive action of degree 3795

A Steiner system $S(5, 8, 24)$ is a collection of 759 8-element subsets known as *octads* of a 24-element set, Ω say, such that any 5-element subset of Ω is contained in precisely one octad. It turns out that such a system is unique up to relabelling, and the group of permutations of Ω which preserves such a system is a copy of the Mathieu group M_{24} which acts 5-transitively on the points of Ω. We let $\Omega = P_1(23)$, the 24-point projective line, and choose the Steiner system so that it is preserved by the projective special linear group $L_2(23) = \langle \alpha : s \mapsto s+1, \gamma : s \mapsto -1/s \rangle$. Let $P(\Omega)$ denote the power set of Ω regarded as a 24-dimensional vector space over the field GF_2, then the 759 octads span a 12-dimensional subspace, the *binary Golay code* $\mathcal{C}$, which contains the empty set ϕ, 759 octads, 2576 12-element subsets called *dodecads*, 759 16-*ads*, which are the complements of octads, and the whole set Ω. The stabilizer in M_{24} of an octad (and its complementary 16-ad) is a subgroup of shape $2^4 : A_8$, in which the elementary abelian normal subgroup of order 16 fixes every point of the octad. The stabilizer of a dodecad is the smaller Mathieu group M_{12}. From the above, the complement of a dodecad must be another dodedad; the stabilizer of such a pair of complementary dodecads, known collectively as a *duum* (as in *two umb*ral dodecads), has shape $M_{12} : 2$, the automorphism group of M_{12}.

Now, a 16-ad can be partitioned into two disjoint octads in 15 ways, and so Ω can be partitioned into three mutually disjoint octads in $759 \times$

$15/3 = 3795$ ways; such a partition is known as a *trio* of octads and is denoted by $U \cdot V \cdot W$, where U, V and W are the octads. The group M_{24} acts transitively on such partitions, and the stabilizer in M_{24} of such a trio is a maximal subgroup of shape $2^6 : (L_3(2) \times S_3)$, which of course has index 3795 in M_{24}. From the above assertions about $\mathcal{C}$, it is clear that the symmetric difference of two octads which intersect in four points must be another octad, and so we see that the 24 points of Ω can be partitioned into six complementary *tetrads* (4-element subsets) such that the union of any two of them is an octad. Such a partition is called a *sextet* and, since a sextet is determined by any one of its six tetrads, 5-transitivity on Ω ensures that M_{24} acts transitively on the $\binom{24}{4}/6 = 1771$ sextets. The stabilizer of one such is a maximal subgroup of shape $2^6 : 3 \dot{} S_6$. A sextet with tetrads a, b, c, d, e, f is denoted $abcdef$; the visible $2^6{:}3$ of the sextet group stabilizes all the tetrads in the sextet and the quotient S_6 acts naturally upon them.

If the vector space $P(\Omega)$ is factored by the subspace $\mathcal{C}$, we obtain the 12-dimensional *cocode* $\mathcal{C}^*$, whose elements (modulo $\mathcal{C}$) are the empty set ϕ, 24 *monads* (1-element subsets), 276 *duads* (2-element subsets), 2024 *triads* (3-element subsets) and the 1771 sextets. Since the symmetric difference of two even subsets necessarily has even cardinality itself, the duads and sextets, together with the empty set, form an 11-dimensional subspace which is clearly an irreducible 2-modular module for M_{24}. A duad $\{a, b\}$ is usually denoted ab.

A sextet $abcdef$ is said to *refine* a trio $A \cdot B \cdot C$ if (after suitable reordering) we have $A = a \dot{\cup} b$, $B = c \dot{\cup} d$ and $C = e \dot{\cup} f$. A trio is said to *coarsen* a sextet if the sextet refines the trio. A given sextet refines just 15 trios; conversely, a trio has just seven refinements to a sextet. These seven sextets, together with the empty set, form a 3-dimensional subspace of $\mathcal{C}^*$.

Now fix a trio $T = A \cdot B \cdot C$. Up to ordering the octads in the trios, the intersection array a trio makes with T is one of the following:

$$
\begin{array}{ccccc}
\begin{matrix} 8 & 0 & 0 \\ 0 & 8 & 0 \\ 0 & 0 & 8 \end{matrix} &
\begin{matrix} 8 & 0 & 0 \\ 0 & 4 & 4 \\ 0 & 4 & 4 \end{matrix} &
\begin{matrix} 0 & 4 & 4 \\ 4 & 0 & 4 \\ 4 & 4 & 0 \end{matrix} &
\begin{matrix} 0 & 4 & 4 \\ 4 & 2 & 2 \\ 4 & 2 & 2 \end{matrix} \ \ \text{and} &
\begin{matrix} 4 & 2 & 2 \\ 2 & 4 & 2 \\ 2 & 2 & 4 \end{matrix}
\end{array}
$$

where these account for 1, 42, 56, 1008 and 2688 trios, respectively. The stabilizer of T acts transitively on the trios that have a particular intersection array with T. We label these intersection arrays I_0, I_1, I_2, I_3 and I_4, respectively. The collapsed coset graph of valence 42 corresponding to this permutation action is given in Figure 5.22. For further information about such graphs, the reader is referred to ref. [1].

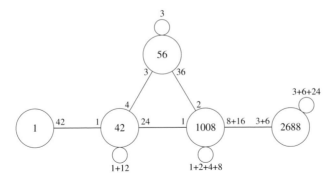

Figure 5.22. The trio graph.

The intersection arrays which give the number of points in which the tetrads of one sextet intersect the tetrads of another, up to re-ordering the tetrads in the two sextets, are given as follows:

$$
\begin{pmatrix}
4 & 0 & 0 & 0 & 0 & 0 \\
0 & 4 & 0 & 0 & 0 & 0 \\
0 & 0 & 4 & 0 & 0 & 0 \\
0 & 0 & 0 & 4 & 0 & 0 \\
0 & 0 & 0 & 0 & 4 & 0 \\
0 & 0 & 0 & 0 & 0 & 4
\end{pmatrix}
\quad
\begin{pmatrix}
2 & 2 & 0 & 0 & 0 & 0 \\
2 & 2 & 0 & 0 & 0 & 0 \\
0 & 0 & 2 & 2 & 0 & 0 \\
0 & 0 & 2 & 2 & 0 & 0 \\
0 & 0 & 0 & 0 & 2 & 2 \\
0 & 0 & 0 & 0 & 2 & 2
\end{pmatrix}
\quad
\begin{pmatrix}
3 & 1 & 0 & 0 & 0 & 0 \\
1 & 3 & 0 & 0 & 0 & 0 \\
0 & 0 & 1 & 1 & 1 & 1 \\
0 & 0 & 1 & 1 & 1 & 1 \\
0 & 0 & 1 & 1 & 1 & 1 \\
0 & 0 & 1 & 1 & 1 & 1
\end{pmatrix}
\quad
\begin{pmatrix}
2 & 0 & 0 & 0 & 1 & 1 \\
0 & 2 & 0 & 0 & 1 & 1 \\
0 & 0 & 2 & 0 & 1 & 1 \\
0 & 0 & 0 & 2 & 1 & 1 \\
1 & 1 & 1 & 1 & 0 & 0 \\
1 & 1 & 1 & 1 & 0 & 0
\end{pmatrix}
$$

For a given fixed sextet S, the numbers of sextets having intersection array with S as shown above are 1, 90, 240 and 1440, respectively. The stabilizer of S acts transitively on sextets having a given intersection array with S; we shall label the above arrays J_0, J_1, J_2 and J_3 in the order in which they appear. The collapsed coset graph in which sextets are joined if their intersection arrays are of type J_1 is given in Figure 5.23.

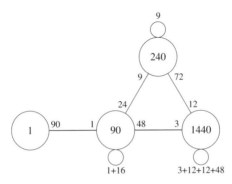

Figure 5.23. The sextet graph.

The Miracle Octad Generator and the hexacode

The Miracle Octad Generator (MOG) [29] is an arrangement of the 24 points of Ω into a 4×6 array in which the octads assume a particularly recognizable form, so it is easy to read them off. Naturally, the six columns of the MOG will form a sextet and the pairing of columns $12 \cdot 34 \cdot 56$ will form a trio. The three octads in this trio are known as the *bricks* of the MOG. The hexacode $\mathcal{H}$ [24] is a 3-dimensional quaternary code of length 6 introduced by John Conway whose codewords give an algebraic notation for the binary codewords of $\mathcal{C}$ as given in the MOG; see Figure 5.24. The reader should be warned that the MOG used in ref. [24] is the mirror image of the original, which was first published in ref. [29]. Equivalently, the two versions are obtained from one another by interchanging the last two columns.

Explicitly, if $\{0, 1, \omega, \bar{\omega}\} = K \cong \mathbb{F}_4$, then

$$\mathcal{H} = \langle (1, 1, 1, 1, 0, 0,), (0, 0, 1, 1, 1, 1), (\bar{\omega}, \omega, \bar{\omega}, \omega, \bar{\omega}, \omega) \rangle$$
$$= \{(0, 0, 0, 0, 0, 0), \ (0, 0, 1, 1, 1, 1)(9 \text{ such}), \ ((\bar{\omega}, \omega, \bar{\omega}, \omega, \bar{\omega}, \omega)(12 \text{ such}),$$
$$(\bar{\omega}, \omega, 0, 1, 0, 1)(36 \text{ such}), \ (1, 1, \omega, \omega, \bar{\omega}, \bar{\omega})(6 \text{ such})\},$$

where multiplication by powers of ω are of course allowed, and an S_4 of permutations of the columns corresponding to

$$S_4 \cong \langle (1\ 3\ 5)(2\ 4\ 6), (1\ 2)(3\ 4), (1\ 3)(2\ 4) \rangle$$

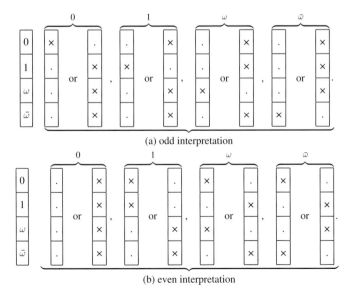

(a) odd interpretation

(b) even interpretation

Figure 5.24. Correspondence between hexacodewords and subsets of ω: (a) odd interpretation; (b) even interpretation.

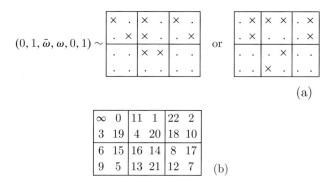

$$(0, 1, \bar{\omega}, \omega, 0, 1) \sim$$ or (a)

∞	0	11	1	22	2
3	19	4	20	18	10
6	15	16	14	8	17
9	5	13	21	12	7

(b)

Figure 5.25. (a) Interpreting hexacodewords and (b) the MOG numbering.

preserves the code. Each hexacodeword has an *odd* and an *even* interpretation, as shown in Figure 5.25, where the positions in which 1 appears are marked with an X symbol. Each interpretation corresponds to 2^5 binary codewords in $\mathcal{C}$, giving the $2^6 \times 2 \times 2^5 = 2^{12}$ binary codewords of $\mathcal{C}$. The rows of the MOG are labelled in descending order, with the elements of K as shown in Figure 5.24; thus the top row is labelled 0. Let $h = (h_1, h_2, \ldots, h_6) \in \mathcal{H}$. Then in the odd interpretation, if $h_i = \lambda \in K$ we place 1 in the λ position in the ith column and zeros in the other three positions, or we may complement this and place 0 in the λth position and 1s in the other three positions. We do this for each of the six values of i and may complement freely so long as *the number of 1s in the top row is odd*. So there are 2^5 choices.

In the even interpretation, if $h_i = \lambda \neq 0$ we place 1 in the 0th and λth positions and zeros in the other two, or, as before, we may complement. If $h_i = 0$, then we place 0 in all four positions or 1 in all four positions. This time we may complement freely so long as *the number of 1s in the top row is even*.

Examples of the odd and even interpretations, together with the MOG numbering, are shown in Figure 5.25. Figure 5.25(b) shows the standard labelling of the 24 points of Ω with the projective line $P_1(23)$ such that all permutations of $L_2(23)$ are in M_{24}.

The hexacode also provides a very useful notation for the elements of the elementary abelian normal subgroup of order 2^6 in the sextet group, consisting of those elements which fix every tetrad (in this case the columns of the MOG). An entry $h_i = \lambda$ means that the affine element $x \mapsto x + \lambda$ should be placed in the ith column. Thus 0 means all points in that column are fixed; 1 means 'interchange first and second, third and fourth entries in that column'; ω means 'interchange first and third, and second and fourth entries'; and $\bar{\omega}$ means 'interchange first and fourth, and second and third entries'. Note that this implies that 45 of the non-trivial elements are of shape $1^8.2^8$, whilst 18 are fixed-point-free involutions. These 18 form the involutions of six disjoint Klein fourgroups, such as

$\{(\bar{\omega}, \omega, \bar{\omega}, \omega, \bar{\omega}, \omega), (1, \bar{\omega}, 1, \bar{\omega}, 1, \bar{\omega}), (\omega, 1, \omega, 1, \omega, 1)\}$, which are of course permuted by the factor group S_6 of the sextet group. The actions on the six fourgroups and the six tetrads are non-permutation identical.

Finding elements of M_{24}

During the sequel we shall on occasions need to produce elements of M_{24} which stabilize certain configurations. The following question arises: what is the minimal test one can apply to guarantee that a permutation of the points of Ω does in fact preserve the Steiner system and so lie in M_{24}. This question was answered by Curtis [30], where a set of eight octads together with a further set of three octads were given with the properties that: (i) all 11 together with the whole set Ω give a basis for $\mathcal{C}$ and (ii) provided a permutation takes each of the eight octads to an octad, then it is in M_{24}. But $\mathcal{C}$ is self-dual, i.e. $\mathcal{C} = \mathcal{C}^{\perp}$, and so an 8-element subset of Ω is an octad if, and only if, it is orthogonal to a basis of $\mathcal{C}$. Thus a permutation is in M_{24} if, and only if, the image of each octad in our set of 8 intersects all 11 octads evenly. It is further shown in ref. [30] that this test is the best possible of its type.

Elements of certain cycle-shapes can, however, be readily written down using the MOG. In particular, given any octad F and any two points of Ω, i and j say, not in F, then there is a unique involution $F_{ij} \in M_{24}$ of cycle shape $1^8.2^8$ fixing F pointwise and interchanging i and j. The transpositions of a fixed-point-free involution of M_{24} may be paired in a unique manner to form the tetrads of a sextet; thus the centralizer of such an element lies in a copy of the sextet group. An element in this class also fixes a number of dodecads (2^5 to be precise) which have two points in each tetrad of that sextet; it is thus determined by such a dodecad. If S is a sextet and D is a dodecad with two points in each of its tetrads, then we denote this element of shape 2^{12} by S^D.

Furthermore, the group M_{24} has two orbits on subsets of size 6: *special hexads* S_6 which are contained in an octad, and *umbral hexads* U_6 which are not. Given an umbral hexad U, we may omit its points one by one and complete the resulting pentad to an octad. This process yields six triples which must partition the 18 remaining points. We may give one of these triples an arbitrary sense, and then there exists a unique element of M_{24} of cycle shape $1^6.3^6$ which fixes U pointwise and rotates this triple in the sense we have chosen.

When in the following we need to produce elements of M_{24} with particular properties, we shall restrict ourselves wherever possible to these three classes, which can be readily checked by hand.

5.7.1   The Janko group J_4

This section is based on unpublished work by J. Bolt, J. N. Bray and R. T. Curtis.

Janko discovered his fourth and largest group in 1976; it was thus the last of the sporadic simple groups to appear. It arose through his determination of finite simple groups having centralizer of an involution of a particular shape, in this case $2^{1+12} : 3 \cdot M_{22}$, which is to say an extra special group of order 2^{13} extended by the triple cover of the Mathieu group M$_{22}$. As the title of his famous paper [61] makes clear, the group contains the largest Mathieu group M$_{24}$ and what was, at the time, considered to be the full covering group of M$_{22}$, i.e. $6 \cdot M_{22}$. In fact, since that time M$_{22}$ has been found to possess a Schur multiplier of order 12 (see ref. [72]) and to be the only sporadic simple group with a quadruple cover.

The group J$_4$ has been investigated in detail by a number of authors, including Benson [10] and Lempken [66], the second of whom constructed its lowest dimensional complex representation of degree 1333. However, the group is best studied computationally by way of its remarkable 112-dimensional representation over the field $\mathbb{F}_2$, discovered by Norton [74]. More recently, Ivanov and Meierfrankenfeld [54] have used the amalgam method to produce a geometric construction of J$_4$ in 1333-dimensional complex space. A description of the methods employed can be found in ref. [52]. A wealth of information about this remarkable group can be found in ref. [53].

In what follows, we define J$_4$ as an image of the progenitor

$$2^{\star 3795} : M_{24},$$

where the 3795 symmetric generators correspond to the trios. This approach builds on Bolt's Ph.D. thesis [11], and eventually makes use of the double coset enumeration program of Bray and the author [17]. Norton's remarkable 112-dimensional representation is used to verify that the relations we use do, in fact, hold in J$_4$, thus giving a lower bound for the order of the group we have defined; the double coset enumeration gives an upper bound. All intermediate results which underlie the enumeration are, however, proved by hand. For those familiar with J$_4$, we would point out at this early stage that, in fact, it has two classes of subgroups isomorphic to M$_{24}$, both of which are contained in the maximal $2^{11} : M_{24}$, where the elementary abelian normal subgroup is the irreducible 11-dimensional submodule of $\mathcal{C}^{\star}$, the cocode of the binary Golay code. Our copy of M$_{24}$ belongs to the class that contains both *2A* and *2B* involutions of J$_4$; a representative of the other class of M$_{24}$ contains no involutions of J$_4$-class *2B*. The names for particular conjugacy classes of elements in the various groups follow the ATLAS [25].

Verifying relations holding in J$_4$

In what follows, we shall use Lemma 3.3 to identify relations by which it might be sensible to factor our chosen progenitor. We shall then need to verify that these relations do in fact hold in J$_4$. To do this we work in the 112-dimensional representation over $\mathbb{F}_2$.

Firstly, we start with the 112×112 matrices over $\mathbb{F}_2$ generating J$_4$ that are given in the www-ATLAS [75]. (These are derived from the original ones produced by Norton and colleagues) The generators a and b satisfy: a is in class $2A$ (of J_4), b is in class $4A$, $o(ab) = 37$ and $o(abab^2) = 10$. These conditions determine the pair (a, b) up to Aut J_4-conjugacy; as such, the pair (a, b) is considered to be a pair of *standard generators* for J_4. We then define the following:

```
x:=((a*b*a*b^2)^5)^((a*b)^4);
y:=((a*b^2)^4)^((a*b)^2*(b*a)^18);
z:=(x*y)^2*(x*y^2)^2*(x*y)^3;


N:=sub<G|x,y>;
// Control subgroup isomorphic to M24.
N1:=sub<G|x,z>;
// Standard trio subgroup.


c:=a^(a*b*a*b^3*(a*b*a*b^2)^3);
t:=(c*z)^15;
// t is the symmetric generator t1.


g1:=(x*y*x*y*x*y*x*y^2*x*y*x*y^2)^((x*y)^6);
g2:=x*y;
g3:=(x,y*x*y*x*y)^((x*y)^6*(x*y^2)^6*(x*y)^3*(x*y^2)
^6);
gg1:=[g1^i:i in [0..14]];
gg2:=[g2^i:i in [0..22]];
gg3:=[g3^i:i in [0..10]];


T:=[t^(gg1[i]*gg2[j]*gg3[k]):k in [1..11],j in [1..23],i
in [1..15]];
// All symmetric generators.
E:=Eigenspace(x,1) meet Eigenspace(y,1);
v:=E.1;
V:=[v^u:u in T];
```

So we have found generators x and y of a copy N of M_{24} within J_4 in terms of standard generators; we have obtained words in x and y for a copy of the trio group in M_{24}, and we have found an element t in J_4 which commutes with this trio group. Thus t has (at most) 3795 images under conjugation by N.

The progenitor and additional relations

As our control subgroup we shall take the Mathieu group M_{24} acting on the 3795 trios; thus we shall take as progenitor

$$P = 2^{*3795} : M_{24}.$$

We must now consider by what additional relations we should factor this infinite group. Since we are aiming to construct the Janko group J_4, we can use its 112-dimensional representation to confirm that it is indeed an image of this progenitor, and to verify that the relations we produce do hold. However, it is of interest to observe how naturally the two relations we need emerge simply by considering M_{24}.

Given two trios with intersection array I_1, the (trio-wise) stabilizer of them both fixes a unique further trio. Namely, if $A \cdot B \cdot C$ and $A \cdot D \cdot E$ are the two trios, the further fixed trio is $A \cdot (B \triangle D) \cdot (C \triangle E)$. Thus if the six tetrads of a sextet are denoted by $\{1, 2, \ldots, 6\}$, then these three trios are $12 \cdot 34 \cdot 56$, $12 \cdot 35 \cdot 46$ and $12 \cdot 36 \cdot 45$, which are stabilized within M_{24} by a subgroup $2^6 : (S_3 \times V_4)$. This last subgroup has trivial centralizer in $M_{24} \cong N$, and so by Lemma 3.3 the only element of N that can be written in terms of these three trios is the identity. Accordingly, we set the product of the three of them equal to the identity, and the first relation by which we quotient out is given by

$$t_{A \cdot B \cdot C} t_{A \cdot D \cdot E} t_{A \cdot (B \triangle D) \cdot (C \triangle E)} = 1.$$

An immediate consequence of this is that $t_{A \cdot B \cdot C} t_{A \cdot D \cdot E}$ is an involution and so $t_{A \cdot B \cdot C}$ and $t_{A \cdot D \cdot E}$ commute.

NOTATION If $T = A \cdot B \cdot C$ and $U = A \cdot D \cdot E$ are trios with intersection array I_1, we define $T + U$ to be the trio $A \cdot B \triangle D \cdot C \triangle E$. Note that $T + U = U + T$, $T + (T + U) = U$ and $U + (T + U) = T$, and also that $T + U$ has intersection array I_1 with both T and U. We might as well also define $T + T = 0$, where 0 is a symbol satisfying $0 + T = T + 0 = T$ for all trios T. (Of course, we have not defined trio 'addition' for all pairs of trios T and U.) We justify this notation by noting that when we make $P(\Omega)$ into an $\mathbb{F}_2$-vector space the addition is the symmetric difference and that two of the octads of $T + U$ are formed by taking symmetric differences of octads of T and U.

For the second relation, we consider two trios whose intersection array is I_3. These may be taken to be the top two trios in Figure 5.26. The (trio-wise) stabilizer of them both has order 2^6 and fixes a further three trios, which are also shown in the diagram. We have joined two trios if their intersection array is I_3; two unjoined trios have intersection array I_1.

The centralizer of the group of order 2^6 stabilizing these five trios is, in fact, its centre, which is a Klein fourgroup. So Lemma 3.3 tells us that there are just three non-trivial elements of N which could be written as words in these five symmetric generators. Now the set stabilizer of these five trios has order 2^9; it acts on them as the obvious dihedral group D_8 of symmetries and fixes just one element of the aforementioned Klein fourgroup (namely ν_2 below, which fixes all five trios) whilst interchanging the other two. So, if any non-trivial element of N can be written in terms of these five trios, ν_2 can. If we label the trios round the edge A, B, C and D in order

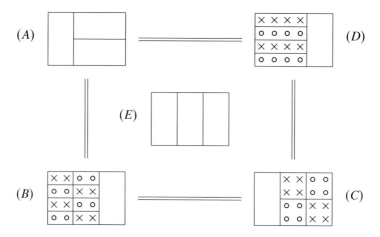

Figure 5.26. The five trios fixed by the stabilizer of two trios having intersection array I_3.

(either clockwise or anticlockwise), with the central one labelled E, then the shortest relation which does not lead to collapse is $t_A t_B t_A t_D = \nu_2$, where

$$\nu_2 = \begin{array}{|c|c|}\hline & \\\hline & \\\hline & \\\hline & \\\hline\end{array}.$$

Naturally, we use our 112-dimensional representation to check that this relation holds in J_4. Note, however, that the two relations we have described have arisen quite spontaneously from a consideration of the action of M_{24} on trios. The full D_8 of symmetries of the diagram is realized within $N_N(N_{AB}) \cong N_{AB}.D_8$, a group of order $512 = 2^9$, and so the eight (potentially distinct) possibilities for the relation $t_A t_B t_A t_D = \nu_2$ are, in fact, equivalent. Inverting (both sides of) this relation yields $t_D t_A t_B t_A = \nu_2$ (this is also achieved by conjugating by t_D). We also note that $N_{AB} = N_{BA} = N_{BC} = N_{CD} = N_{DA}$, and that ν_2 is the central involution in $N_N(N_{AB})$.

Note also that $A + C = B + D = E$, so $t_C, t_D \in \langle t_A, t_B, t_E \rangle$ and also $\nu_2 = t_A t_B t_A t_D \in \langle t_A, t_B, t_E \rangle$. Also $t_A t_B t_A t_B = t_A t_B t_A t_D t_E = \nu_2 t_E = t_E \nu_2$, an involution; so $t_A t_B$ has order 4. So we have shown Lemma 5.3.

LEMMA 5.3 *If the trios T and U have intersection arrays I_0, I_1 or I_3, then $t_T t_U$ has order 1, 2 or 4, respectively. Moreover $\langle t_A, t_B, t_C, t_D, t_E, \nu_2 \rangle = \langle t_A, t_B, t_E \rangle \cong D_8 \times 2$.*

The symmetric presentation

The aim of the rest of this chapter is to demonstrate the following theorem.

THEOREM **5.12** We have

$$G = \frac{2^{\star 3795} : M_{24}}{t_T t_U t_{T+U} = 1;\ t_A t_B t_A t_D = v_2} \cong J_4,$$

where T and U have intersection array I_1, and A, B, D and v_2 (and also C and E) are as in Figure 5.26.

At this stage, we should, of course, like to use our double coset enumerator to find all double cosets of the form NwN in the usual manner, where $N \cong M_{24}$. Unfortunately this is well out of range as M_{24} has index of order 10^{11} in J_4. However, John Bray has modified the enumerator so that it can handle double cosets of the form HwN for $H \neq N$ and is particularly proficient if $N \leq H$. In order to exploit this we shall deduce from our relations that the group we have defined contains an elementary abelian subgroup of order 2^{11} which is normalized by $N \cong M_{24}$; we shall then use $H = 2^{11} : M_{24}$ in our enumeration.

Note It will turn out that the symmetric generators are in J_4-class $2A$, with the products $t_T t_U$ residing in J_4-classes $1A$, $2A$, $2B$, $4B$ and $5A$, respectively, where T and U have intersection arrays in the order given above. Of course, we shall not use this information when deriving consequences of our relations.

A maximal subgroup containing N

Our next step is to construct from our 3795 symmetric generators 1771 new involutions which will correspond to sextets. We first need to investigate further the relationship between sextets and trios.

The 15 trios coarsening a sextet

The 15 trios coarsening the sextet $D = abcdef$ take the form $ab \cdot cd \cdot ef$ and thus correspond to the synthemes of $\{a, b, c, d, e, f\}$. Denote the set of these 15 trios by $\mathrm{coa}(D)$. We shall now study the group $\langle t_T : T \in \mathrm{coa}(D) \rangle$. Only a quotient S_6 of the sextet stabilizer $N_D \cong 2^6{:}3{\cdot}S_6$ acts non-trivially on $\{t_T : T \in \mathrm{coa}(D)\}$. Applying the outer automorphism of S_6, which interchanges synthemes and duads, allows us (for now only) to label these 15 symmetric generators as u_{ij} using unordered pairs of $\{1, 2, 3, 4, 5, 6\}$. The extra relation $t_{ab \cdot cd \cdot ef} t_{ab \cdot ce \cdot df} t_{ab \cdot cf \cdot de} = 1$ then becomes $u_{12} u_{34} u_{56} = 1$ (and its conjugates under S_6). We aim to show the following:

$$\frac{2^{\star 15} : S_6}{u_{12} u_{34} u_{56} = 1} \cong 2^5 : S_6 \cong 2^{1+4} : S_6,$$

where the notation 2^{1+4} indicates that the elementary abelian group of order 2^5 possesses an invariant subgroup of order 2, which is thus central in the whole group. Let $\mathcal{U} = \{u_{ij} : 1 \leqslant i < j \leqslant 6\}$. It is obvious that u_{12} and u_{34} commute. Now,

$$u_{12}^{u_{13}} = u_{13}u_{12}u_{13} = u_{13}u_{45}u_{36}u_{13} = u_{26}u_{36}u_{25}u_{46}$$

$$= u_{26}u_{14}u_{46} = u_{35}u_{46} = u_{12},$$

and so u_{12} and u_{13} commute. Thus $\langle \mathcal{U} \rangle$ is abelian, indeed an elementary abelian 2-group. Let z_{abc} be the involution $u_{ab}u_{bc}u_{ac}$, where the commutativity of $\langle \mathcal{U} \rangle$ ensures that this is independent of the ordering of a, b and c. We have

$$z_{123}z_{124} = u_{12}u_{23}u_{13}u_{12}u_{24}u_{14} = u_{23}u_{14}u_{13}u_{24} = u_{56}u_{56} = 1,$$

and so $z_{abc} = z_{abd} = z_{ade} = z_{def}$, i.e. $z = z_{abc}$ is independent of $\{a, b, c\}$, and so the involution z is centralized by the whole of S_6. We now claim that $\langle \mathcal{U} \rangle = \langle u_{12}, u_{13}, u_{14}, u_{15}, z \rangle$, for (using also the symmetry under $\text{Sym}_{\{2,3,4,5\}}$) we can easily calculate that $u_{23} = u_{12}u_{13}z$, $u_{26} = u_{13}u_{14}u_{15}z$ and $u_{16} = u_{12}u_{13}u_{14}u_{15}$. To determine the submodule structure, we look at the action of S_6 on non-zero vectors of $\langle \mathcal{U} \rangle$ (they have sizes $1 + 15 + 15$), and we easily determine that the orbits of size 15 (they have representatives u_{12} and $u_{12}z$) generate the whole of $\langle \mathcal{U} \rangle$ and that the orbit of size 1 generates $\langle z \rangle$.

Converting back to our trio notation, we find that the following:

$$t_{ab \cdot cd \cdot ef} t_{ad \cdot cf \cdot eb} t_{af \cdot cb \cdot ed} = s_D,$$

where s_D (the element z above) is an involution centralized by the whole of the sextet group $N_D \cong 2^6{:}3^{\cdot}S_6$. We refer to the s_D as *sextet elements* (or involutions). Note that s_D is independent of the three trio elements (where the trios are all coarsenings of D) used to define it. We have also shown the following.

LEMMA 5.4 The group $\langle t_T : T \in \text{coa}(S) \rangle$ is elementary abelian of order 2^5 and has a unique involution (which we call s_D) that commutes with the whole of $N_D \cong 2^6{:}3^{\cdot}S_6$. (There is no reason *a priori* why s_D should be non-trivial, but in our matrix group it is actually an involution.)

In the course of proving the above result, we have also demonstrated the following lemma.

LEMMA 5.5 If the trios T and U have intersection array I_2, then t_T and t_U commute.

The seven sextets refining a trio

For a trio T, define $\text{ref}(T)$ to be the set of seven sextets refining a trio. Now that we have defined the sextet elements s_D, we wish to investigate the group $\langle s_X : X \in \text{ref}(T) \rangle$ for a particular (fixed) trio T. If T is the MOG trio, then the seven refinements have identical tetrads in each of the three bricks.

$$O = \boxed{\begin{smallmatrix} \cdot & \cdot \\ \cdot & \cdot \\ \cdot & \cdot \\ \cdot & \cdot \end{smallmatrix}} \ ; A = \boxed{\begin{smallmatrix} \cdot & \cdot \\ \cdot & \cdot \\ \times & \times \\ \times & \times \end{smallmatrix}} \ ; B = \boxed{\begin{smallmatrix} \times & \times \\ \cdot & \cdot \\ \times & \times \\ \times & \times \end{smallmatrix}} \ ; C = \boxed{\begin{smallmatrix} \times & \times \\ \times & \times \\ \times & \times \\ \cdot & \cdot \end{smallmatrix}} \ ;$$

$$D = \boxed{\begin{smallmatrix} \times & \cdot \\ \times & \cdot \\ \times & \cdot \\ \times & \cdot \end{smallmatrix}} \ ; E = \boxed{\begin{smallmatrix} \cdot & \times \\ \cdot & \times \\ \times & \cdot \\ \times & \cdot \end{smallmatrix}} \ ; F = \boxed{\begin{smallmatrix} \cdot & \times \\ \times & \cdot \\ \cdot & \times \\ \times & \cdot \end{smallmatrix}} \ ; G = \boxed{\begin{smallmatrix} \cdot & \times \\ \times & \cdot \\ \times & \cdot \\ \cdot & \times \end{smallmatrix}} \ .$$

A useful mnemonic might be: Adjacent, Broken, Central, Descending, Echelon, Flagged, Gibbous.

Note that s_D thus denotes the sextet element corresponding to the columns of the MOG. In $\mathcal{C}^*$, the cocode of the Golay code $\mathcal{C}$, these seven sextets certainly comprise the non-zero vectors of a 3-dimensional subspace, and it will be our first objective to show that our two relations imply that the subgroup of G generated by the seven sextet elements corresponding to these seven sextets is also elementary abelian of order 2^3. It is sufficient to show Lemma 5.6.

LEMMA 5.6 *If A, B and C are the three sextets defined above, then we have* $s_A s_B \, s_C = 1.$

Proof The subgroup of M_{24} fixing the three sextets A, B and C is a subgroup L of index 7 in the trio group of shape $2^6 : (S_3 \times S_4)$. We shall now identify an orbit of this group on trios and thus obtain a subprogenitor, which we shall investigate. In fact, there are 36 octads of the form $[0, X \text{ or } X', X \text{ or } X']$, where X is one of the three tetrads A, B, C shown above as tetrads of sextets refining the MOG trio, and X' denotes the complement of X. The notation indicates the intersection of the octad with each brick of the MOG. There are then 18 trios which contain one of the three octads of the MOG trio, and two of the above 36 octads. We denote, for instance,

$$A_1^+ = \{[0' \ 0 \ 0], [0 \ A \ A], [0 \ A' \ A']\} = \boxed{\begin{smallmatrix} \times & \times & \circ & \circ & \circ & \circ \\ \times & \times & \circ & \circ & \circ & \circ \\ \times & \times & \cdot & \cdot & \cdot & \cdot \\ \times & \times & \cdot & \cdot & \cdot & \cdot \end{smallmatrix}} \ ,$$

$$A_1^- = \{[0' \ 0 \ 0], [0 \ A \ A'], [0 \ A' \ A]\} = \boxed{\begin{smallmatrix} \times & \times & \circ & \circ & \cdot & \cdot \\ \times & \times & \circ & \circ & \cdot & \cdot \\ \times & \times & \cdot & \cdot & \circ & \circ \\ \times & \times & \cdot & \cdot & \circ & \circ \end{smallmatrix}} \ .$$

Thus in a symbol $X_i^{\pm}$ the X denotes one of A, B or C, the subscript denotes which brick of the MOG trio is to be present, and the sign indicates that the pattern in the other two MOG octads is repeated $(+)$ or complemented $(-)$. There are thus $3 \times 3 \times 2 = 18$ such trios, and they clearly form an orbit under the action of the group L; we thus have a subprogenitor

$$P_L = 2^{\star 18} : (2^6 : (S_3 \times S_4)),$$

which inherits a number of relations from Theorem 5.12:

(i) $A_1^+ A_1^- = t_E$;

(ii) $A_1^+ B_1^+ C_1^+ = 1$;

(iii) $A_1^+ A_2^+ = A_2^+ A_1^+$ (derived from $A_1^+ A_2^+ A_3^- = s_A$);

(iv) $A_1^+ B_3^+ A_1^+ B_3^- = \nu_2$, of Theorem 5.12.

Of course, all conjugates of these relations under L also hold. We shall work in $H_L = \langle X_i^{\pm} \mid X \in \{A, B, C\}, i \in \{1, 2, 3\}\rangle$, the group generated by the 18 symmetric generators, and in our calculations we use the fact that

$$K = \left\langle t_E, \nu_1 = \quad , \nu_2 = \quad , \nu_3 = \quad \right\rangle \cong 2^3$$

lies in its centre. Note that relation (iv) and conjugation by elements of L imply that $(X_i^{\pm} Y_j^{\pm})^2 = \nu_k t_E$, for $X, Y \in \{A, B, C\}$, $X \neq Y$, i, j, k distinct. Using these relations, we have the following:

$$
\begin{aligned}
s_A s_B &= A_1^+ \ A_2^+ \ A_3^- \ B_2^- \ B_2^- \ B_1^- = A_1^+ \ A_2^+ \ C_3^+ \ B_2^- \ B_1^- \\
&= A_1^+ \ A_2^+ \ B_2^+ \ C_3^+ \ B_1^- \ \nu_1 = A_1^+ \ C_2^+ \ C_3^+ \ B_1^- \ \nu_1 \\
&= A_1^+ \ C_2^+ \ B_1^+ \ C_3^+ \ \nu_2 \nu_1 = A_1^+ \ B_1^- \ C_2^+ \ C_3^+ \nu_3 \nu_2 \nu_1 \\
&= C_1^- \ C_2^+ \ C_3^+ = s_C,
\end{aligned}
$$

as required.   □

The subgroup $2^{11} : M_{24}$

Now that we have shown that the product of two sextet elements whose associated sextets have intersection array J_1 is indeed the sextet element corresponding to the sextet which is the sum in $\mathcal{C}^*$ of those two sextets, we are in a position to prove that

$$\frac{2^{\star 1771} : M_{24}}{s_A s_B s_C = 1} \cong 2^{11} : M_{24}.$$

We may proceed using our double coset enumerator [17] to enumerate the (N, N)-double cosets where $N \cong M_{24}$. We find that there are three double cosets, containing 1, 1771 and 276 single cosets, respectively, and thus that there are $2048 = 2^{11}$ single cosets. But the above progenitor certainly does map onto a group of shape $2^{11} : M_{24}$, in which the 2^{11} when considered as an M_{24}-module is the even part of the Golay cocode, and in which the additional relation does hold. Coset enumeration shows that the image has the right order, and so this surjection is an isomorphism. Taking the group M_{24} as permutations on 24 letters, appropriate computer input and output in MAGMA [19] is as follows:

```
> sext:=sub<m24|Stabilizer(m24,{24,3,6,9}),
> Stabilizer(m24,{23,19,15,5})>;
> f,N,k:=CosetAction(m24,sext);
> oo:=Orbits(Stabilizer(N,1));
> [#oo[i]:i in [1..#oo]];
[ 1, 90, 240, 1440 ]
> Fix(Stabilizer(N,[1,Random(oo[2])]));
{ 1, 135, 315 }
```

The sextet stabilizer has been obtained as the subgroup generated by the stabilizers of each of two tetrads whose union is an octad, in this case $\{24, 3, 6, 9, 23, 19, 15, 5\}$. For computational purposes, we have replaced 0 of the 24-point projective line by 23, and ∞ by 24. The next few commands identify an *even line* of sextets, so the three sextets labelled 1, 135 and 315 here correspond to an image of A, B and C:

```
> RR:=[<[1,135,315],Id(N)>];
> HH:=[N];
> CT:=DCEnum(N,RR,HH:Print:=5,Grain:=100);

Index: 2048 === Rank: 3 === Edges: 7
> CT[4];
[
    [],
    [ 1 ],
    [ 1, 5 ]
]
> CT[7];
[ 1, 1771, 276 ]
>
```

The single relation by which we have factored the progenitor corresponds to

$$s_1 \, s_{135} \, s_{315} = 1.$$

The output tells us in CT[4] that there are just three double cosets of form NwN, namely N, Ns_1N and Ns_1s_5N, and in CT[7] that these three contain 1, 1771 and 276 single cosets of N, respectively.

The above progenitor is what is called a *universal representation group* for the point-line incidence system $\mathcal{G} = (P, L)$, where P is a set of points and L a set of lines with three points per line (see, for example ref. [2]). Explicitly, we define the following:

$$R(\mathcal{G}) = \langle t_i, i \in P \mid t_i^2 = 1, \ t_i t_j t_k = 1 \text{ for } \{i, j, k\} \in L \rangle,$$

so we require that the three points on a line correspond to the three non-trivial elements of a Klein fourgroup. It is shown in ref. [2] that the universal representation group in our example, where the points are the sextets and the lines are sets such as $\{A, B, C\}$, is abelian, and Ronan and Smith [76] show that as a vector space over $\mathbb{F}_2$ it has dimension 11.

Since the identification of this group is crucial to the enumeration which follows, and to make the present argument self-contained, we demonstrate how this result can be proved by hand using the techniques of symmetric generation. We start by showing the following.

LEMMA **5.7**

$$\frac{2^{\cdot 35} : A_8}{\dfrac{abcd}{efgh} \cdot \dfrac{abef}{cdgh} \cdot \dfrac{abgh}{cdef} = 1} \cong 2^6 : A_8,$$

an elementary abelian group of order 64 extended by the alternating group on eight letters. Here A_8 is viewed as a permutation group acting on the 35 partitions of eight letters into two fours, and there are 35 symmetric generators corresponding to these partitions.

Proof Our additional relation implies that any two symmetric generators whose tetrads intersect evenly commute; we must investigate a pair of symmetric generators whose tetrads intersect oddly. But we have

$$x = \frac{abcd}{efgh} \cdot \frac{ebcd}{afgh} = \frac{abcd}{efgh} \cdot \frac{abef}{cdgh} \cdot \frac{abef}{cdgh} \cdot \frac{ebcd}{afgh} = \frac{abgh}{cdef} \cdot \frac{afcd}{begh},$$

which is centralized by

$$\langle (b\ c\ d), (c\ d)(gh), (fgh), (ae)(bf)(cg)(dh), (bgh) \rangle \cong S_6$$

with orbits $\{a, e\}$ and $\{b, c, d, f, g, h\}$. In particular, x corresponds to the unordered pair $\{a, e\}$ of the eight letters and so there are at most 28 such elements. Multiplication of x by any symmetric generator corresponding to a partition with a and e on opposite sides results in a single symmetric generator and, since $(b\ f)(c\ g\ d\ h)$ centralizes x, we have

$$\frac{abcd}{efgh} \cdot \frac{ebcd}{afgh} = \left(\frac{abcd}{efgh} \cdot \frac{ebcd}{afgh} \right)^{(b\ f)(c\ g\ d\ h)} = \frac{afgh}{ebcd} \cdot \frac{abcd}{efgh}.$$

Thus the generators all commute with one another, and the group they generate has order $1 + 35 + 28 = 2^6$.   □

We are now in a position to prove Theorem 5.13 by hand.

Theorem 5.13

$$G = \frac{2^{\bullet 1771} : M_{24}}{s_A s_B s_C = 1} \cong 2^{11} : M_{24},$$

an elementary abelian group of order 2048 extended by M_{24}.

Proof Firstly, note that, for a fixed octad O, the 35 sextets that have a tetrad in O satisfy the relation of Lemma 5.7 and so generate an elementary abelian group of order 2^6. We call such a subspace the *octad space* of the octad O and denote it by U_O. Now any two sextets which intersect as J_1 or J_2, as in the labelled arrays on p. 179, lie together in some octad space and so commute. The product of two J_2-related sextet elements is a 'duad-type' element in that octad space. We see that

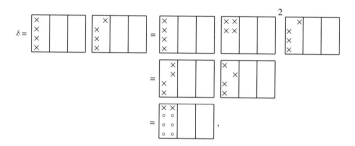

where in each case four ×s indicate the sextet element defined by that tetrad; when necessary below we shall complete all six tetrads of a sextet element. In the final line we indicate the 'duad-type' element and, with the os, the octad space to which it belongs. We wish to show that this duad-type element is independent of the octad space used to define it. Certainly it commutes with the subgroup of the relevant octad stabilizer fixing the duad, which is to say a group of shape $2^4 : S_6$. Moreover, we have the following:

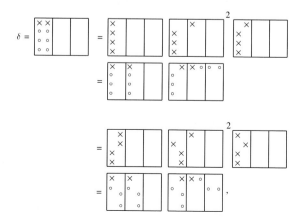

which shows that

$$\mu = \quad .$$

It is of course our objective to show that this element μ is in reality the identity element. We see that μ is fixed by

$$H = \left\langle \quad , \quad , \quad , \quad \right\rangle .$$

Note that each of these elements is of one of the types described at the beginning of this section and so may be readily shown to lie in our copy of M_{24}. We now argue that $H = N \cong M_{24}$. We note that H is visibly transitive on the 24 points, and so must be contained in one of the transitive maximal subgroups of M_{24} or be the whole group. The possibilities are $H \leq L_2(7)$, $L_2(23)$, the trio stabilizer, the sextet stabilizer or the duum stabilizer $M_{12} : 2$. But the first three of these generators of H generate a subgroup with suborbits of lengths $1 + 1 + 22$, so it contains elements of order 11; this group cannot fix a duum (it would have to have blocks of imprimitivity of size 12), and so H can only be a copy of $L_2(23)$ or M_{24}. However, involutions in $L_2(23)$ are fixed-point-free, and H contains involutions of cycle shape $1^8.2^8$. So the only possibility is that $H = N$.

This is a powerful fact, which can now be used to complete our proof that $\mu = 1$. For instance, let

$$\rho = \quad \in H = N \cong M_{24}.$$

Applying this element to μ and cancelling, we obtain the following:

$$\mu = \quad ,$$

and so

$$\delta_1 = \begin{array}{|c c|c|} \hline \times & \times & \\ \circ & \circ & \\ \circ & \circ & \\ \circ & \circ & \\ \hline \end{array}$$

is fixed by ρ as well as a subgroup of shape $2^4 : S_6$ fixing a duad and an octad containing it, which we saw previously. The maximality of this subgroup in $M_{22} : 2$ shows that δ is fixed by the whole of $M_{22} : 2$ and is thus independent of the octad space used to define it; it thus depends only on the duad and has just $\binom{24}{2} = 276$ images under the action of M_{24}. Now, any two duads lie together in some octad space and so they commute. Moreover, any sextet is the product of two duads in some octad space, and so the group generated by the duads equals the group generated by the sextets, and this group is abelian. But any duad and any sextet lie together in some octad space, and their product is either a duad or another sextet. So the set of duads and sextets (together with the identity) is closed under mulitplication and forms an elementary abelian group of order $1 + 1771 + 276 = 2^{11}$ as required. □

So let $H = \langle N, t_S \rangle$, where S is a sextet. The above shows that $H \cong 2^{11} : M_{24}$. We now use our double coset enumerator [17] to enumerate the (H, N)-double cosets within our symmetric presentation. We must first obtain M_{24} as permutations on 3795 letters:

```
> m24<a,b>:=PermutationGroup<24|(1, 2, 3, 4, 5, 6, 7, 8, 9,
 10,11,12, 13, 14, 15, 16, 17, 18, 19, 20, 21, 22, 23),(1, 11)
 (3, 19) (4, 20)(5, 9)(6, 15)(13, 21)(14, 16)(23, 24)>;

> #m24;
244823040

> oct:=Stabilizer(m24,{ 3, 5, 6, 9, 15, 19, 23, 24 });
> triopw:=Stabilizer(oct,{ 1, 4, 11, 13, 14, 16, 20, 21 });
> trio:=sub<m24|triopw,m24!(1, 2, 23)(3, 4, 18)(5, 21, 7)
(6, 16,8)(9, 13, 12)(10, 19, 20)(11, 22, 24)(14,17, 15),
m24!(1, 23)(3, 4)(5, 21)(6, 16)(9, 13)(11, 24)(14, 15)
(19, 20)>;

> Index(m24,trio);
3795
```

The group M_{24} is entered as permutations on 24 letters. The subgroup oct is the stabilizer of an octad, and triopw is the stabilizer of each of three octads comprising a trio. We then adjoin permutations permuting those

three octads to obtain the full trio stabilizer `trio` and verify that it has
index 3795 in M_{24}:

```
> f,N,k:=CosetAction(m24,trio);

> st1:=f(trio);
> for i in [1..3795] do
for> orb:=i^st1;
for> if #orb eq 1008 then
for|if> ss:=i;
for|if> break;
for|if> end if;
for> end for;
> xx:=Fix(Stabilizer(st1,ss));
> xx;
{ 1, 5, 823, 1240, 3115 }
> nxx:=Stabilizer(N,xx);
> znxx:=Centre(nxx);
> #znxx;
2
> 1^nxx;
GSet{ 1, 5, 823, 3115 }
```

We take the action of M_{24} on the trios and obtain a trio `ss` in the 1008-orbit
of the stabilizer of the original trio, which is labelled 1. The set of five trios
`xx` corresponds to Figure 5.9; the set stabilizer of `xx` acts as D_8 on `xx`, and
so we can identify that trio 1240 is the 'central' trio corresponding to E. Now
the element ν_2 of Theorem 5.12 is central in this stabilizer, and so we are
almost ready to write down our relations. We finally require the following:

```
> Fix(Stabilizer(st1,1240));
{ 1, 823, 1240 }
```

We find that the vertex opposite $A = 1$ in Figure 5.9 is labelled 823, and
so, without loss of generality, we have the correspondence $[A, B, C, D, E] =$
$[t_1, t_5, t_{823}, t_{3115}, t_{1240}]$. Thus our two defining relations are (i) $t_1 t_{1240} t_{823} = 1$
and (ii) $t_1 t_5 t_1 t_{3115} = \nu_2$, the non-trivial element in the centre of the stabilizer
of `xx`:

```
> for j in [1..3795] do
for> orb:=j^st1;
for> if #orb eq 56 then
for|if> tt:=j;
for|if> break;
for|if> end if;
for> end for;
> Fix(Stabilizer(st1,tt));
{ 1, 110, 2165 }
```

It remains to extend the subgroup N to H, where $H \cong 2^{11} : M_{24}$. All we need is three trios whose product is a sextet-type element; i.e. a trio **tt** in the 56-orbit of the stabilizer of trio 1, and the unique third trio which is fixed by the stabilizer of 1 and **tt**. The previous calculation shows us that $\{1, 110, 2165\}$ is such a set of trios. We now have the following:

```
> RR:=[<[1,5,1,3115],cnxx.1>,<[1,1240,823],Id(N)>];
> HH:=[*N,<[1,110,2165],Id(N)>*];
> CT:=DCEnum(N,RR,HH:Print:=5,Grain:=100)
```

The two relations are fed into **RR** and the subgroup H is entered as **HH**; an enumeration of the double cosets of form HwN is then carried out. As before, we store representatives for w in **CT[4]**, thus, for example, the fifth double coset here denotes $Ht_1 t_5 t_3 N$, and the number of single cosets each contains is recorded in **CT[7]**. The enumeration took just over two minutes on a 3.2 GHz Pentium 4 PC with 1 GB of memory. A word of warning: the coset representatives in **CT[4]** are not necessarily the shortest possible words in the t_i, but are essentially the first representatives found by the program. This yields the following output:

```
Index: 173067389 === Rank: 20 === Edges: 3121 ===
Status: Early closed === Time:131.062

> CT[4];
[
  [],
  [ 1 ],
  [ 1, 5 ],
  [ 1, 2 ],
  [ 1, 5, 3 ],
  [ 1, 5, 23 ],
  [ 1, 5, 23, 38 ],
  [ 1, 2, 33 ],
  [ 1, 2, 40 ],
  [ 1, 2, 32 ],
  [ 1, 2, 70 ],
  [ 1, 5, 3, 2077 ],
  [ 1, 5, 6, 380 ],
  [ 1, 5, 63, 6 ],
  [ 1, 5, 23, 38, 1034 ],
  [ 1, 5, 23, 38, 276 ],
  [ 1, 2, 219 ],
  [ 1, 2, 32, 2949 ],
  [ 1, 2, 70, 302 ],
  [ 1, 5, 23, 38, 634 ]
]
> CT[7];
```

[1, 3795, 318780, 5100480, 81607680, 1275120, 15301440,
40803840, 1912680,2550240, 7650720, 12241152, 2550240,
967680, 53130, 478170, 212520, 1771, 26565, 11385]

We find that there are (at most) 20 double cosets and $173\,067\,389 = 11^2 \times 29 \times 31 \times 37 \times 43$ single cosets, which gives an upper bound on the order of the group generated by our 112-dimensional matrices (and the group defined by our symmetric presentation) of $173\,067\,389 \cdot 2^{11} \cdot |M_{24}|$. But a random search in our 112-dimensional matrix group easily locates elements of orders 29, 31, 37 and 43. In order to show that 11^2 divides the index of H in G, we seek a subgroup of order 11^3 in our matrix group by using random searches to locate $U_3(11)$ (or $U_3(11){:}2$) and working inside this to locate some subgroup lying in between 11_+^{1+2} and $11_+^{1+2}{:}(5 \times 8{:}_32)$, from which we easily get the desired subgroup. Since H has order $2^{21} \times 3^3 \times 5 \times 7 \times 11 \times 23$, we see that the index $173\,067\,389$ is exactly the index of H in G, and so G has the same order as J_4 (since we know that elements of order 2 that are meant to be in the normal 2^{11} of H still have order 2 in our matrix group, and so H does not collapse to a proper image).

We now check primitivity of the above action by showing that H is maximal in G. But we know a set of (H, N)-double coset representatives, and since $N \le H$, the (H, H)-double coset representatives can be chosen from among these. We need only check that $\langle H, g \rangle = G$ whenever g is an (H, H)-double coset representative (other than the case when g is obviously in H). This is easily done with the enumerator (one adds these double coset representatives to the generators for H and sets the enumerator running again). We now show the following.

THEOREM **5.14** *G is simple.*

Proof Let K be a non-trivial proper normal subgroup of G. Then in the already mentioned $173\,067\,389$-point action, G is primitive and so K is transitive. The point-stabilizer can be taken to be H, and we must have $(K \cap H) \lhd H$, and so $K \cap H \cong 1$, 2^{11} or $K \cap H = H$. If $K \cap H = H$, then $K = G$, contradicting the fact that K is proper. If $K \cap H = 1$, then $|K| = 173\,067\,389$, the minimum possible order of a non-trivial normal subgroup of G, so K is characteristically simple and hence simple as its order is not a proper power. But this is impossible as K has odd order. Lastly we consider the case when $K \cap H \cong 2^{11}$. Since we have now dealt with the case $K \cap H = 1$, K is now minimal normal and therefore (characteristically) simple. Now let $P \cong 2^{11}$ be a Sylow 2-subgroup of K (we can take $P = K \cap H$). Now, $H \le N_G(P)$ and P is not normal in G (since $173\,067\,389 \nmid |P| = 2048$), so, by the maximality of H, we get $H = N_G(P)$. So $N_K(P) = K \cap N_G(P) = K \cap H = P$, and the Burnside normal p-complement theorem with $p = 2$ shows that K is not simple, a contradiction. Thus the theorem is proved. □

An ordinary presentation of J_4

An ordinary presentation of J_4 deduced from our symmetric presentation will take the following form:

$$\langle x, y, t \mid R(x, y), t^2 = 1 = [t, w_1(x, y)] = [t, w_2(x, y)]$$

$$= u_1(x, y, t) = u_2(x, y, t)\rangle,$$

where $R(x, y)$ are relations in x and y which ensure that $\langle x, y \rangle \cong M_{24}$, w_1 and w_2 are words in x and y such that $\langle w_1, w_2 \rangle \cong 2^6 : (S_3 \times L_3(2))$, and u_1 and u_2 are relations in x, y and t which correspond to our two additional relations. Explicitly we have the following:

$$\langle x, y, t \mid x^2 = y^3 = (xy)^{23} = [x, y]^{12} = [x, yxy]^5 = (xy(xyxy^{-1})^3)^4$$

$$= (xyxyxy^{-1})^3(xyxy^{-1}xy^{-1})^3 = t^2 = [t, yxy(xy^{-1})^2(xy)^3]$$

$$= [t, x] = (yt^{yxy^{-1}xyxy^{-1}x})^3 = ((yxyxyxy)^3 tt^{(xy)^3y(xy)^6y})^2 = 1 \rangle.$$

The restriction of this presentation to $\langle x, y \rangle$ (by removing the relations involving t) yields a presentation of M_{24}. This has been demonstrated by enumerating the $10\,644\,480$ cosets of $\langle xy \rangle$, which in MAGMA with Hard:=true set can be done in the space required to store 11×10^6 cosets defining fewer than 12.5×10^6 cosets in total. We note that

$$\langle x, y, (yt^{(yx)^4y^{-1}x})^3 \rangle \cong 2^{11} : M_{24}.$$

It is not possible to enumerate the $173\,067\,389$ cosets of this subgroup on machines with 32-bit addressing architecture. (The minimum table size needed is $4 \times 4 \times 173\,067\,389 = 2\,769\,078\,224$ bytes, i.e. about $2.58\,\text{GB}$, and we do not even know if a coset enumerator can successfully perform such an enumeration in a manner that defines few redundant cosets.)

The symmetric generators lie in class $2A$, and we conclude this section by giving words in x, y and t which generate $C_G(t) \cong 2^{1+12} : 3 \cdot M_{22}.2$. Thus,

$$C_G(t) = \langle t, x, yxy(xy^{-1})^2(xy)^3, t^{xyxy^{-1}xy}, t^{xyxy^{-1}} \rangle$$

$$= \langle yxy(xy^{-1})^2(xy)^3, xt^{xyxy^{-1}} \rangle.$$

Single relator symmetric presentation

We conclude by mentioning that Bray has shown that our first additional relation, namely $t_T t_U t_{T+U} = 1$, is essentially redundant. In fact, Theorem 5.15 follows.

THEOREM 5.15

$$G = \frac{2^{\star 3795} : M_{24}}{t_A t_B t_A t_D = \nu_2} \cong J_4 \times 2,$$

where A, B, D and ν_2 are as in Figure 5.9.

Proof Note that G has a subgroup of index 2, namely the set of elements πw, where $\pi \in N$ and w is a word in the symmetric generators of even length, and so cannot be simple. We shall show that the element $t_T t_U t_{T+U}$, which has order 1 in J_4, is an involution in the centre of G. For convenience of notation, we shall denote the symmetric generator t_F, for F a trio, simply by F. Thus we have the following relations:

$$ABAD = BCBA = CDCB = DADC$$
$$= ADAB = BABC = CBCD = DCDA = \nu_2$$

and, inverting,

$$DABA = ABCB = BCDC = CDAD$$
$$= BADA = CBAB = DCBC = ADCD = \nu_2.$$

So $C = BAB\nu_2$, and $D = ABA\nu_2$, and we have $C, D \in \langle A, B, \nu_2 \rangle$. Moreover,

$$\nu_2 = DADC = ABAAABABAB\nu_2^3 = (AB)^4 \nu_2,$$

and so $(AB)^4 = 1$. Moreover, $AC = \nu_2 ABAB$ and so $CA = \nu_2 BABA = AC$; thus A and C commute. Similarly B and D commute, and E (having an octad in common with each of the trios A, B, C and D) commutes with all of them. In particular, ACE is an involution, independent of the ordering of A, C and E. But $ACE = ABABE\nu_2$ and $BDE = BABAE\nu_2 = ACE$; so the element $ACE = z$, say, is fixed by any permutation of M_{24} which fixes $\{A, C, E\}$ and by any permutation of M_{24} which fixes $\{B, D, E\}$. One readily verifies that

$$\langle \text{Stab}_{M_{24}}(\{A, C, E\}), \text{Stab}_{M_{24}}(\{B, D, E\}) \rangle = N \cong M_{24}.$$

Thus z is centralized by N and by t_E, and so by the whole of G. Factoring out the subgroup $\langle z \rangle$ gives the group J_4, as in Theorem 5.12, and so $G \cong 2 \times J_4$ as required. $\qquad\square$

5.7.2 The Conway group ·O

The Leech lattice Λ was discovered by Leech [65] in 1965 in connection with the packing of non-overlapping identical spheres into 24-dimensional space $\mathbb{R}^{24}$ so that their centres lie at the lattice points; see ref. [26]. Its construction relies heavily on the rich combinatorial structure underlying the Mathieu group M_{24}. The full group of symmetries of Λ is, of course, infinite as it contains all translations by a lattice vector. Leech himself considered the subgroup consisting of all symmetries fixing the origin O; he had enough geometric evidence to predict the order of this group to within a factor of 2, but could not prove the existence of all the symmetries he anticipated. It was John McKay who told John Conway about Λ – and the rest, as they say, is history. In two elegant papers (see refs. [20] and [22]) Conway produced a beautifully simple additional symmetry of Λ and found the order of the group it generates together with the monomial group of permutations and sign changes used in the construction of Λ. He proved that this is the full group of symmetries of Λ (fixing O), showed that it is perfect with centre of order 2, and that the quotient by its centre is simple. He called the group $\cdot$O to signify that it was the stabilizer of O in the full group of symmetries of Λ, and he extended the notation to $\cdot$2 and $\cdot$3, the stabilizers of vectors of type 2 and type 3, respectively. The symbol $\cdot$1 was then used to denote the quotient $\cdot$O$/\langle\pm1\rangle$.

In this section we use the methods of symmetric generation to define $\cdot$O directly from M_{24} by considering a homomorphic image of an infinite group, which we denote by

$$P = 2^{\star\binom{24}{4}} : M_{24}.$$

Here $2^{\star\binom{24}{4}}$ denotes a free product of $\binom{24}{4}$ cyclic groups of order 2, corresponding to the tetrads of the 24-point set on which M_{24} acts. This free product is extended by M_{24} itself to form a semi-direct product in which M_{24} acts in the natural manner on tetrads. Lemma 3.3, which first appeared in ref. [33], is then used to yield a relation, first shown to be sufficient to define $\cdot$O by Bray, by which we can factor P without leading to total collapse. Since P is a semi-direct product, every element of it can be written as πw, where π is an element of M_{24} and w is a word in the $\binom{24}{4}$ involutory generators of the free product. The relator by which we factor takes a particularly simple form, with the length of w being just 3; thus, the corresponding relation has the form $\nu = t_T t_U t_V$, where $\nu \in M_{24}$ and T, U, and V are tetrads.

Having defined this quotient G, we use the double coset enumerator of Bray and Curtis [17] to demonstrate that it is indeed a group of the required order. We then seek a faithful representation of minimal degree and, unsurprisingly, come up with dimension 24. Embedding G in $O_{24}(\mathbb{R})$ is readily accomplished, and it turns out that the involutory generators corresponding to the tetrads are simply the negatives of Conway's original elements. The Leech lattice follows, of course, by letting this orthogonal group act on the vectors in the standard basis of $\mathbb{R}^{24}$.

The progenitor for $\cdot$O *and the additional relation*

As has been mentioned above, the Mathieu group M_{24} acts quintuply transitively on 24 letters and so permutes the $\binom{24}{4}$ tetrads transitively. This action is not, however, primitive as the six tetrads which together comprise a sextet form a block of imprimitivity. We shall consider the progenitor

$$P = 2^{\star\binom{24}{4}} : M_{24}.$$

Thus, a typical symmetric generator will be denoted by t_T, where T is a tetrad of the 24 points of Ω. Whereas the rank of the symmetric group S_{24} acting on tetrads is just 5, depending only on the number of points in which a tetrad intersects the fixed tetrad T, the tetrad stabilizer in M_{24}, which has shape $2^6 : (3 \times A_5) : 2$, has 14 orbits on tetrads. We shall be concerned with pairs of tetrads which lie together in a common octad of the system and which intersect one another in two points. Having fixed the tetrad T, there are clearly $\binom{4}{2} \cdot 5 \cdot \binom{4}{2} = 180$ possibilities for a tetrad U to intersect it in this manner. The stabilizer in M_{24} of both T and U thus has order $2^6 \cdot 3 \cdot 120/180 = 2^7$ and shape $2^4 : 2^3$. Indeed, the stabilizer of an octad has shape $2^4 : A_8$, where the elementary abelian 2^4 fixes every point of the octad, and we have simply fixed a partition of the eight points of the octad into pairs (and fixed each of those pairs). To be explicit, we let T be the tetrad consisting of the top two points in each of the first two columns of the MOG, and we let U be the first and third points in each of the first two columns. In what follows, we shall often denote the element t_T by ×s in the four positions of T as displayed in the MOG diagram. Thus,

$$
t_T = \begin{array}{|cc|c|c|}
\hline \times & \times & . & . \\
\times & \times & . & . \\
\hline . & . & . & . \\
. & . & . & . \\
\hline
\end{array}, \quad
t_U = \begin{array}{|cc|c|c|}
\hline \times & \times & . & . \\
. & . & . & . \\
\hline \times & \times & . & . \\
. & . & . & . \\
\hline
\end{array}, \quad
\begin{array}{|cc|c|c|}
\hline a & a & & \\
b & b & & \\
\hline c & c & & \\
d & d & & \\
\hline
\end{array}.
$$

Now, the tetrads T and U determine a pairing of of the first brick into duads a, b, c and d. The stabilizer of T and U, $\mathrm{Stab}_{M_{24}}(TU)$, must fix each of these duads and so commute with the symmetric generators $\mathcal{X} = \{t_{ab} = t_T, t_{ac} = t_U, t_{ad}, t_{bc}, t_{bd}, t_{cd}\}$. Lemma 3.3 says that the only elements of $N \cong M_{24}$ which can be written in terms of the elements in $\mathcal{X}$ without causing collapse must lie in the centralizer in M_{24} of $\mathrm{Stab}_{M_{24}}(TU)$, and we have

$$
C_{M_{24}}(\mathrm{Stab}_{M_{24}}(TU)) = Z(\mathrm{Stab}_{M_{24}}(TU)) = \left\langle \nu = \begin{array}{|c|cc|c|}\hline & & & \\ & \bullet\!-\!\bullet & \bullet\!-\!\bullet & \\ & \vdots & \vdots & \\ & \bullet\!-\!\bullet & \bullet\!-\!\bullet & \\ & \bullet\!-\!\bullet & \bullet\!-\!\bullet & \\\hline\end{array} \right\rangle \cong C_2.
$$

We now seek to write $\nu = w(t_T, t_U, \dots)$, a word in the elements of $\mathcal{X}$ of shortest possible length without causing collapse. But $l(w) = 1$ would mean that $\nu = t_T$ commutes with the stabilizer of T, which it does not; $l(w) = 2$ would mean that $\nu = t_T t_U$, but this means that $\nu = t_{ab}t_{ac} = t_{ac}t_{ad} = t_{ad}t_{ab}$, and on multiplying these three relations together we obtain $\nu = 1$. So the minimum length for w is 3, and we have a relation $\nu = xyz$. Firstly, note that each of x, y and z commutes with ν and so $xy = z\nu$ is of order 2; thus x and y commute with one another, and similarly all three elements x, y and z commute with one another. In particular, the three elements x, y and z must be distinct if we are to avoid collapse. Suppose now that two of the tetrads are complementary within the octad $\{a, b, c, d\}$, so that, without loss of generality, we have $t_{ab}t_{cd}t_{ac} = \nu$. But an element such as

$$
\rho = \begin{array}{|c|c|cc|}
\hline & & \rule{0.6em}{0.4pt} & \\
\hline \mid\ \mid & \mid\ \mid & \times & \\
\hline
\end{array} \in M_{24}
$$

commutes with ν and acts as $(a)(b)(c\ d)$ on the pairs; conjugating our relation by this element, we have $t_{ab}t_{cd}t_{ac} = t_{ab}t_{cd}t_{ad}$, and so $t_{ac} = t_{ad}$, leading to a contradiction. There are thus just two possibilities for a relation of length 3:

(i) $t_{ab}t_{ac}t_{bc} = \nu$;

(ii) $t_{ab}t_{ac}t_{ad} = \nu$.

In order to see that case (i) fails, we proceed as follows. The orbits of M_{24} on the subsets of Ω were calculated by Todd and displayed in a convenient diagram by Conway [23]. The 8-element subsets fall into three orbits, which are denoted by S_8, T_8 and U_8 and referred to as *special* (the octads themselves), *transverse* and *umbral*. Umbral 8-element subsets fall uniquely into four duads in such a way that the removal of any one of the duads leaves a special hexad (i.e. a hexad which is contained in an octad). An explicit example is given by the pairs $\{a, b, c, e\}$:

$$
\begin{array}{|c c|c c|c|}
\hline
\times & \times & \times & \times & \\
\hline
\times & \times & & & \\
\hline
\times & \times & & & \\
\hline
\end{array}
\ ,\qquad
\begin{array}{|c c|c c|c|}
\hline
a & a & e & e & \\
\hline
b & b & & & \\
\hline
c & c & & & \\
\hline
\end{array}
\ .
$$

We work in the elementary abelian group of order 8 generated by $\{t_{ab}, t_{ac}, t_{ae}\}$. Then

$$
1 = t_{ab}^2 t_{ac}^2 t_{ae}^2 = t_{ab}t_{ac}\cdot t_{ac}t_{ae}\cdot t_{ae}t_{ab}
$$

$$
= t_{bc}\ \boxed{}\quad t_{ce}\ \boxed{}\quad t_{be}\ \boxed{}
$$

$$
= t_{be}\ \boxed{}\quad t_{be}\ \boxed{}
$$

$$
= \ \boxed{}\ ,
$$

which is a contradiction. In the following section, we shall explore the consequences of the relation in case (ii), namely that $t_{ab}t_{ac}t_{ad} = \nu$.

Investigation of the image of our progenitor

The rest of this section will be devoted to showing Theorem 5.16 holds.

THEOREM 5.16

$$G = \frac{2^{\binom{24}{4}} : M_{24}}{t_{ab}t_{ac}t_{ad} = \nu} \cong \cdot O,$$

where the duads a, b, c, d partition an octad, and $C_{M_{24}}(\mathrm{Stab}_{M_{24}}(a, b, c, d)) = \langle \nu \rangle$.

From Corollary 3.1, we see that G is perfect. In order to proceed, it will be useful to introduce names for certain frequently used tetrads. If T is a tetrad, then the associated symmetric generator will, of course, be denoted by t_T. Thus the subsets of the first brick (first two columns) of the MOG shown on p. 189 give us the elements $\{t_A, t_B, \ldots, t_G\}$.

In fact, the seven sextets defined by these seven tetrads are precisely the set of sextets which refine the MOG trio, the six tetrads in each case being the same pattern (and its complement) repeated in each brick of the MOG. If T is one such tetrad, then we denote by T' its complement in the octad shown, which we shall denote by O. Thus we should have $t_A = t_{cd}$ and $t_{A'} = t_{ab}$ in the notation of this section. Our relation then becomes $t_{A'}t_{B'}t_{C'} = \nu$. We have already seen that, if T and U are two tetrads which intersect in two points and which lie together in the same octad, then t_T and t_U commute; thus, since t_A commutes with $t_{B'}, t_{C'}$ and ν, t_A commutes with $t_{A'}$. So two symmetric generators whose tetrads are disjoint and whose union is an octad also commute. In fact, Lemma 5.8 follows.

LEMMA 5.8 The involution $t_A t_{A'} = t_{A'} t_A = t_X t_{X'}$ for X any tetrad in the octad O; and so $t_A t_{A'} = \epsilon_O$, say, an involution which commutes with the octad stabilizer of shape $2^4 : A_8$.

    Proof The element $t_A t_{A'}$ certainly commutes with the subgroup H of the octad stabilizer which fixes a partition of the eight points into two tetrads, namely a subgroup of M_{24} of shape $2^4 : (A_4 \times A_4) : 2$, which is maximal in the octad stabilizer $2^4 : A_8$. But, using the relation twice, we have

$$t_A t_{A'} = t_A t_B^2 t_{A'} = t_{C'} \, \nu \, t_C \, \nu = t_{C'} t_C,$$

which is fixed by

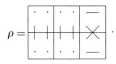

$$\rho =$$

But ρ does not preserve the partition of O into two tetrads, and so $\langle H, \rho \rangle \cong 2^4 : A_8$. $\square$

In this manner, we obtain just 759 *octad-type* involutions of the form ϵ_O for $O \in \mathcal{C}_8$, the set of octads in the Golay code. We shall show that $E = \langle \epsilon_O \mid O \in \mathcal{C}_8 \rangle \cong 2^{12}$, an elementary abelian group of order 2^{12}. Before proceeding, let us suppose that O and O' are octads which intersect in four points; then

$O \setminus O' = T_1$, $O' \setminus O = T_2$ and $O \cap O' = T_3$ are three tetrads of a sextet. By the above, we have

$$\epsilon_O \epsilon_{O'} = t_{T_1} t_{T_3} t_{T_2} t_{T_3} = t_{T_1} t_{T_2} = \epsilon_{O+O'},$$

where $O + O'$ denotes the symmetric difference of the octads O and O'; in this respect, ϵ_O and $\epsilon_{O'}$ behave like octads in the Golay code. When we have occasion to display a particular symmetric generator t_T in the MOG, we shall place $\times$ in the positions of the tetrad T; when we wish to display an element ϵ_O, we shall insert o in the positions of the octad O.

Elementary abelian subgroup of order 2^{12}

We may readily convert this problem into the language of progenitors, as we have 759 involutory generators corresponding to the octads, subject to the above relation and permuted by M_{24}. We are attempting to show the following lemma.

LEMMA **5.9**

$$\frac{2^{\star 759} : M_{24}}{\epsilon_O \epsilon_U \epsilon_{O+U} = 1} = 2^{12} : M_{24},$$

where O and U are two octads which intersect in four points, and $O + U$ denotes their symmetric difference.

This is an ideal opportunity to illustrate the double coset enumerator of Bray and the author [17]. We need the action of M_{24} on 759 points, and may obtain this using MAGMA as follows:

```
> g:=Sym(24);
> m24:=sub<g|g!(1,2,3,4,5,6,7,8,9,10,11,12,13,14,15,
16,17,18,19,20,21,22,23),
> g!(1, 11)(3, 19)(4, 20)(5, 9)(6, 15)(13, 21)(14, 16)
(23, 24)>;
> #m24;
244823040
> oct:=Stabilizer(m24,{24,23,3,6,9,19,15,5});
> f,nn,k:=CosetAction(m24,oct);
> Degree(nn);
759
```

We have input two permutations of degree 24 which generate M_{24} and then asked for the action of this group on the cosets of the stabilizer of an octad. We must now feed in our additional relation, and to do this we need an octad which intersects the first one in four points. This means it must lie in the 280-orbit of the stabilizer of the first octad:

```
> oo:=Orbits(Stabilizer(nn,1));
> [#oo[i]:i in [1..#oo]];
```

```
[ 1, 30, 280, 448 ]
> r:=Random(oo[3]);
> r;
52
> Fix(Stabilizer(nn,[1,52]));
{ 1, 52, 367 }
```

The octad labelled 52 is a random member of the 280-orbit. The only other octad fixed by the stabilizer of octads 1 and 52 is the one labelled 367, and so the relation must be that the product of these three octads is the identity:

```
> RR:=[<[1,52,367],Id(nn)>];
> HH:=[nn];
> CT:=DCEnum(nn,RR,HH:Print:=5, Grain:=100);

Index: 4096 === Rank: 5 === Edges: 13 === Time: 2.169

> CT[4];
[
    [],
    [ 1 ],
    [ 1, 2 ],
    [ 1, 32 ],
    [ 1, 32, 752 ]
]
> CT[7];
[ 1, 759, 2576, 759, 1 ]
>
```

This relation is fed into RR, and the double coset enumerator tells us that there are five double cosets with the expected sizes. With the current labelling of octads, they are N, $N\epsilon_1 N$, $N\epsilon_1\epsilon_2 N$, $N\epsilon_1\epsilon_{32}N$ and $N\epsilon_1\epsilon_{32}\epsilon_{752}N$. The subgroup clearly maps onto the binary Golay code $\mathcal{C}$, and, since it has order 2^{12}, it is isomorphic to $\mathcal{C}$.

The subgroup generated by the 759 involutory generators in the above progenitor is a universal representation group, as on p. 191. It is shown in ref. [2] that the universal representation group in our example, where the points are the octads and the lines are sets such as $\{O, U, O+U\}$, where O and U are octads which intersect in four points, is elementary abelian of order 2^{12}. This fact has been proved above by our double coset enumeration, but, since it is of the utmost importance, we choose to prove it again by hand.

Proof by hand We are aiming to show that

$$E = \langle \epsilon_O \mid O \in \mathcal{C}_8 \rangle \cong 2^{12}$$

is an elementary abelian group of order 2^{12}. We shall show that any two of the generators commute with one another, and that the number of elements in the group they generate is at most 2^{12}. Since the octad-type vectors in the Golay code certainly satisfy our additional relation, we shall have proved the result. The relation tells us that if two octads O and U intersect in four points, then ϵ_O and ϵ_U commute and have product ϵ_{O+U}, another generator.

Suppose now that $O, U \in \mathcal{C}_8$ with $O \cap U = \phi$. Then, if T_1, T_2, T_3, T_4 are tetrads of any one of the seven sextets refining the trio $O : U : O + U + \Omega$ chosen so that $O = T_1 + T_2$, $U = T_3 + T_4$, we have

$$\epsilon_O\epsilon_U = t_{T_1}t_{T_2}t_{T_3}t_{T_4} = t_{T_1}t_{T_3}t_{T_2}t_{T_4} = \epsilon_{T_1+T_3}\epsilon_{T_2+T_4}.$$

There are $7 \times 2 \times 2 = 28$ choices for the octad $T_1 + T_3$, which, together with O and U, yield all 30 octads in the 16-ad $O + U$. So,

$$\epsilon_O\epsilon_U = \epsilon_V\epsilon_W$$

for any $V, W \in \mathcal{C}_8$ with $O + U = V + W$, and we may write $\epsilon_O\epsilon_U = \epsilon_{O+U}$, an element which is only dependent on the 16-ad $O + U$. If $U : V : W$ is a trio, then $\epsilon_U\epsilon_V\epsilon_W$ commutes with the whole of the trio stabilizer, since the three elements commute with one another. But if $T_1, T_2, T_3, T_4, T_5, T_6$ is any one of the seven refinements of this trio, then

$$\epsilon_U\epsilon_V\epsilon_W = t_{T_1}t_{T_2}t_{T_3}t_{T_4}t_{T_5}t_{T_6},$$

which commutes with the whole of the sextet stabilizer, since these six symmetric generators commute with one another. Together these two subgroups generate the whole of M_{24}, and so

$$\epsilon_U\epsilon_V\epsilon_W = \epsilon_{U'}\epsilon_{V'}\epsilon_{W'} = \epsilon_\Omega,$$

say, for any trio $U' : V' : W'$. Clearly, $\epsilon_\Omega\epsilon_O = \epsilon_{\Omega+O}$ for any $O \in \mathcal{C}_8$.

We must now consider the case $O, U \in \mathcal{C}_8$ with $|O \cap U| = 2$:

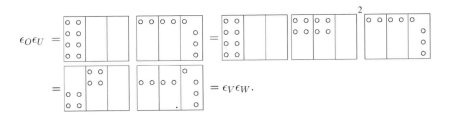

The element $\epsilon_O\epsilon_U$ clearly commutes with a subgroup of M_{24} isomorphic to the symmetric group S_6 preserving the partition of the 24 points into subsets of sizes $6 + 6 + 2 + 10$ determined by $O \setminus U, U \setminus O, O \cap U, \Omega - (O \cup U)$. But the foregoing calculation shows that it also commutes with the corresponding copy of S_6 preserving $V - W, W - V, V \cap W, \Omega - (V \cup W)$. Together, these two copies of S_6 generate a subgroup of M_{24} isomorphic to the Mathieu group M_{12} acting on $O + U = V + W$. So $\epsilon_O\epsilon_U$ depends only on the dodecad $O + U$, and we may write $\epsilon_O\epsilon_U = \epsilon_{O+U}$. In particular, we see that $\epsilon_O\epsilon_U = \epsilon_U\epsilon_O$. We obtain just 2576 new elements of E in this manner. If $O \in \mathcal{C}_8$ and $C \in \mathcal{C}_{16}$, then if $O \cap C = \phi$ we have $\epsilon_O\epsilon_C = \epsilon_\Omega$ as above. Otherwise we have $|O \cap C| = 4$

or 6, and in either case we can find octads V and W so that $C = V \cup W$ and $|O \cap V| = 4$. Then $\epsilon_O \epsilon_C = \epsilon_O \epsilon_V \epsilon_W = \epsilon_{O+V} \epsilon_W$, a case we have already considered. It remains to show that if $O \in \mathcal{C}_8$, $C \in \mathcal{C}_{12}$, then $\epsilon_O \epsilon_C$ is in the set of elements already produced. Now $|O \cap C| = 6, 4$ or 2. In the first case, $\epsilon_O \epsilon_C = \epsilon_{C+O}$, as we have already seen. In the second case, we may choose V to be any octad containing the intersection $O \cap C$ and two further points of C; then, $\epsilon_O \epsilon_C = \epsilon_O \epsilon_V \epsilon_W = \epsilon_{O+C} \epsilon_W$, which again we have already considered. To deal with the final case, when $|O \cap C| = 2$, recall that any pair of points outside a dodecad determines a partition of the dodecad into two hexads such that either hexad together with the pair is an octad. Choose the pair to lie in $O \setminus C$; since octads must intersect one another evenly, one of the hexads contains $O \cap C$ and the other is disjoint from O. We may write $\epsilon_C = \epsilon_V \epsilon_W$, where V and W are octads intersecting in our chosen pair, such that $|O \cap V| = 4$. Then $\epsilon_O \epsilon_C = \epsilon_O \epsilon_V \epsilon_W = \epsilon_{O+V} \epsilon_W$, which we have already considered. Thus our set consisting of the identity, 759 octad-type elements, 759 16-ad type elements, the element ϵ_Ω and 2576 dodecad-type elements is closed under multiplication, and so forms an elementary abelian group of order $4096 = 2^{12}$.   □

Note that the element ϵ_Ω defined above commutes with M_{24} and with the symmetric generators t_T. It is thus central in the image group G, and in any faithful irreducible representation of G it will be represented by $-I$, a scalar matrix with -1 down the diagonal.

Mechanical enumeration of the cosets of H in G

The subgroup $N \cong M_{24}$ is too small for us to enumerate all double cosets of the form NwN. However, we have now constructed, by machine and by hand, a subgroup $H \cong 2^{12} : M_{24}$, and Bray has modified our double coset enumerator to cope with enumeration of double cosets of the form HwN. Indeed, it is particularly adept at performing this kind of enumeration if $N \le H$, as is the case here. We must first obtain M_{24} as a permutation group of degree $\binom{24}{4} = 10\,626$ acting on the cosets of the stabilizer of a tetrad, in this case $\{24(= \infty), 3, 6, 9\}$. We now seek a second tetrad which intersects the first tetrad in two points and whose union with the first tetrad lies in an octad. Since there are just $\binom{4}{2} \times 5 \times \binom{4}{2} = 180$ such tetrads, we immediately see that we require the fifth orbit of the stabilizer of the first tetrad. We choose a random tetrad in this orbit and are given the tetrad labelled 8696:

```
> s24:=Sym(24);
> m24:=sub<s24|
s24!(1,2,3,4,5,6,7,8,9,10,11,12,13,14,15,16,17,18,19,20,
21,22,23),
s24!(24,23)(3,19)(6,15)(9,5)(11,1)(4,20)(16,14)(13,21)>;
> #m24;
244823040
> xx:={24,3,6,9};
```

```
> sxx:=Stabilizer(m24,xx);
> f,nn,k:=CosetAction(m24,sxx);
> st1:=Stabilizer(nn,1);
> oo:=Orbits(st1);
> [#oo[i]:i in [1..#oo]];
[ 1, 5, 80, 80, 180, 320, 320, 360, 640, 960, 960, 1920,
1920, 2880 ]
> r:=Random(oo[5]);r;
8696
```

The stabilizer of two such tetrads is a subgroup of the octad stabilizer $2^4 : A_8$ of shape $2^4 : 2^3$ with orbits on the 24 points of lengths $2+2+2+2+16$. This clearly fixes six tetrads, and we must find out which of the four possibilities completes the word of our relation. Consideration of the orders of the stabilizers of pairs of these six tetrads soon reduces the possibilities to two, one of which leads to collapse, as on p. 201. It turns out that $[1, 8696, 4203]$ is the word in the symmetric generators we require, and so we set it equal to the unique non-trivial element in the centralizer of the stabilizer of these three tetrads. The union of the tetrads labelled 1 and 325 is an octad, and so the subgroup N, together with the element $t_1 t_{325}$ (which is written here as $< [1, 325], Id(N) >$), generate $2^{12} : M_{24}$:

```
> Fix(Stabilizer(nn,[1,8696]));
{ 1, 325, 887, 4203, 5193, 8696 }
> cAB:=Centralizer(nn,Stabilizer(nn,[1,8696]));
> #cAB;
2
> RR:=[<[1,8696,4203],cAB.1>];
> HH:=[*nn,<[1,325],Id(nn)>*];
> CT1:=DCEnum(nn,RR,HH:Print:=5,Grain:=100);
Dealing with subgroup.
Pushing relations at subgroup.
Main part of enumeration.
Index: 8292375 === Rank: 19 === Edges: 1043 === Time:
505.07
> CT1[4];
[
     [],
     [ 1 ],
     [ 1, 2 ],
     [ 1, 2, 24 ],
     [ 1, 4 ],
     [ 1, 17 ],
     [ 1, 7 ],
     [ 1, 2, 61 ],
     [ 1, 2, 8 ],
     [ 1, 2, 17 ],
```

```
      [ 1, 2, 59 ],
      [ 1, 2, 14 ],
      [ 1, 2, 117 ],
      [ 1, 2, 204 ],
      [ 1, 2, 7 ],
      [ 1, 2, 1 ],
      [ 1, 2, 259 ],
      [ 1, 2, 1212 ],
      [ 1, 2, 17, 1642 ]
]
> CT1[7];
[ 1, 1771, 637560, 2040192, 26565, 637560, 21252, 2266880,
370944, 728640, 91080, 566720, 91080, 425040, 42504, 759,
340032, 1771, 2024 ]
```

The output shows us that there are 19 double cosets of the form HwN, where $H \cong 2^{12} : M_{24}$ and $N \cong M_{24}$; the index of H in G is $8\,292\,375$ (note the misprint in the first edition of the ATLAS). Canonical double coset representatives are given, and we see that the graph obtained by joining a coset of H to those cosets obtained by multiplication by a symmetric generator has diameter 4.

Alternatively, we may enumerate double cosets of the form KwN, where $K \cong \mathrm{Co}_2$, the stabilizer of a type 2 vector in the Leech lattice. Explicitly,

$$K = \langle \mathrm{Stab}_N(\infty), t_T \mid \infty \in T \rangle.$$

Note that this is, in a sense, preferable as the index is much smaller; however, there is more work entailed as we have not yet identified Co_2. Thus we have the following:

```
> m23:=Stabilizer(m24,24);
> #m23;
10200960
> HH:=[*f(m23),<[1],Id(nn)>*];
> CT:=DCEnum(nn,RR,HH:Print:=5,Grain:=100);
Dealing with subgroup.
Pushing relations at subgroup.
Main part of enumeration.
Index: 196560 === Rank: 16 === Edges: 178 === Time: 282.661

> CT[4];
[
    [],
    [ 3 ],
    [ 3, 2 ],
    [ 3, 2, 1197 ],
    [ 3, 26 ],
```

```
    [ 3, 5 ],
    [ 3, 2, 4 ],
    [ 3, 2, 1 ],
    [ 3, 53 ],
    [ 3, 2, 14 ],
    [ 3, 2, 90 ],
    [ 3, 2, 9188 ],
    [ 3, 2, 417 ],
    [ 3, 2, 2554 ],
    [ 3, 2, 540 ],
    [ 3, 2, 90, 340 ]
]
>CT[7];
[ 24, 6072, 53130, 30912, 21252, 12144, 30912, 21252,
759, 12144,6072,759,276, 276, 552, 24 ]
```

In this case, the index is 196 560 (note that K does not contain ϵ_Ω, which negates a 2-vector), and the number of double cosets is 16.

A representation of the group G

The lowest dimension in which M_{24} can be represented faithfully as matrices over the complex numbers $\mathbb{C}$ is 23, and since G is perfect and the element ϵ_Ω is to be represented by $-I_n$ (which has determinant $(-1)^n$), any faithful representation of G must have even degree. So the lowest dimension in which we could represent G faithfully is 24. Certainly $E = 2^{12} : M_{24}$ acts monomially in this dimension, with M_{24} acting as permutations and the elements ϵ_C for $C \in \mathcal{C}$ acting as sign changes on the $\mathcal{C}$-set C. Let ρ denote such a representation of G. We are led to seeking elements $\rho(t_T)$, for T a tetrad of the 24 points, which

(i) commute with the tetrad stabilizer of M_{24};

(ii) commute with elements ϵ_O, where O is an octad which is the union of two tetrads in the sextet defined by T;

(iii) have order 2; and

(iv) satisfy the additional relation.

Condition (i) requires the matrix representing $\rho(t_T)$ to have the following form:

$$
\rho(t_T) = \begin{pmatrix}
aI+bJ & eJ & eJ & eJ & eJ & eJ \\
fJ & cI+dJ & gJ & gJ & gJ & gJ \\
fJ & gJ & cI+dJ & gJ & gJ & gJ \\
fJ & gJ & gJ & cI+dJ & gJ & gJ \\
fJ & gJ & gJ & gJ & cI+dJ & gJ \\
fJ & gJ & gJ & gJ & gJ & cI+dJ
\end{pmatrix},
$$

where the 24×24 matrix has been partitioned into blocks corresponding to the sextet defined by the tetrad (which itself corresponds to the block in the first row and the first column); I denotes the 4×4 identity matrix and J denotes the 4×4 all-1s matrix. But this must commute with sign changes on the first two tetrads; so negating the first two columns must have the same effect as negating the first two rows. Thus $e = f = g = 0$, and the matrix takes the following form:

$$\rho(t_T) = \mathrm{diag}(aI + bJ, (cI + dJ)^5).$$

Condition (iii) requires

$$(aI + bJ)^2 = (cI + dJ)^2 = I_4 = a^2I + (2ab + 4b^2)J = c^2I + (2cd + 4d^2)J.$$

So we have $a = \pm 1$ and either $b = 0$ or $a = -2b$, and similarly $c = \pm 1$ and $d = 0$ or $c = -2d$. But if T_1, T_2 are two tetrads of the same sextet, then $t_{T_1} t_{T_2} = \epsilon_{T_1 + T_2}$, and so

$$(aI + bJ)(cI + dJ) = acI + (ad + bc + 4bd)J = -I.$$

So $ac = -1$ and, if either of b or d is zero, then so is the other. Both being zero would mean that $\rho(t_T)$ is a diagonal matrix, and so the additional relation of condition (iv) could not possibly hold. So neither is zero, and $a = \pm 1, c = -a, b = -a/2, d = a/2$, and it remains to determine the sign. In order to do this, we first restrict our attention to the action on the octad which is the union of the three tetrads in the additional relation, and we order the rows and columns to correspond to the pairings a, b, c, d. We let

$$L = \begin{pmatrix} -1 & 1 \\ 1 & -1 \end{pmatrix}, J = \begin{pmatrix} 1 & 1 \\ 1 & 1 \end{pmatrix},$$

so that $L^2 = -2L, J^2 = 2J, LJ = JL = 0$. Our relation requires the following:

$$\pm \frac{1}{8} \left(\begin{array}{cc|cc} L & J & & \\ J & L & & \\ \hline & & -L & -J \\ & & -J & -L \end{array} \right) \left(\begin{array}{cc|cc} L & & J & \\ -L & & & -J \\ \hline & J & & L \\ & -J & & -L \end{array} \right) \left(\begin{array}{cc|cc} L & & & J \\ & -L & -J & \\ \hline & -J & -L & \\ J & & & L \end{array} \right)$$

$$\pm \frac{1}{8} \left(\begin{array}{cc|cc} L & J & & \\ J & L & & \\ \hline & & -L & -J \\ & & -J & -L \end{array} \right) \left(\begin{array}{cc|cc} -2L & -2J & & \\ -2J & -2L & & \\ \hline & & 2L & 2J \\ & & 2J & 2L \end{array} \right) = \pm \frac{1}{8} \left(\begin{array}{cc|cc} -8I & & & \\ & -8I & & \\ \hline & & -8I & \\ & & & -8I \end{array} \right) = I_8,$$

and so we require the negative sign. To complete the verification that the additional relation holds, and that this product does indeed produce the required permutation of M_{24}, we restrict our attention to the second brick of the MOG when we obtain

$$\frac{1}{8}\left(\begin{array}{cc|cc} L & J & & \\ J & L & & \\ \hline & & L & J \\ & & J & L \end{array}\right)\left(\begin{array}{cc|cc} L & & J & \\ & L & & J \\ \hline J & & L & \\ & J & & L \end{array}\right)\left(\begin{array}{cc|cc} L & & & J \\ & L & J & \\ \hline & J & L & \\ J & & & L \end{array}\right)$$

$$=\frac{1}{8}\left(\begin{array}{cc|cc} L & J & & \\ J & L & & \\ \hline & & L & J \\ & & J & L \end{array}\right)\left(\begin{array}{cc|cc} -2L & 2J & & \\ 2J & -2L & & \\ \hline & & -2L & 2J \\ & & 2J & -2L \end{array}\right)$$

$$=\frac{1}{2}\left(\begin{array}{cccc} L+J & & & \\ & L+J & & \\ & & L+J & \\ & & & L+J \end{array}\right)=\left(\begin{array}{cc|cc|cc} 1 & & & & & \\ & 1 & & & & \\ \hline & & 1 & & & \\ & & & 1 & & \\ \hline & & & & 1 & \\ & & & & & 1 \end{array}\right),$$

as required. Note that the element t_T is in fact $-\xi_T$, where ξ_T is the element produced in Conway (ref. [23], p. 237) to show that the Leech lattice is preserved by more than the monomial group $H = 2^{12} : M_{24}$. Observe moreover that $\rho(t_T)$ is an orthogonal matrix, and so the group $\rho(G)$ preserves lengths of vectors and angles between them. In the notation of ref. [23], we let $\{v_i \mid i \in \Omega\}$ be an orthonormal basis for a 24-dimensional space over $\mathbb{R}$, and for $X \subset \Omega$ we let v_X denote

$$v_X = \sum_{i \in X} v_i.$$

For T a tetrad of points in Ω, we let $\{T = T_0, T_1, \ldots, T_5\}$ be the sextet defined by T. Then the element t_T acts as follows:

$$t_T = -\xi_T : v_i \mapsto \begin{cases} v_i - \frac{1}{2}v_T & \text{for } i \in T = T_0 \\ \frac{1}{2}v_{T_i} - v_i & \text{for } i \in T_i, i \neq 0, \end{cases}$$

so, as described in ref. [23], t_T is best applied to a vector in $\mathbb{R}^{24}$:

for each tetrad T_i work out one half the sum of the entries in T_i and subtract it from each of the four entries; then negate on every entry except those in $T = T_0$.

The Leech lattice Λ

In order to obtain the Leech lattice Λ, we simply apply the group we have constructed to the standard basis vectors and consider the $\mathbb{Z}$-lattice

Table 5.6. The shortest vectors in the Leech lattice

Shape	Calculation	Number
$(4^2.0^{22})$	$\binom{24}{2}.2^2$	1104
$(2^8.0^{16})$	759.2^7	97 152
(-3.1^{23})	24.2^{12}	98 304
	Total	196 560

spanned by the set of images. More specifically, in order to avoid fractions, we normalize by applying the group to the vectors $8v_i$. Let Λ denote this lattice. If T denotes the first column of the MOG, we have

$$t_T = \begin{array}{|c c|c c|c c|} \hline \times & . & . & . & . & . \\ \times & . & . & . & . & . \\ \hline \times & . & . & . & . & . \\ \times & . & . & . & . & . \\ \hline \end{array} : \begin{array}{|c c|c c|c c|} \hline 8 & . & . & . & . & . \\ . & . & . & . & . & . \\ \hline . & . & . & . & . & . \\ . & . & . & . & . & . \\ \hline \end{array} \mapsto \begin{array}{|c c|c c|c c|} \hline 4 & . & . & . & . & . \\ -4 & . & . & . & . & . \\ \hline -4 & . & . & . & . & . \\ -4 & . & . & . & . & . \\ \hline \end{array} .$$

So, under the permutations of the quintuply transitive M_{24} and the sign changes of E, every vector of the shape $(\pm 4)^4.0^{20}$ is in Λ. In particular, we have $(4, 4, 4, 4, 0, 0^{19}) + (0, -4, -4, -4, -4, 0^{19}) = (4, 0^3, -4, 0^{19}) \in \Lambda$, and so every vector of the form $((\pm 4)^2.0^{22})$ is in Λ. Moreover, we see that

$$t_T : \begin{array}{|c c|c c|c c|} \hline 4 & 4 & . & . & . & . \\ . & . & . & . & . & . \\ \hline . & . & . & . & . & . \\ . & . & . & . & . & . \\ \hline \end{array} \mapsto \begin{array}{|c c|c c|c c|} \hline 2 & -2 & . & . & . & . \\ -2 & 2 & . & . & . & . \\ \hline -2 & 2 & . & . & . & . \\ -2 & 2 & . & . & . & . \\ \hline \end{array} ;$$

$$\begin{array}{|c c|c c|c c|} \hline 0 & 2 & 2 & 2 & 2 & 2 \\ 2 & . & . & . & . & . \\ \hline 2 & . & . & . & . & . \\ 2 & . & . & . & . & . \\ \hline \end{array} \mapsto \begin{array}{|c c|c c|c c|} \hline -3 & -1 & -1 & -1 & -1 & -1 \\ -1 & 1 & 1 & 1 & 1 & 1 \\ \hline -1 & 1 & 1 & 1 & 1 & 1 \\ -1 & 1 & 1 & 1 & 1 & 1 \\ \hline \end{array} .$$

The first image shows that Λ contains every vector of the form $((\pm 2)^8.0^{16})$, where the non-zero entries are in the positions of an octad of the Steiner system preserved by our copy of M_{24} and the number of minus signs is even (since $\mathcal{C}$-sets intersect one another evenly). The second shows that every vector of the form $(-3, 1^{23})$ followed by a sign change on a $\mathcal{C}$-set is also in Λ. We may readily check that this set of vectors, which have normalized length 2×16, is closed under the action of t_T and hence of G; it is normally denoted by Λ_2 (see Table 5.6).

Clearly G acts as a permutation group on the $196\,560/2 = 98\,280$ pairs consisting of a type 2 vector and its negative; the stabilizer of such a pair has just three non-trivial orbits on the other pairs where the orbit in which a particular pair lies depends only on the angles its vectors make with the fixed vectors. The permutation character of this action is $\chi_1 + \chi_3 + \chi_6 + \chi_{10}$, of degrees 1, 299, 17 250, 80 730 respectively, as listed in the ATLAS [25].

In ref. [21], Theorem 5, Conway gives a beautifully simple characterization of the vectors of Λ, normalized as above.

THEOREM **5.17** [Conway] The integral vector $x = (x_1, x_2, \ldots, x_{24})$ is in Λ if, and only if,

(i) the x_i all have the same parity;

(ii) the set of i, where x_i takes any given value (modulo 4), is a $\mathcal{C}$-set; and

(iii) $\sum x_i \equiv 0$ or 4 (modulo 8) according as $x_i \equiv 0$ or 1 (modulo 2).

It is readily checked that the above list of 2-vectors contains all vectors of (normalized) length 32 having these properties, and it is clear that these properties are enjoyed by all integral combinations of them.

EXERCISE **5.2**

(1) Using Conway's description of the (normalized) vectors of the Leech lattice (see Theorem 5.17), write down the shapes of all vectors in Λ_3, that is to say all vectors with norm equal to 3×16. Verify that these are the orbits under the action of the monomial group $N \cong 2^{12} : M_{24}$ consisting of permutations of the basis vectors and sign changes on codewords of the Golay code. Calculate the number of vectors of each type, and use tetrad-type elements ξ_T to show that these fuse into one orbit in $\cdot O$; hence work out the index of $\cdot 3$ or Co_3, the stabilizer of a 3-vector in $\cdot O$.

(2) Let $u_1 = (5.1^{23}) \in \Lambda_3$. Show that u can be expressed as the sum of two vectors in Λ_2 in precisely 276 ways, which fall into two orbits of lengths 23 and 253 under the action of permutations fixing u. Now take $u_2 = ((-3)^3.1^{21}) \in \Lambda_3$ and identify the 276 ways in which it can be decomposed into the sum of two 2-vectors. Show that the orbits under the action of the group isomorphic to $M_{21} : S_3$ fixing u_2 cannot fuse to an orbit of size 23, and so Co_3 must act transitively on this set of 276 objects. Finally, let $u_3 = (2^{12}.0^{12}) \in \Lambda_3$. Write $u_3 = w_{31} + w_{32}$, where $w_{31}, w_{32} \in \Lambda$, in such a way that you can produce an element of $\cdot O$ which fixes u_3 and interchanges w_{31} and w_{32}. Deduce that the subgroups of index 276 in $\cdot 3$ themselves have subgroups of index 2. (This subgroup of index 2 is in fact the McLaughlin simple group; see Section 5.7.3).

(3) We have seen that $\cdot 3$ acts transitively on 276 letters; proceed as follows to show that this action is in fact doubly transitive. Let $u_1 = (5.1.1^{22}) \in \Lambda_3$ and $v_1 = (4^2.0^{22}) \in \Lambda_2$. Note that these vectors are preserved by a subgroup isomorphic to M_{22}. Now write down all 2-vectors which

have inner product 24 with u_1 and 16 with v_1 and record the orbit lengths when your copy of M_{22} acts on these 275 vectors. Now let $u_2 = ((-3)^3.1^{21}) \in \Lambda_3$ and $v_2 = (4^20.0^{21}) \in \Lambda_2$ and record the orbit lengths of the subgroup of shape $M_{21} : 2$, fixing u_2 and v_2 on the similarly defined 275 2-vectors. Conclude that the stabilizer of a 3-vector and a 2-vector having inner product 24 with one another acts transitively on the 275 2-vectors defined as above.

(4) Use the methods of the above examples to show that ·O acts transitively on triangles of type 233. The stabilizer of such a triangle is the automorphism group of the Higman–Sims group, $G \cong HS{:}2$, where the outer automorphism interchanges the two edges of type 3. Give an example of such a 233-triangle in which the two 3-vectors are interchanged by either a permutation of M_{24} or a sign change on a codeword of the Golay code. Taking $u_1 = (5.1.1^{22}) \in \Lambda_3$ and $v_1 = (4.-4.0^{22}) \in \Lambda_2$, together with some other choice of 233-triangle, show that G acts as a rank 3 permutation group on the 100 2-vectors having inner product 24 with u_1 and 16 with v_1. What are the suborbit lengths?

Show, moreover, that there are 176 pairs of 2-vectors having inner product 24 with u and 8 with v (the two vectors in a pair summing to u), and a further 176 pairs of 2-vectors having inner product 24 with $u-v$ and eight with v (the two vectors in a pair summing to $u-v$). So the outer automorphism of HS:2 interchanges the two sets of 176 pairs (the two *halves*). Show further that the set of inner products between two pairs in the same half is constant, and that for a fixed pair in one half the inner products with a pair from the other half is $\{0, 0, 16, 16\}$ in 50 cases and $\{8, 8, 8, 8\}$ in the remaining 126 cases.[6] Show that $G' \cong HS$ acts doubly transitively on each of the halves.

A presentation of the Conway group · O

It is of interest to deduce an ordinary presentation of ·O from our symmetric presentation. We first need a presentation for the control subgroup M_{24}, and we choose one based on that given in ref. [33], p. 390, which defines M_{24} as an image of the progenitor

$$2^{\star 7} : L_3(2).$$

Consider first

$$\langle x, y, t \mid x^7 = t^2 = y^2 = (xy)^3 = [x, y]^4 = [y^x, t] = 1, y = [t, x^2]^3 \rangle.$$

We see that $L = \langle x, y \rangle \cong L_3(2)$ or the trivial group. If it is the former, which must be the case since we can find permutations of 24 letters satisfying all these relations, then, without loss of generality, we may calculate within L

---

[6] This is a realization of Graham Higman's geometry with 176 *points* and 176 *quadrics*; each quadric contains 50 points and each point belongs to 50 quadrics. See Section 5.3.

by letting $x \sim (0\ 1\ 2\ 3\ 4\ 5\ 6)$, $y \sim (3\ 4)(5\ 1)$. Moreover $\langle t, t^{x^2} \rangle \cong D_{12}$ or an image thereof, and centralizes y. Thus t commutes with

$$\langle y, y^{x^{-2}}, y^x \rangle = \langle (3\ 4)(5\ 1), (1\ 2)(3\ 6), (4\ 5)(6\ 2) \rangle \cong S_4,$$

and so $|t^L| = 7$. Labelling $t^{x^i} = t_i$ for $i = 0, 1, \ldots, 6$, we see that elements of L must permute the t_i by conjugation just as the above permutations representing x and y permute their subscripts. Thus $\langle x, y, t \rangle = \langle x, t \rangle$ is a homomorphic image of the progenitor $2^{*7} : L_3(2)$. Following Curtis, Hammas and Bray (ref. [36], Table 8, p. 33), we factor this by the following additional relators:

$$(yt^{x^{-1}}t^x)^4 \sim ((3\ 4)(5\ 1)t_6 t_1)^4$$
$$(yxt)^{11} \sim ((0\ 1\ 6)(2\ 3\ 5)t_0)^{11}.$$

To verify the claim made in ref. [36] that this is a presentation for M_{24}, we observe that restricting the given relations to the four symmetric generators $\{t_0, t_3, t_6, t_5\}$ yields $(0\ 3) = (t_5 t_6)^3$ and $((0\ 3\ 6)t_0)^{11} = 1$. Thus we have the symmetric presentation given by

$$J = \frac{2^{*4} : S_4}{(3\ 4) = (s_1 s_2)^3, ((1\ 2\ 3)s_1)^{11}},$$

which is equivalent to the following presentation:

$$\langle u, v, s \mid u^4 = v^2 = (uv)^3 = s^2 = (uvs)^{11} = 1, v = [s, u]^3 \rangle.$$

Either the symmetric or the ordinary presentation is readily shown to define the projective special linear group $L_2(23)$.

We have now identified a sufficiently large subgroup, and we can perform a coset enumeration of

$$M = \langle x, t \mid x^7 = t^2 = y^2 = (xy)^3 = [x, y]^4 = [y^x, t] = 1, y = [t, x^2]^3,$$
$$(yt^{x^{-1}}t^x)^4 = (yxt)^{11} = 1 \rangle$$

over the subgroup $\langle t, t^{x^3}, t^{x^6}, t^{x^5} \rangle \cong L_2(23)$ to obtain index 40 320 and so $|M| = 244\,823\,040$. In fact, the subgroup $\langle x, ytxy \rangle \cong M_{23}$, and so the group M can be obtained acting on 24 letters.

In order to extend this to a presentation of $\cdot O$, we must adjoin a generator s, say, which commutes with the stabilizer in M of a tetrad of the 24 points, and we must then factor out the additional relation. The clue is to consider an element of M_{24} of order 12 and cycle shape 2.6.4.12. Let σ be such an element; then σ^3 has shape $2.2^3.4.4^3$, written so as to reveal the cycles of σ. Now the 2-orbit of σ, together with each of the other 2-orbits of σ^3 in turn, may be taken to be the three tetrads in our additional relation, when the element v is simply σ^6. So our additional relation may be taken to be $(\sigma^2 s)^3 = 1$, where s is the symmetric generator corresponding to any one of these three tetrads. The above will be clarified by exhibiting such an element in the MOG-diagram. Thus,

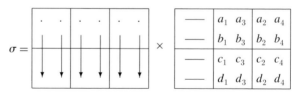

is an element of the required shape; it is expressed as the product of an element of order 3 which fixes the columns and the top row whilst rotating the other three rows downwards, and an element of order 4 which fixes the rows and acts on the six columns as the permutation $(1\,2)(3\,5\,4\,6)$. The three tetrads are those labelled A', B' and C' on p. 204, and the element $\sigma^6 = \nu$ as on p. 202. We find that the element $\sigma = xyt^x$ is in this conjugacy class.

We must now seek elements in M_{24} which generate the stabilizer of a tetrad which contains the 2-cycle of σ and is fixed by σ^3. Noting that any element of M_{24} can be written in the form πw, where $\pi \in L_3(2)$ and w is a word in the seven symmetric generators, we find computationally that

$$u_1 = \quad (1\,6\,5)(2\,4\,3)t_5t_1t_2t_0t_6t_5 \quad = x^3yx^3t^{x^{-2}}t^xt^{x^2}tt^{x^{-1}}t^{x^{-2}},$$
$$u_2 = \qquad (2\,6)(4\,5)t_4t_1t_0t_1t_5 \qquad = (y^xt)^{t^xt^{x^{-2}}},$$

will suffice so we adjoin s such that $s^2 = [s, u_1] = [s, u_2] = 1$ and have now defined the progenitor [7]

$$2^{\cdot\binom{24}{4}} : M_{24}.$$

It remains to factor this by the additional relation which now takes the simple form $(\sigma^2 s)^3 = 1$. Our double coset enumeration has proved that

$$\cdot O = \langle x, y, t, s \| x^7 = t^2 = y^2 = (xy)^3 = [x, y]^4 = [y^x, t] = 1, y = [t, x^2]^3,$$
$$(yt^{x^{-1}}t^x)^4 = (yxt)^{11} = s^2 = [s, x^3yx^3t^{x^{-2}}t^xt^{x^2}tt^{x^{-1}}t^{x^{-2}}]$$
$$= [s, (y^xt)^{t^xt^{x^{-2}}}] = ((xyt^x)^2s)^3 = 1\rangle.$$

5.7.3   Some related groups: ·2, ·3 and McL

As was mentioned in Section 5.7.2, the Conway group $\cdot O$ acts transitively on Λ_2, the set of 196 560 vectors of type 2. One such vector is given by

$$u = \begin{array}{|cc|cc|cc|}
\hline
-3 & 1 & 1 & 1 & 1 & 1 \\
1 & 1 & 1 & 1 & 1 & 1 \\
\hline
1 & 1 & 1 & 1 & 1 & 1 \\
1 & 1 & 1 & 1 & 1 & 1 \\
\hline
\end{array},$$

---

[7] It turns out that the element x^3yx maps the tetrad corresponding to s to another tetrad in the same sextet; so ss^{x^3yx} lies in the copy of the Golay code normalized by our copy of M_{24}. Thus, $\langle x, y, t, [s, x^3yx]\rangle \cong 2^{12} : M_{24}$.

which is visibly fixed by a subgroup isomorphic to M_{23}. Moreover, this vector is fixed by our tetradic elements t_T, for T a tetrad of the 24 points permuted by M_{24}, so long as T contains the coordinate on which u has coefficient -3. For recall that t_T is simply $-\xi_T$, the negative of the original Conway element. This element acts on a vector as follows: for each tetrad of the sextet defined by T take the average of the entries of that tetrad and subtract this from each of the entries in that tetrad. Finally negate on each tetrad except T itself. So if

$$\xi_T = \begin{array}{|c|c|c|} \hline \times & & \\ \hline \times & & \\ \hline \times & & \\ \hline \times & & \\ \hline \end{array} \quad ,$$

then the columns are the tetrads; on the first column the average is zero, so we subtract nothing and do not change the sign; on the other columns we subtract 2 from every entry and then negate, thus fixing vector u. From these remarks we see that the single relation which is used to define $\cdot O$ is inherited by $\cdot 2$, the subgroup of $\cdot O$ fixing a type 2 vector, and, since the subgroup M_{23} is maximal in $\cdot O$, we see that $\cdot 2$ is a homomorphic image of the analogous presentation and, in fact, we have the following.

THEOREM 5.18

$$\frac{2^{\star\binom{23}{3}} : M_{23}}{t_{ab} t_{ac} t_{ad} = \nu} \cong \cdot 2,$$

where a, b, c and d are the pairs as in Section 5.7.2 and ν is the permutation of M_{24} shown in that section and below:

$$\begin{array}{|c c|c|} \hline a & a & \\ \hline b & b & \\ \hline c & c & \\ \hline d & d & \\ \hline \end{array}$$

and

$$\nu = \begin{array}{|c|c|c|} \hline & \bullet\!-\!\bullet & \bullet\!-\!\bullet \\ \hline & \bullet\!-\!\bullet & \bullet\!\mid\!\bullet \\ \hline & \bullet\!-\!\bullet & \bullet\!-\!\bullet \\ \hline & \bullet\!-\!\bullet & \bullet\!-\!\bullet \\ \hline \end{array}$$

To use the double coset enumerator to verify that this does in fact define $\cdot 2$, we must first obtain M_{23} acting on the 1771 unordered triples of the 23 points. Accordingly, we input two permutations of 23 points which generate M_{23} and ask for the action of this group on the cosets of the stabilizer of a triple of points:

```
> s23:=Sym(23);
> a:=s23!(1,2,3,4,5,6,7,8,9,10,11,12,13,14,15,16,17,
          18,19,20,21,22,23);
```

```
> b:=s23!(11,1)(4,20)(16,14)(13,21)(22,2)(18,10)
(8,17)(12,7);
> m23:=sub<s23|a,b>;
> #m23;
10200960
> xx:={3,6,9};
> sxx:=Stabilizer(m23,xx);
> f,nn,k:=CosetAction(m23,sxx);
> Degree(nn);
1771
```

Our second triple yy must have a point in common with this fixed triple xx, and must be such that its union with xx lies in a *heptad* (seven points making an octad with the point fixed by our copy of M_{23}). There are thus $3 \times 5 \times \binom{4}{2} = 90$ possible triples, so we look for the 90-orbit of the stabilizer of xx. Now the stabilizer of both xx and yy fixes just one other triple, zz say, namely the point of intersection of xx and yy together with the remaining two points of the heptad. Moreover,

$$C_{M_{23}}(\mathrm{Stab}_{M_{23}}(xx, yy, zz)) \cong C_2,$$

and so a generator, ν say, for this cyclic group is the only non-trivial element of M_{23} which could be written in terms xx, yy and zz without causing collapse. Lemma 3.3 thus leads us directly to factoring by the relation $xx.yy.zz = \nu$, where we have taken $\{xx, yy, zz\}$ to be the triples labelled $\{1, 874, 1389\}$:

```
> oo:=orbits(Stabilizer(nn,1));
> [#Set(oo[i]):i in [1..#oo]];
[ 1, 20, 60, 90, 160, 480, 480, 480 ]
> rr:=Random(Set(oo[4]));
> Fix(Stabilizer(nn,[1,rr]));
{ 1, 874, 1389 }
> RR:=[<[1,874,1389],Centralizer(nn,Stabilizer
(nn,[1,rr])).1>];
> CT:=DCEnum(nn,RR,[nn]:Grain:=100);
> CT[1];
4147200
> CT[4];
[
    [],
    [ 1 ],
    [ 1, 3 ],
    [ 1, 3, 206 ],
    [ 1, 10 ],
    [ 1, 7 ],
    [ 1, 3, 36 ],
```

```
    [ 1, 3, 9 ],
    [ 1, 3, 10 ],
    [ 1, 3, 7 ],
    [ 1, 3, 122 ],
    [ 1, 3, 4 ],
    [ 1, 3, 148 ],
    [ 1, 10, 74, 5 ],
    [ 1, 5 ],
    [ 1, 7, 38 ],
    [ 1, 104 ],
    [ 1, 3, 170 ],
    [ 1, 3, 9, 11, 727 ],
    [ 1, 3, 163 ],
    [ 1, 3, 104 ],
    [ 1, 3, 1 ],
    [ 1, 3, 148, 79 ],
    [ 1, 10, 5 ],
    [ 1, 7, 213 ],
    [ 1, 3, 7, 276 ],
    [ 1, 7, 242 ]
]
> CT[7];
[ 1, 1771, 17710, 212520, 212520, 212520, 53130, 28336,
510048,425040, 30360, 425040, 212520, 70840, 5313, 850080,
1771,85008,170016, 141680, 7590, 506, 283360, 60720, 85008,
15456,28336 ]
>
```

As in previous examples, CT[1] tells us the index of $N =\cong M_{23}$ in the group we have defined, namely 4 147 200, as required; CT[4] gives CDCRs for the 27 double cosets; and CT[7] gives the number of single cosets each of them contains. The usual word of warning is required: the CDCRs are not necessarily the shortest words possible for that double coset; they are simply the first the program discovered.

EXERCISE 5.3

(1) Prove as follows that the group Co_2 (or ·2, to use Conway's terminology) acts transitively on the set of 2-vectors having inner product 8 with the fixed vector, and that ·O thus acts transitively on triangles of type 223 (having two edges in Λ_2 and the third edge in Λ_3).[8] Let the fixed vector be $(4^2.0^{22})$, which is visibly preserved by a subgroup

---

[8] The stabilizer of such a triangle (or 2-dimensional sublattice) is isomorphic to the McLaughlin group McL; its outer automorphism has the effect of fixing the edge in Λ_3 and interchanging the two 2-vectors. Note that this exercise can be used to show that ·O acts transitively on Λ_3.

$H \cong 2^{10} : M_{22}$, and work out the orbits of H on suitable 2-vectors. (You should find orbits of lengths $22\,528 + 2048 + 22\,528$.) Now take the fixed vector to be $(-3, 1^{23})$, which is preserved by $K \cong M_{23}$, and in particular is fixed by a permutation π of order 23. Such a permutation can fix no further 2-vectors and so every orbit of K on permissible 2-vectors has length divisible by 23. Deduce that the three orbits of H above must fuse into a single orbit under the action of $\cdot 2$.

(2) Use a similar approach to show that $\cdot O$ acts transitively on triangles of type 222 and of type 224. Work out the order of the stabilizer of each such triangle.

The next related group we should like to deal with is the McLaughlin group McL, which was investigated by Bradley in ref. [12]. This time we take the Mathieu group M_{22} as control subgroup and let the symmetric generators correspond to the 672 dodecads which contain a given one of the two fixed points but not the other. Now the subgroup of M_{22} fixing such a dodecad is isomorphic to $L_2(11)$ (take M_{12} fixing a pair of disjoint dodecads and fix a point in each dodecad), and it has an orbit of 165 such dodecads which intersect the fixed dodecad in eight points. As usual, we invoke Lemma 3.3 and note that the stabilizer of two such dodecads has order $660/165 = 4$; it is thus a Sylow 2-subgroup of $L_2(11)$, which is a self-centralizing Klein fourgroup. So the only non-trivial elements of M_{22} which could be written in terms of the two symmetric generators corresponding to these two dodecads lie in this copy of V_4. Since they commute with the two symmetric generators in question, a possible relation would have to have form $(t_i t_j)^k = \nu$. Consistent with our practice of always choosing the shortest possible word in the symmetric generators which does not lead to collapse, we choose $k = 2$ and factor by

$$
\left(
\begin{array}{|c|c|c|} \hline
x & x & x \\ \hline
  x &   x &   x \\ \hline
x & x & x \\ \hline
x & x & x \\ \hline
\end{array}
\;
\begin{array}{|c|c|c|c|} \hline
x & x\;x & x\;x \\ \hline
x & & \\ \hline
x\;x & x & x \\ \hline
& x & x \\ \hline
\end{array}
\right)^2 = \nu,
$$

where the copy of M_{22} we are using fixes the two leftmost points in the top row (corresponding to ∞ and 0), and ν bodily interchanges the last two bricks of the MOG while fixing the eight points in the first brick. This is essentially the same relation as that given in Bradley [12] (p. 86). With this interpretation we have the following.

THEOREM 5.19

$$
\frac{2^{\cdot 672} : M_{22}}{(t_1 t_2)^2 = \nu} \cong \mathrm{McL} : 2.
$$

To verify this using the double coset enumerator, we first obtain M_{22} by fixing $\infty = 24$ and $0 = 23$ in M_{24}. We then choose xx to be a dodecad containing ∞ but not 0 and take the action of M$_{22}$ on the cosets of its stabilizer to obtain the permutation action of degree 672:

```
> s24:=Sym(24);
> a:=s24!(1, 2, 3, 4, 5, 6, 7, 8, 9, 10, 11, 12, 13, 14, 15,
   16, 17, 18, 19, 20, 21, 22, 23);
> b:=s24!(1, 11)(3, 19)(4, 20)(5, 9)(6, 15)(13, 21)
   (14, 16) (23, 24);
> m24:=sub<s24|a,b>;
> m22:=Stabilizer(m24,[23,24]);

> xx := { 5, 7, 10, 11, 14, 15, 17, 19, 20, 21, 22 };
> stxx:=Stabilizer(m22,xx);
> f,nn,k:=CosetAction(m22,stxx);
> Degree(nn);
672
```

We must now locate a point, *ss* say, in the 165-orbit of this stabilizer and factor by a relation which puts the word $[1, ss, 1, ss]$ equal to one of the three involutions in the Klein fourgroup. A word of warning is required here: two of the three involutions work (and are interchanged by an outer automorphism), but the third does not:

```
> oo:=orbits(Stabilizer(nn,1));
> [#oo[i]:i in [1..#oo]];
[ 1, 55, 55, 66, 165, 330 ]
> ss:=Random(Set(oo[5]));

> RR:=[<[1,ss,1,ss],Stabilizer(nn,[1,ss]).2>];
> CT:=DCEnum(nn,RR,[nn]:Print:=1,Grain:=100);

Index: 4050 === Rank: 8 === Edges: 38 === Time: 2.484
> CT[4];
[
   [],
   [ 1 ],
   [ 1, 2 ],
   [ 1, 4 ],
   [ 1, 4, 268 ],
   [ 1, 2, 44 ],
   [ 1, 2, 289 ],
   [ 1, 28 ]
]
> CT[7];
[ 1, 672, 462, 1232, 22, 1155, 176, 330 ]
```

These orbits can be seen in the Leech lattice as follows. As mentioned above, the McLaughlin group stabilizes a triangle of type 223 in Λ; the

Table 5.7. The orbits of M_{22} on $(2025+2025)$
2-vectors

Shape	Calculation	Number
$(5.1.1^{22})$		
$(4.4.0^{22})$		
$(1.-3.1^{22})$		
$(4.-4.0^{22})$	1	1
$(2.-2.-2.2^5.0^{16})$	77.6	462
$(0.0.2^8.0^{14})$	330	330
$(1.-1.3.-1^6.1^{15})$	176.7	1232
$(4.0-4.0^{21})$	22	22
$(2.2.-2^2.2^4.0^{16})$	$77.\binom{6}{2}$	1155
$(3.1.-1^{11}.1^{11})$	672	672
$(1.3.-1^7.1^{15})$	176	176

2-dimensional sublattice spanned by the edges of this triangle can be extended to a 3-dimensional sublattice whose additional 'short'[9] vectors have types 2, 2, 3 and 4. As 223-triangle we may take $u = (5, 1, 1^{22})$, $v = (4, 4, 0^{22})$, $w = (1, -3, 1^{22})$, each of which is fixed by M_{22}; to extend it to a 3-dimensional sublattice of the required type, we must seek vectors in Λ_2 which have inner product 16 with u, inner product 16 with one of v and w and 0 with the other. Thus, as orbits under M_{22}, we obtain Table 5.7.

The *calculations* make use of the fact that there are 77 octads through a given two points, 176 octads through one of the two points but not the other, and 330 octads disjoint from the two points. Moreover, there are 672 dodecads through one of the two points but not the other.

These vectors fall into two sets of size 2025 interchanged by the outer automorphism of McL, which also interchanges v and w whilst fixing u. Of course, our symmetric generators lie in McL : 2 \ McL and so have this effect.

The most natural action of McL:2, however, is on the 275 2-vectors which have inner product 16 with each of v and $-w$. Explicitly, see Table 5.8.

From the point of view of M_{22}, these are simply the 22 points, the 77 hexads (octads through the two fixed points) and the 176 heptads (octads through one of the two fixed points but not the other). We can define a graph of valence 112 on these 275 vertices, where two vertices are joined if the inner product of the corresponding vectors is 8. So, if P is a point, h is a hexad and k is a heptad: P joins h if, and only if, $P \notin h$; P joins k if, and only if, $P \in k$; h_1 joins h_2 if, and only if, $h_1 \cap h_2 = \phi$; k_1 joins k_2 if, and only if, $|k_1 \cap k_2| = 1$; and h joins k if, and only if, $|h \cap k| = 3$. A diagram of this graph appears in Figure 5.27.

---

[9] Every vector in the Leech lattice is congruent modulo 2Λ to a unique vector in Λ_2, a unique vector in Λ_3 or a set of 24 mutually orthogonal vectors of Λ_4 together with their negatives. Modulo 2Λ, these 2^{12} vectors form a vector space of dimension 12 over $\mathbb{Z}_2$.

Table 5.8. The orbits of M_{22} on
275 2-vectors

Shape	Number
$(4.\ 4.\ 0^{22})$	
$(-1.\ 3.\ -1^{22}$	
$(0.\ 4.\ -4.\ 0^{21})$	22
$(2.\ 2.\ -2.^6\ 0^{16})$	77
$(1.\ 3.\ 1^7.\ -1^{15}$	176

Our double coset enumerator has shown that the group defined in Theorem 5.19 has maximal order $4050 \times |M_{22}|$; and that if the stabilizer of a 223-triangle satisfies the additional relation, then this really is its order. We can do this by realizing our progenitor as symmetries of the 275 vertices of the graph in Figure 5.27, and showing that the additional relation is satisfied. This could be done by hand, but in the following exercises we indicate how it can be done using MAGMA.

EXERCISE 5.4

(1) First input M_{22} as permutations on 22 letters and, using the labelling shown in Section 5.7, fix a hexad and a heptad. Obtain the 77 and 176 point actions of M_{22} by letting it act on the cosets of the stabilizers of this hexad and heptad, respectively:

```
> m22:=sub<Sym(22)|(1,2,4,8,16,9,18,13,3,6,12)
  (5,10,20,17,11,22,21,19,15,7,14),
> (11,1)(4,20)(16,14)(13,21)(22,2)(18,10)(8,17)(12,7)>;
> #m22;
443520
```

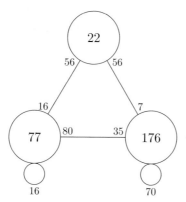

Figure 5.27. The rank 3 graph preserved by McL:2.

```
> hex:={3,6,9,19,15,5};
> sthex:=Stabilizer(m22,hex);
> f77,g77,k77:=CosetAction(m22,sthex);
> Degree(g77);
77
> hept:={3,6,9,11,4,16,13};
> sthept:=Stabilizer(m22,hept);
> f176,g176,k176:=CosetAction(m22,sthept);
> Degree(g176);
176
```

We now embed M_{22} as m22p275 in the symmetric group on 275 letters, with orbits of lengths 22, 77 and 176. The following function f converts a permutation of M_{22} acting on 22 letters into a permutation on 275 letters:

```
> s275:=Sym(275);
> f:=func<uu|s275!([i^uu:i in [1..22]]
>   cat [j^f77(uu)+22:j in [1..77]]
>   cat [k^f176(uu)+99:k in [1..176]])>;
> Order(sub<s275|f(m22.1),f(m22.2)>);
443520
> m22p275:=sub<s275|f(m22.1),f(m22.2)>;
```

We must now find a subgroup of M_{22} isomorphic to $L_2(11)$ and find an involution in S_{275} which commutes with it. It turns out that the centralizer in S_{275} of this copy of $L_2(11)$ has order 48, but if we pick out the involutions which fix 11 points we find there are just three possibilities: two of these generate the alternating group A_{275}, and the third, t_1 say, gives us the McLaughlin group. Under conjugation by m22p275, t_1 will have 672 images. Complete your verification by showing that if t_2 is in the 165 orbit of our $L_2(11)$ then $(t_1.t_2)^2$ is an element of order 2 in m22p275:

```
> l11:=Stabilizer(m22,{1,2,4,8,16,9,18,13,3,6,12});
> l11p275:=sub<s275|f(l11.1),f(l11.2),f(l11.3),
f(l11.4)>;
> #l11p275;
660
> cl11p275:=Centralizer(s275,l11p275);
> #cl11p275l;
48
```

5.8   The Fischer groups

Not surprisingly, given their definition in terms of a special conjugacy class of involutions known as *3-transpositions* (see below), the three Fischer groups may be defined in a concise and attractive manner by way of symmetric presentations. In this section we shall give such a symmetric presentation for two of these groups, but refer the reader who requires a fuller explanation to refs. [55] and [56], where many other related results may be found. Although we shall not give full proofs of the assertions made, we shall introduce, with explanations, a different approach to verifying their validity.

The clue lies in Example 4.2 in Section 4.4 on the group of the 28 bitangents. We took as control subgroup the symmetric group S_8 in its degree 35 action on the partitions of the eight letters into two tetrads. Thus our progenitor took the form $P = 2^{\star 35} : S_8$, and the symmetric generators could be denoted by symbols such as

$$\begin{bmatrix} 1234 \\ \hline 5678 \end{bmatrix}.$$

We factored this by two relations:

$$(1\ 2)(3\ 4)(5\ 6)(7\ 8) = \begin{bmatrix} 1234 \\ \hline 5678 \end{bmatrix}\begin{bmatrix} 1256 \\ \hline 3478 \end{bmatrix}\begin{bmatrix} 1278 \\ \hline 3456 \end{bmatrix}$$

and

$$\left((1\ 5) \begin{bmatrix} 1234 \\ \hline 5678 \end{bmatrix} \right)^3 = 1,$$

and obtained the symplectic group $S_6(2)$.

This prompts us to take as control subgroup S_{12} in its action on the partitions of the 12 letters into three tetrads. Since the number of such partitions is given by $\frac{1}{6}\binom{12}{4}\binom{8}{4} = 5775$, we are considering the progenitor $2^{\star 5775} : S_{12}$. We now factorize by the analogous relations to those used for $S_6(2)$ and find the result given in Theorem 5.20.

THEOREM 5.20

$$\frac{2^{\star 5775} : S_{12}}{\begin{bmatrix} 1234 \\ \hline 5678 \\ \hline 90xy \end{bmatrix}\begin{bmatrix} 1256 \\ \hline 3478 \\ \hline 90xy \end{bmatrix}\begin{bmatrix} 1278 \\ \hline 3456 \\ \hline 90xy \end{bmatrix} = (1\ 2)(3\ 4)(5\ 6)(7\ 8)} \cong Fi_{23},$$

$$\left((1\ 5) \begin{bmatrix} 1234 \\ \hline 5678 \\ \hline 90xy \end{bmatrix} \right)^3 = 1,$$

the second largest Fischer group.

Naturally we refer to these symmetric generators as *trifid maps*. At the end of this section we record what happens if one or other of these two relations is removed,[10] but in the meantime we indicate how the double coset enumerator may be used to verify this claim. We first require the

---

[10] Recall that in the $S_6(2)$ case removal of either of the two additional relations resulted in $2 \times S_6(2)$.

stabilizer in S_{12} of a partition into three tetrads, namely a subgroup of shape $S_4 \wr S_3$. This enables us to obtain S_{12} in its action on 5775 points:

```
> s12:=Sym(12);
> hh:=Stabilizer(s12,[{1,2,3,4},{5,6,7,8},{9,10,11,
12}]);
> hhb:=sub<s12|hh,s12!(1,5)(2,6)(3,7)(4,8),
> s12!(1,5,9)(2,6,10)(3,7,11)(4,8,12)>;
> Index(s12,hhb);
5775
> f,nn,kk:=CosetAction(s12,hhb);
```

We must now input the two additional relations, and to do this identify the three partitions mentioned in the first relation as being those labelled $\{1, 471, 1114\}$. Finally, in order to put the second relation in canonical form, we need the labels of two partitions interchanged by the transposition (1 5):

```
> Fix(sub<nn|f(s12!(1,2)),f(s12!(3,4)),f(s12!(5,6)),
f(s12!(7,8)),
> f(s12!(9,10)),f(s12!(10,11,12))>);
{ 1, 471, 1114 }
> 1^f(s12!(1,5));
23
> RR:=[<[1,471,1114],f(s12!(1,2)(3,4)(5,6)(7,8))>,
>          <[1,23,1],f(s12!(1,5))>];
> CT:=DCEnum(nn,RR,
>[*f(s12!(1,2,3,4,5,6,7,8,9,10)),f(s12!(1,2)),<[1],
Id(nn)>*]:
>  Print:=5,Grain:=300);

Index: 31671 === Rank: 4 === Edges: 22 ===  Time: 10.625

>
> CT[4];
[
    [],
    [ 5 ],
    [ 5, 25 ],
    [ 5, 25, 76 ]
]
> CT[7];
[ 66, 5775, 20790, 5040 ]
```

The enumeration is carried out over a subgroup which centralizes the transposition (11 12), and we obtain index $31\,671$. So the image group possesses a conjugacy class of involutions of length dividing $31\,671$. Indeed, we are suggesting that in the image our control subgroup S_{12} acts on this class of involutions with three orbits of lengths 66 (its own transpositions), 5775 (the symmetric generators), 20 790 and 5040. Arguments which allow us to make this claim, given that we know that Fi_{23} is a homomorphic image, are outlined later in this section.

Proceeding in the manner suggested by this result, we consider the symmetric group S_{16} as control group acting on the partitions of the 16 letters into four tetrads (the *quadrifid maps*), factored by the same analogous relations. Rather disappointingly, though not surprisingly since the largest symmetric group to appear in a sporadic group is S_{12}, we find that, with $n = 16!/4!(4!)^4 = 2\,627\,625$, Theorem 5.21 results.

THEOREM 5.21

$$G = \frac{2^{\star n} : \mathrm{S}_{16}}{\begin{bmatrix} 1234 \\ 5678 \\ 90uv \\ wxyz \end{bmatrix}\begin{bmatrix} 1256 \\ 3478 \\ 90uv \\ wxyz \end{bmatrix}\begin{bmatrix} 1278 \\ 3456 \\ 90uv \\ wxyz \end{bmatrix} = (1\ 2)(3\ 4)(5\ 6)(7\ 8)} \cong \langle 1 \rangle,$$

$$\left((1\ 5) \begin{bmatrix} 1234 \\ 5678 \\ 90uv \\ wxyz \end{bmatrix} \right)^3 = 1,$$

the trivial group.

This last fact is a relatively straightforward consequence of the following lemma.

LEMMA 5.10

$$G = \frac{2^{\star 36} : \mathrm{S}_6(2)}{(\mathrm{B}t_\star)^3 = 1} \cong \langle 1 \rangle,$$

where B denotes one of the bifid maps, say 1234/5678, and $t_\star$ denotes the symmetric generator fixed by the copy of S_8 which served as the control subgroup in Section 4.4.

Proof Recall that the 36 objects being permuted by $\mathrm{S}_6(2)$ may be taken to be the symbol $\star$ and the 35 bifid maps, so our set of symmetric generators is given by

$$\mathcal{T} = \{t_\star\} \cup \{t_\mathrm{B} \mid \mathrm{B} \text{ a bifid map}\}.$$

Careful though: the bifid maps do not act on one another by conjugation (as in the 63-point action) but by *right multiplication* of cosets. Thus the action $\mathcal{T} \times N \mapsto \mathcal{T}$ is given by

$$\left(\frac{1256}{1278}, \frac{1234}{5678} \right) \mapsto \frac{1278}{3456}; \left(\frac{5234}{1678}, \frac{1234}{5678} \right) \mapsto \frac{5234}{1678}.$$

Now let

$$c = t_C, d = t_D, e = t_E,$$

where

$$C = \frac{1234}{5678}, D = \frac{1256}{3478} \text{ and } E = \frac{1278}{3456};$$

note that $cde = (1\ 2)(3\ 4)(5\ 6)(7\ 8)$. Then $(t_\star c)^3 = 1$, and so

$$t_\star c = (t_\star c)^4 = [t_\star, t_\star^c] = [t_\star, t_\star^{de}] = [t_\star^e, t_\star^d]^e = [e'^\star, d'^\star]^e = [e, d]^{t_\star e} = 1.$$

We have used the fact that c, d and e commute with one another, and that, if u and v are involutions in a group such that $(uv)^3 = 1$, then $u^v = v^u$.

But this says that $t_\star = c = t_C$, where C was any bifid map; thus, in particular, $c = d$ and the simple group $N = S_6(2)$ collapses to the identity. But then the additional relation becomes $t_\star^3 = 1$, and so $t_\star = 1$ and $G = 1$. □

This lemma is proved in much more generality in ref. [56], but this will suffice for our purposes.

Proof of Theorem 5.17 We consider the subgroup of G generated by all the quadrifid maps which have $90uv$ and $wxyz$ as two of their rows. There are of course 35 of these and they must generate a homomorphic image of $S_6(2)$. But the transposition $(9\ w)$ of our S_{16} clearly commutes with a maximal subgroup of this group isomorphic to S_8 acting on $\{1, 2, \ldots, 8\}$, but does not commute with the whole $S_6(2)$. We thus have a subprogenitor $2^{\star 36} : S_6(2)$, where the 36 symmetric generators are $(9\ w)$ and its images under conjugation by this copy of $S_6(2)$, factored by a relation equivalent to that in Lemma 5.10. But this yields the trivial group, and so some of our quadrifid maps are mapped to the identity. Thus all the quadrifid maps are mapped to the identity and so is the transposition $(1\ 5)$. Thus the control subgroup also maps to the identity and $G = \langle 1 \rangle$.   □

In an unpublished manuscript, Bray shows the following.

THEOREM 5.22

$$G = \cfrac{2^{\star 462} : S_{12}}{\left[\begin{array}{c}123456\\7890xy\end{array}\right]\left[\begin{array}{c}123789\\4560xy\end{array}\right]\left[\begin{array}{c}123456\\7890xy\end{array}\right]\left[\begin{array}{c}1230xy\\456789\end{array}\right] = 1} \cong O_{10}^-(2) : 2,$$

$$\left((1\ 7)\left[\begin{array}{c}123456\\7890xy\end{array}\right]\right)^3 = 1,$$

where the symmetric generators now correspond to partitions of the 12 letters into two sixes.

Call these *bihex maps* and note that the 66 transpositions in the S_{12} are conjugate in G to the 462 bihex maps, forming a conjugacy class of $66 + 462 = 528$ involutions in the outer half of $O_{10}^-(2)$, in a directly analogous manner to the $S_6(2)$ case.

This suggests a different approach to working out the order of a symmetrically presented group G: instead of trying to find the index of the control subgroup N (or some other subgroup) in G, we count the involutions in a conjugacy class of G. Firstly, note the following elementary result.

LEMMA **5.11** Suppose that $G = \langle X \rangle$, where X is a conjugacy class of elements in G. Let $\psi : G \to H$ be a homomorphism of G onto a group H and assume that ψ restricted to X is a bijection. Then $\ker(\psi) \leq Z(G)$.

Proof Let $k \in \ker(\psi)$ and let $x \in X$. Then

$$\psi(x^k) = \psi(x)^{\psi(k)} = \psi(x),$$

and so $x = x^k$ as ψ is a bijection on the conjugacy class X. Hence $k \in C_G(x)$, and, since this holds for all $x \in X$ and X generates G, we have $k \in Z(G)$, as required. $\qquad\square$

Since the progenitors under consideration are (essentially) perfect and we know that they map onto specific *target groups*, we may then appeal to our knowledge of possible perfect central extensions of these target groups (their *Schur multipliers*).

In counting involutions we make use of a result of Virotte-Ducharme [85], which we have translated into the language of progenitors and slightly generalized. Before stating this result, we clarify some definitions: a set of involutions $\mathcal{D}$ in a group G is said to be a *G-closed set* of *3-transpositions* provided that $\mathcal{D}$ is a union of conjugacy classes of G and that products xy for $x, y \in \mathcal{D}$ have order at most 3. If, in addition, $\mathcal{D}$ generates G, then we say that G is a *3-transposition group* with respect to $\mathcal{D}$. Note that here $\mathcal{D}$ need not be a single conjugacy class.

THEOREM **5.23** Let $P = 2^{\star n} : N$ and let $t = t_0 \in \mathcal{T}$, the set of symmetric generators of the progenitor P. Furthermore, let $N_0 = \mathrm{Stab}_N(t)$, with $|N : N_0| = |\mathcal{T}| = n$, and suppose that N is a 3-transposition group with respect to $\mathcal{C}$ such that

$$N = N_0 \cup N_0 \mathcal{C} \cup N_0 \mathcal{C} \mathcal{C}. \tag{5.5}$$

Let $\mathcal{R} = \{(tc)^3 \mid c \in \mathcal{C} \setminus N_0\}$ and set $G = P/\mathcal{R}$. Let $\mathcal{D}$ denote the image of the set $\mathcal{C} \cup \mathcal{T}$. Then G is a 3-transposition group with respect to $\mathcal{D} \setminus \{1\}$. Moreover,

(i) if G is a faithful image of P, then $|\mathcal{D}| = |\mathcal{C}| + |\mathcal{T}|$;

(ii) if N acts faithfully on the cosets of N_0, or equivalently on $\mathcal{T}$, then $\mathcal{D}$ is a conjugacy class of G.

Before proving this theorem, it is worth noting that both our examples $2^{\star 35} : S_8 \mapsto S_6(2)$ and $2^{\star 462} : S_{12} \mapsto O_{10}^-(2) : 2$ satisfy its conclusions. Thus in the first case $|\mathcal{D}| = |\mathcal{C}| + |\mathcal{T}| = \binom{8}{2} + 35 = 63$; and in the second $|\mathcal{D}| = |\mathcal{C}| + |\mathcal{T}| = \binom{12}{2} + 462 = 528$. In the first case, Equation (5.5) is equivalent to requiring that a given bifid map can be conjugated to any other by a product of at most two transpositions of S_8. This is clearly the case, so Theorem 5.23 can be applied as it stands. However, in the second case we would require that a given bihex map can be conjugated into any other by a product of at most two transpositions. But two bihex maps whose hexads intersect one another in three letters require at least three transpositions to conjugate one into the other, so the theorem cannot be applied as it stands. On the other hand, if we let

$$h = \begin{bmatrix} 123456 \\ 7890xy \end{bmatrix}$$

then $h^{(6\ y)} = (6\ y)^h$, and so conjugation of a transposition of S_{12} by a non-commuting bihex map results in a bihex map. Moreover,

$$h\, h^{(5\ x)(6\ y)} = (h^{(6\ y)}h^{(5\ x)})^{(6\ y)} = ((6\ y)^h(5\ x)^h)^{(6\ y)} = ((6\ y)(5\ x))^{h(6\ y)},$$

of order 2. So if the hexads of two such elements intersect evenly (i.e. two in one and four in the other) then the elements commute. Finally note that Bray's first relation shows that a bihex map conjugated by another bihex map whose hexads cut it 3.3 results in a third bihex map. So the union of the set of transpositions and the set of bihex maps *is* a conjugacy class in the image group as claimed above.

Proof of Theorem 5.5 We shall work throughout in the image group G, and so expressions for elements are to be interpreted as their images after quotienting by $\langle \mathcal{R}^P \rangle$. The definition implies that $[t, x] = 1$ if $x \in \mathcal{C} \cap N_0$, and that $(tx)^3 = 1$ if $x \in \mathcal{C} \setminus N_0$. If $t \neq s \in \mathcal{T}$ then $s = t^g$ for some $g \in G$. Now, Equation (5.5) implies that we may write $g = mc$ or $g = mcd$, where $m \in N_0$ and $c, d \in \mathcal{C}$. If $d \in N_0$ then $mcd = mdc^d = m'e$, where $m' \in N_0$ and $e = c^d \in \mathcal{C}$; so we may assume that neither c nor d lies in N_0. So to show that G is a 3-transposition group with respect to $\mathcal{C} \cup \mathcal{T}$, it remains to show that all products of the form $ts = tt^c$ or tt^{cd} have order 1, 2 or 3. But $tt^c = ct$ of order dividing 3, and

$$tt^{cd} = (t^d t^c)^d = (d^t c^t)^d = (dc)^{td},$$

which has the same order as dc, namely 2 or 3 by hypothesis. Thus tx has order less than or equal to 3 for all $x \in \mathcal{D}$, and the transitivity of N on $\mathcal{T}$ shows that this is true with t replaced by any element of $\mathcal{T}$. Finally we must show that $\mathcal{D}$ is a conjugacy class of G. It will suffice to show that $\mathcal{D}^t = \mathcal{D}$ as $G = \langle N, t \rangle$ and $\mathcal{D}$ is certainly preserved by N. But $c^t = c$ or $c^t = t^c \in \mathcal{T}$ for all $c \in \mathcal{C}$; so it remains to show that $r^t \in \mathcal{D}$ for all $r \in \mathcal{T}$. But, as above, $r = t^c$ or $r = t^{cd}$ for some $c, d \in \mathcal{C} \setminus N_0$. Then $(t^c)^t = c$ and

$$(t^{cd})^t = (t^c)^{dt} = c^{tdt} = c^{dtd} = (c^d)^{td} = e^{td} = (e^t)^d = e^d \in \mathcal{C} \text{ or } t^{ed} \in \mathcal{T},$$

where $e = c^d \in \mathcal{C}$.   □

At this stage we use the group $O_{10}^-(2) : 2$, for which a symmetric presentation was given in Theorem 5.18, as control subgroup acting on the cosets of our previous control subgroup S_{12} with degree $104\,448$.

THEOREM **5.24**

$$G = \frac{2^{\star 104448} : (O_{10}^-(2) : 2)}{\left(\left[\frac{123456}{7890xy}\right]t\right)^3} \cong 3\dot{}\,Fi_{24}.$$

The proof of this result involves diagram geometries and relies on the results of Ronan [77] and Ivanov [51], which together imply that the only geometries with a particular diagram (known as $\mathcal{G}_{24}$) and which satisfy a certain transitivity condition are those attached to the groups $3\dot{}\,Fi_{24}$ and Fi_{24}. We do not have space to describe this theory here, but refer the reader to ref. [56], where there is a brief summary.

We conclude this section by stating as promised what happens if one or other of the additional relations in Theorem 5.16 is removed. If the second relation is removed, then the result is reminiscent of the $S_6(2)$ case and we obtain the following theorem.

THEOREM **5.25**

$$\frac{2^{\star 5775} : S_{12}}{\begin{bmatrix} 1234 \\ 5678 \\ 90xy \end{bmatrix}\begin{bmatrix} 1256 \\ 3478 \\ 90xy \end{bmatrix}\begin{bmatrix} 1278 \\ 3456 \\ 90xy \end{bmatrix} = (1\ 2)(3\ 4)(5\ 6)(7\ 8)} \cong 2 \times Fi_{23},$$

However, if instead the first relation is removed, then John Bray has proved the following remarkable result.

THEOREM **5.26** [Bray]

$$\frac{2^{\star 5775} : S_{12}}{\left((1\ 5)\begin{bmatrix} 1234 \\ 5678 \\ 90xy \end{bmatrix}\right)^3} \cong Fi_{23} \times O_{10}^+(3) : 2.$$

5.9 Transitive extensions and the O'Nan group

In all the groups investigated so far in Part II, our procedure has been to start with an interesting transitive permutation group N of degree n, usually but not necessarily primitive, and form the progenitor $2^{\star n} : N$. We then employ various methods, in particular applying Lemma 3.3, to produce suitable additional relations by which to factor this progenitor. The structure of the control subgroup N will of course be critical, and many

of our most pleasing results have been obtained when the centralizer of a
2-point stabilizer is cyclic of order 2 with a generator interchanging the two
fixed points.

It is natural to ask what happens if N possesses a transitive extension of
degree $n+1$. To be precise, suppose we have obtained an interesting finite
group G as follows:

$$\frac{2^{\star n} : N}{\pi_1 w_1, \ldots, \pi_r w_r} \cong G,$$

with symmetric generators $\mathcal{T} = \{t_1, \ldots, t_n\}$, $\pi \in N$ and $w_i = w_i(t_1, \ldots, t_n)$,
a word in the t_i. Then, supposing that M is a doubly transitive permutation
group of degree $n+1$ with point stabilizer $M_0 \cong N$, what will happen if we
form

$$\frac{2^{\star n+1} : M}{\pi_1 w_1, \ldots, \pi_r w_r},$$

with symmetric generators $\hat{\mathcal{T}} = \{t_0, t_1, \ldots, t_n\}$? The simple answer is that
this straightforward approach to extension is invariably doomed to fail-
ure for the simple reason that these additional relations often fail to satisfy Lemma 3.3 applied to the extended progenitor. The image group will
collapse dramatically.

Perhaps the best way of seeing what goes wrong is by way of a familiar
example. So, consider the progenitor $P = 2^{\star 11} : L_2(11)$, when $N \cong L_2(11)$, the
point stabilizer $N_1 \cong A_5$, and the 2-point stabilizer[11] is $N_{12} \cong S_3$. Lemma 3.3
says that any permutation in N which can be written in terms of t_1 and t_2
without causing collapse must lie in $C_N(N_{12})$, and in this case we have that
$C_2 \cong C_N(N_{12}) = \langle \pi_{12} \rangle$, say. Accordingly, in the investigation which led to
a symmetric presentation for the Janko group J_1 we factored by a relation
$\pi_{12} = t_1 t_2 t_1 t_2 t_1$.

Now, $L_2(11)$ in its action on 11 letters possesses a transitive extension,
namely the Mathieu group M_{11} acting triply transitively on 12 letters. So we
have $M \cong M_{11}$, $M_0 \cong L_2(11)$ and $M_{01} \cong A_5$. But A_5 centralizes no non-trivial
element, and so the previous relation would infringe Lemma 3.3.

All is not lost, however! It turns out that the way to proceed makes
use of the fact that any pair of non-commuting involutions, a and b, say,
generate a dihedral group of order $2k$ which has trivial centre, if k is odd, and
centre of order 2, namely $\langle (ab)^{k/2} \rangle$, if k is even. So we form the progenitor
$\hat{P} = 2^{\star(n+1)} : M$ with symmetric generating set $\mathcal{S} = \{s_0, s_1, \ldots, s_n\}$, where
we are supposing that $\langle s_i, s_j \rangle \cong D_8$ for $i \neq j$.[12] We now set $t_i = (s_0 s_i)^2$ for
$i = 1, \ldots, n$, and note that $\hat{P}$ contains a sub-progenitor $P = 2^{\star n} : N$ whose
symmetric generators are the t_i. We may now factor $\hat{P}$ by the relations
$\pi_i w_i$ used to define G above (written as words in the t_i) to ensure that

---

[11] Note that we have continued to take the set of letters acted on by N to be $\{1, \ldots, 11\}$,
so as to be consistent with the previous paragraph; however, we should normally take $L_2(11)$
acting on $\mathbb{Z}_{11}$, the integers modulo 11.
[12] Note that we could replace D_8 by D_{4k} for any $k \geq 2$.

the resulting group $\hat{G}$ contains copies of some homomorphic image of G, which we aim to make isomorphic copies of G itself. Note further that, since $t_i \in Z(\langle s_0, s_i \rangle)$, s_0 commutes with $\langle t_1, \ldots, t_n \rangle$.

Applying this process to the J_1 case mentioned above, we start with our copy of

$$\mathrm{L}_2(11) \cong N = \langle (0\ 1\ 2\ 3\ 4\ 5\ 6\ 7\ 8\ 9\ \mathrm{X}),\ (3\ 4)(2\ \mathrm{X})(5\ 9)(6\ 7) \rangle,$$

where X stands for '10'; we then let $\sigma = (3\ 4)(0\ 1\ 8)(2\ 5\ 6\ \mathrm{X}\ 9\ 7) \in N$ and recall that

$$\frac{2^{\bullet 11} : \mathrm{L}_2(11)}{(\sigma t_0)^5} \cong \mathrm{J}_1,$$

the smallest Janko group. The process of transitive extension leads us to form the following group:

$$\frac{2^{\bullet 12} : \mathrm{M}_{11}}{(\sigma(s_\infty s_0)^2)^5, (s_\infty s_0)^4},$$

where the 12 symmetric generators are $\{s_\infty, s_0, \ldots, s_{\mathrm{X}}\}$. Then

$$\langle (s_\infty s_0)^2, \ldots, (s_\infty s_{\mathrm{X}})^2 \rangle = \mathrm{J}_1 \ \text{ or } \ \langle 1 \rangle,$$

the trivial group.[13] Moreover, s_∞ commutes with this subgroup and so, if the former case holds, we have a subgroup of shape $2 \times \mathrm{J}_1$. As readers familiar with the sporadic groups will know, there is indeed a group with a class of involutions whose centralizers have this shape, namely the O'Nan group. It turns out that the above is insufficient to define it, but that Theorem 5.27 holds.

THEOREM 5.27

$$\frac{2^{\bullet 12} : \mathrm{M}_{11}}{(\sigma(s_\infty s_0)^2)^5, (s_\infty s_0)^4, (\sigma^3 s_\infty s_3)^5} \cong \mathrm{O'N} : 2.$$

Note that all relators involve words of even length in the s_i, and so the image must contain a subgroup of index 2 and the (images of) the symmetric generators must lie outside this subgroup.

Motivated by this example, Bolt [11] together with Bray investigated such transitive extensions of progenitors and found that the family of groups of form $\mathrm{L}_5(p) : 2$ arise in this manner.

5.10 Symmetric representation of groups

It is the purpose of this section to demonstrate how construction of a group by way of a symmetric presentation gives rise to an alternative, concise but informative, method of representing its elements, which will prove particularly useful for some of the sporadic simple groups. In general, if we

---

[13] Since J_1 is simple.

wish to multiply and invert elements in a straightforward manner, we must represent them as either permutations or as matrices. These two operations are readily performed on permutations; moreover, the cycle shape of an element immediately yields its order and often its conjugacy class. However, for the larger sporadic groups the lowest degree of a permutation representation is unmanageable (the Monster group is at best a permutation group on 10^{20} letters). Operations on matrices are much more difficult and time-consuming, and basic information about an element is not readily recovered from its matrix representation; even so, Parker's MeatAxe [7] effectively handles very large matrices over finite fields, and much recent progress in computing with matrix groups has been made by Leedham-Green and colleagues. Moreover, some impressively low degree representations of sporadic groups over finite fields have been found by, for instance, Norton, Parker and Wilson (see references in ref. [38]). Group elements can, of course, be expressed as words in any generating set, but even recognizing the identity element can be a formidable task. Again, given a short sequence of letters whose stabilizer is trivial, a permutation is uniquely defined by its image. This gives a remarkably concise notation for elements of the group, but does not readily admit the basic operations.

The approach illustrated here, which exploits the Cayley diagram of G over the control subgroup N, combines conciseness with acceptable ease of manipulation. Inversion is as straightforward as for permutations, and multiplication can be performed manually or mechanically by means of a short recursive algorithm.

The desirability of such a concise representation of group elements is convincingly illustrated by consideration of the Janko group J_1, whose order is only twice that of the Mathieu group M_{12}, a permutation group on 12 letters, but whose lowest permutation degree is 266. Whilst it is true that MAGMA [19] and other group theoretic packages handle permutations of this size with immense ease, recording and transmitting particular elements (other than electronically) is rather inconvenient. So, as in Curtis and Hasan [37], we shall illustrate the method by applying it to J_1.

Since the diameter of the graph in Figure 5.5 is 4 and the edges correspond to symmetric generators, it is clear that every element of G can be represented by an element of $L_2(11)$, that is to say a permutation on 11 letters followed by a word in the symmetric generators of length at most 4. Indeed, the absence of cycles of length less than 5 in the Cayley graph shows that this representation is unique when the length of the word is 2 or less. Clearly such a *symmetric representation* is far more convenient to work with than permutations on 266 letters, provided we are able to multiply and invert elements easily. In fact, it is our main purpose to present computer programs which do precisely this. The usefulness of this approach is demonstrated in the following, where we list *all* involutions of J_1 up to conjugation by permutations in N, in symmetrically represented form. Symmetrically represented generators for a representative of each conjugacy class of maximal subgroups of J_1 are given in Curtis [34].

Table 5.9. *The involutions of* J_1 *under conjugation by* $L_2(11)$

Representative involution		Corresponding edge	Number
0		$(*, 0)$	11
0^1	$= 101 = \pi_{01}.10$	$(1, 01)$	110
	$= (0, 1)(2, 6)(7, X)(3, 4).10$		
0^{10}	$= \pi_{01}$	$(10, 01)$	55
	$= (0, 1)(2, 6)(7, X)(3, 4)$		
0^{18}	$= \pi_{2X}.1$	$(18, 10)$	165
	$= (2, X)(0, 8)(5, 6)(7, 9).1$		
0^{12}	$= \pi_{57}.326$	$(12, 012)$	660
	$= (5, 7)(1, X)(3, 6)(4, 9).326$		
0^{124}	$= \pi_{5X}.39$	$(5X9, X59)$	330
	$= (5, X)(0, 7)(2, 4)(3, 9).39$		
0^{1240}	$= \pi_{5X}.7390$	$(05X2, 5X2)$	132
	$= (5, X)(0, 7)(2, 4)(3, 9).7390$		

The involutions of J_1

Let $\mathcal{J}$ be the unique conjugacy class of involutions in J_1 and let $x \in \mathcal{J}$. Then $C_{J_1}(x) \cong 2 \times A_5$, which is the stabilizer of an edge in the Livingstone graph. This stabilizer is the centralizer of an involution, so we see that there is a one-to-one correspondence between the involutions and the edges of the graph. In Table 5.9 we give a representative for each of the seven orbits of involutions under conjugation by the control subgroup N, together with the corresponding edge and the number of involutions in each orbit. Recall that π_{ij} stands for the element of N given by $t_i t_j t_i t_j t_i$.

The algorithms and programs

As explained above, any element of G can be written, not necessarily uniquely, as the product of a permutation of $N \cong L_2(11)$ followed by a word of length at most 4 in the symmetric generators. In Section 1.5 of Curtis [34], I outlined a procedure for multiplying elements represented in this fashion. It may be useful to see this process being carried out manually.

EXAMPLE

$$\pi_{5X}07.\pi_{01}10 = \pi_{5X}\pi_{01}0^{\pi_{01}}7^{\pi_{01}}10$$

$$= \pi_{5X}\pi_{01}1X10$$

$$= \pi_{5X}\pi_{01}\pi_{1X}1X0$$

$$= (5, X)(0, 7)(2, 4)(3, 9).(0, 1)(2, 6)(7, X)(3, 4).$$

$$(1, X)(8, 0)(9, 3)(6, 4)1X0$$

$$= (0, 1, 8)(2, 9, 6)(5, 7, X)1X0,$$

which is in its canonically shortest form.

In this chapter we have computerized such a procedure in two independent ways. The first, see Section 5.11.1, makes full use of the ease with which MAGMA handles permutations of such low degree. Elements represented as above are transformed into permutations on 266 letters, and any group theoretic function can then be applied before transformation back into the symmetric representation. Thus, for example, the procedure *cenelt* returns generators for the centralizer of a given element symmetrically represented. One can readily write procedures to perform whatever task one chooses in MAGMA, and keep a record of the results in this short form.

The programs given in Section 5.11.2 are rather more interesting, both mathematically and computationally. To multiply two elements we first use *unify* to express

$$\pi u \cdot \sigma v = \pi \sigma \cdot u^\sigma v,$$

where $\pi, \sigma \in N$ and u, v are words in the elements of $\mathcal{T}$, as a single sequence *ss* of length $11 + length(u) + length(v) \leq 19$. The first 11 entries give the permutation $\pi\sigma$ and the remainder represents a word of length ≤ 8 in the elements of $\mathcal{T}$. The procedure *canon* now puts *ss* into its canonically shortest form. No other representations of group elements are used; words in the symmetric generators are simply shortened by application of the relations (and their conjugates under N). Working interactively, the response is immediate.

The first type of procedure is heavily MAGMA-dependent, but can be readily modified for other packages such as GAP. On the other hand, the second type of procedure, although written in MAGMA here, could have been written in any high-level language.[14]

5.11 Appendix to Chapter 5

5.11.1 Program 1

We start with a presentation of J_1, based on the symmetric presentation of Theorem 5.2, in which the three generators x, y and t correspond to the two permutations in Section 5.2.1 (iv) (p. 138) and the symmetric generator t_0, respectively. Thus the MAGMA command CosetAction $(J, \mathrm{sub}\langle J | x, y \rangle)$ gives J_1 in its action on 266 letters. We now form the symmetric generators $\mathcal{T} = \{t_i; i = 0, 1, \ldots, X\}$ as permutations on 266 letters and store them as *ts*, a sequence of length 11. Next we build *cst*, a sequence of length 266 whose terms are sequences of integers, representing words in the symmetric generators. These words form a complete set of coset representatives for $L_2(11)$ in J_1 and correspond to the ordering determined by the MAGMA function CosetAction. Given two elements of J_1 symmetrically represented as (xx, uu) and (yy, vv), the procedure mult uses *ts* and *cst* to return the product (zz, ww). As described in Section 5.10, the procedure cenelt

---

[14] I am indebted to John Cannon for improving the MAGMA code in Program 2 (Section 5.11.2).

is used to return generators for the centralizer of a given element (xx, uu), themselves symmetrically represented:

```
J1fmt := recformat<
/*
Data structure for the symmetric representation of J1.
*/
     J1: GrpPerm,
     L11: GrpPerm,
     cst: SeqEnum,  // of integer sequences
     ts: SeqEnum,   // of elements of J1
     tra1: SetIndx, // of elements of L11
     tra2: SetIndx // of elements of L11
>;

//-----------------------------------------------------------------------
----------
prodim := function(pt, Q, I)
/*
Return the image of pt under permutations Q[I] applied sequentially.
*/
  v := pt;
  for i in I do
     v := v^(Q[i]);
  end for;
  return v;
end function;

//-----------------------------------------------------------------------
----------
symrep := function()
/*
Initialize the data structures for the symmetric representation of
J1.
*/
  J<x, y, t> := Group< x, y, t | x^11 = y^2 = (x*y)^3 =
(x^4*y*x^6*y)^2 = 1,
    t^2 = (t, y) = (t^x, y) = (t^(x^8), y) = (y*(t^(x^3)))^5 =
(x*t)^6 = 1>;

  // Construct the sequence of 11 symmetric generators ts as
permutations
  // on 266 letters.

  f, J1, k := CosetAction(J, sub< J | x, y>);
  ts := [ (t^(x^i)) @ f : i in [1 .. 11] ];

  // Construct representatives cst for the control subgroup N = L(2,
11)
  // as words in the symmetric generators consisting of the empty
```

```
word,
  // 11 words of length one, 110 words of length two, 132 words of
length three,
  // and 12 words of length four.

  S11 := SymmetricGroup(11);
  aa := S11 ! (1, 2, 3, 4, 5, 6, 7, 8, 9, 10, 11); bb := S11 ! (3,
4)(2, 10)(5, 9)(6, 7);
  cc := S11 ! (1, 3, 9, 5, 4)(2, 6, 7, 10, 8); L11 := sub< S11 | aa,
bb, cc >;

  cst := [null : i in [1 .. 266]] where null is [Integers() | ];
  for i := 1 to 11 do
    cst[prodim(1, ts, [i])] := [i];
  end for;
  for i := 1 to 11 do
    for j in {1, 2, 3, 4, 5, 6, 7, 8, 9, 10, 11} diff {i} do
   cst[prodim(1, ts, [i, j])] := [i, j];
    end for;
  end for;

  tra1 := Transversal(L11, sub<L11 | cc>);
  for i := 1 to 132 do
    ss := [3, 6, 1]^tra1[i];
    cst[prodim(1, ts, ss)] := ss;
  end for;

  tra2 := Transversal(L11, sub<L11 | aa, cc>);
  for i := 1 to 12 do
    ss := [1, 10, 3, 11]^tra2[i];
    cst[prodim(1, ts, ss)] := ss;
  end for;

  return rec<J1fmt |
    J1 := J1, L11 := L11, cst := cst, ts := ts, tra1 := tra1, tra2
:= tra2
  >;
end function;

//---------------------------------------------------------------------
----------
mult := function(J1Des, x, y)
/*
Return in its symmetric representation the product of elements x and
y of J1
themselves  symmetrically represented.
*/
    J1 := J1Des'J1; cst := J1Des'cst; ts := J1Des'ts; L11 :=
J1Des'L11;
    rrr:= J1Des'L11 ! x[1]; sss:= J1Des'L11 ! y[1];
```

```
  uu := x[2]^sss; vv := y[2];
  tt := &*[J1|ts[uu[i]]: i in [1 .. #uu]] * &*[J1|ts[vv[i]]: i in
[1 .. #vv]];
  ww := cst[1^tt];
  tt := tt * &*[J1|ts[ww[#ww - k + 1]]: k in [1 .. #ww]];
  zz := L11![rep{j: j in [1..11] | (1^ts[i])^tt eq 1^ts[j]}: i in
[1..11]];
  return <rrr*sss * zz, ww>;
end function;


//---------------------------------------------------------------------
----------
sym2per := function(J1Des, x)
/*
Convert an element x of J1 in the symmetric repesentation into a
permutation
acting on 266 letters. The image of an element of N is determined by
its action
on the eleven cosets whose representatives have length one.
*/
  J1 := J1Des'J1; cst := J1Des'cst; ts := J1Des'ts; L11 :=
J1Des'L11;
    tra1 := J1Des'tra1; tra2 := J1Des'tra2;

    xx := J1Des'L11 ! x[1]; uu := x[2];
    p := [1 : i in [1 .. 266]];
    for i := 1 to 11 do
        p[prodim(1, ts, [i])] := prodim(1, ts, [i]^xx);
    end for;
    for i := 1 to 11 do
        for j in {1, 2, 3, 4, 5, 6, 7, 8, 9, 10, 11} diff {i} do
  p[prodim(1, ts, [i, j])] := prodim(1, ts, [i, j]^xx);
        end for;
    end for;
   for i := 1 to 132 do
    t := [3, 6, 1]^tra1[i];
    p[prodim(1, ts, t)] := prodim(1, ts, t^xx);
   end for;
   for i := 1 to 12 do
    t := [1, 10, 3, 11]^tra2[i];
    p[prodim(1, ts, t)] := prodim(1, ts, t^xx);
   end for;

   return (J1 ! p) * &*[J1|ts[uu[j]]: j in [1 .. #uu]];
end function;


//---------------------------------------------------------------------
----------
per2sym := function(J1Des, p)
/*
```

```
Convert permutation p of J1 on 266 letters into its symmetric
representation.
The image of 1 under p gives the coset representative for Np as a
word ww in
the symmetric generators. Multiplication of p by the symmetric
generators of ww
in reverse order yields a permutation which can be identified with an
element
of N by its action on the 11 cosets of length one.
*/
    J1 := J1Des'J1; cst := J1Des'cst; ts := J1Des'ts; L11 :=
J1Des'L11;
    ww := cst[1^p];
    tt := p * &*[J1|ts[ww[#ww - l + 1]]: l in [1 .. #ww]];
    zz := L11![rep{j: j in [1..11] | (1^ts[i])^tt eq 1^ts[j]}: i in
[1..11]];
    return <zz, ww>;
end function;


//-------------------------------------------------------------------
----------
cenelt := function(J1Des, x)
/*
Construct the centraliser of element x of J1 given in its
symmetric
representation. An example of how all the standard procedures of {\sc
Magma}
can be utilized by: transformation to permutations, application of
the procedure,
transformation back to symmetric representation.
*/
    cent := Centralizer(J1Des'J1, sym2per(J1Des, x));
    return <Order(cent), [per2sym(J1Des, c): c in Generators(cent)]>;
end function;
```

5.11.2 Program 2

In this program we assume detailed knowledge of the *control subroup*
$N \cong L_2(11)$, but use no representation of elements of J_1 other than
their symmetric representation. Firstly, the procedure *unify* uses the
identity $\pi u.\sigma v = \pi\sigma.u^\sigma v$ to combine two symmetrically represented
elements (xx, uu) and (yy, vv) into a single sequence ss of length
$(11 + length(uu) + length(vv))$, which represents a permutation of N followed by a word of length ≤ 8 in the elements of $\mathcal{T}$. The procedure *canon*
then takes such a sequence and reduces it to its shortest form using the
following recursive algorithm. We make use of the relations $t_i t_j t_i t_j t_i = \pi_{ij}$
and $t_i t_j t_k t_i t_j = \sigma_{ijk}$, for $\{i, j, k\}$ a *special triple*, i.e. a triple fixed pointwise
by an involution of N.

The algorithm

Step (I)   If two adjacent symmetric generators are equal, delete them.

Step (II)   If a string $t_i t_j t_i$ appears, replace it by $\pi_{ij} t_i t_j$ and move the permutation π_{ij} over the preceding symmetric generators in the standard manner.

Step (III) If a string $t_i t_j t_k$ appears with $\{i, j, k\}$ a special triple, replace it by $\sigma_{ijk} t_j t_i$ and move the permutation to the left as above.

Having completed the above, if $length(ss) \leq 11 + 3 = 14$, finish. Otherwise we may assume all strings $t_i t_j t_k$ have $\{i, j, k\}$ non-special.

Step (IV)   For each string $t_i t_j t_k t_l$ construct a permutation σ and a 5-cycle ρ such that $\sigma t_i t_j t_k$ commutes with ρ. (In fact, $\sigma = \sigma_{pqr}$, where $p = i^{\pi_{jk}}, q = k^{\pi_{ij}}, r = j^{\pi_{ki}}$.) If l is not fixed by ρ, we replace $t_i t_j t_k$ by $\sigma^{-1}(\sigma t_i t_j t_k)^{\rho^m}$, where m is such that $l \in \{i, j, k\}^{\rho^m}$. If $length(ss) \leq 15$, finish. Otherwise take step (V).

Step (V)   We may now assume that any string of length 5 is one of the 11 special pentads preserved by $L_2(11)$. We use the identity $267X8 = \sigma_{X64} X26$ from ref. [34], p. 304, i.e. for $[i, j, k, l, m]$ an even permutation of a special pentad

$$t_i t_j t_k t_l t_m = \sigma_{ljn} t_l t_i t_j,$$

where $n = l^{\pi_{ij}}$.

After each **Step**, recall *canon*.

```
/*-----------------------------------------------
Define the projective special linear group PSL(2,11) as permutations
of degree
11 and sequences of pairs, special triples and permutations equal to
ijkij.
*/
L11 := PermutationGroup< 11 |(1,2,3,4,5,6,7,8,9,10,11),
(2,10)(3,4)(6,7)(5,9) >;
sg := L11!(1, 11, 8)(2, 7, 9, 10, 6, 5)(3, 4);
trans := Transversal(L11, Stabilizer(L11,{3,4}));
prs := {@ {3,4}^x : x in trans @};
trips := {@ {11,1,8}^x : x in trans @};
sgs := [ sg^x : x in trans ];


/*---------------------------------------------------------------
---------
Given two symmetrically represented elements of J1, where x and y are
permutations of N and u and v are words in the symmetric generators,
return
a single sequence of length 11 + l(u) + l(v) using the above
identity.
```

```
*/
Unify := func< x, u, y, v | [ q[p[i]] : i in [1..#p] ] cat v
                            where p is Eltseq(x) cat u
                            where q is Eltseq(y) >;
```

```
/*------------------------------------------------------------------
---------
Return permutation of N given by the word ijiji in the symmetric
generators.
*/
Pi := func< i, j | sgs[Index(prs, {i, j})]^3 >;
```

```
/*------------------------------------------------------------------
---------
For {i,j,k} a special triple, return permutation of N given by word
ijkij.
*/
Sg := func< i, j, k | i^sgs[rr] eq k select sgs[rr] else sgs[rr]^-1
                     where rr is Index(trips, {i,j,k}) >;
```

```
/*------------------------------------------------------------------
---------
For {i,j,k} a non-special triple, return permutations p and s such
that s.ijk
commutes with the 5-cycle p.
*/
Ntrip := function(i,j,k)
    s := Sg(i^Pi(j,k),k^Pi(i,j),j^Pi(k,i));
    l := Rep(Fix(Pi(i,j)) diff {j^Pi(i,k),i^Pi(j,k)});
    p := L11
!(i,k,k^Pi(i,j),l,i^s)(j^Pi(i,k),i^(s*Pi(i,j)),i^Pi(j,k),j,k^s);
       return p, s;
end function;
```

```
/*------------------------------------------------------------------
---------
For ss a sequence representing a permutation of N followed by a word
in the
symmetric generators, return an equivalent sequence of canonically
shortest length.
*/
Canon := function(ss)

  s := ss;

  // Step I.
  if exists(i){ i : i in [12..#s-1] | s[i] eq s[i+1] } then
    s := $$( s[1..i-1] cat s[i+2..#s] );
```

```
     end if;

  // Step II.
  if exists(i){ i : i in [12..#s-2] | s[i] eq s[i+2] } then
     s := $$( [ p[s[k]] : k in [1..i-1] ] cat s[i..i+1] cat
s[i+3..#s]
            where p is Eltseq(Pi(s[i], s[i+1])) );
  end if;

  // Step III.
  if exists(i){ i : i in [12..#s-2] | Index(trips,
{s[i],s[i+1],s[i+2]}) ne 0 } then
     s := $$([ q[s[k]] : k in [1..i-1] ] cat [s[i+1],s[i]] cat
s[i+3..#s])
            where q is Eltseq(Sg(s[i], s[i+1], s[i+2]));
  end if;

  // Step IV.
  if #s ge 15 and exists(m,i,p,q){<m,i,p,q> : m in [1..5], i in
[12..#s-3] |
            s[i+3] in { s[i+1]^(q^m), s[i+2]^(q^m) }
            where q, p is Ntrip(s[i], s[i+1], s[i+2]) } then
     s := $$( [ t[s[k]] : k in [1..11] ] cat [ s[k]^r : k in
[12..i-1] ]
            cat [ s[i-1+k]^(q^m) : k in [1..3]] cat s[i+3..#s]
)
            where t is Eltseq(r) where r is p^(-1)*p^(q^m);
  end if;

  // Step V.
  if #s ge 16 then
     s := $$([s[l[i]] : i in [1..11]] cat [s[15],s[12],s[13]] cat
s[17..#s])
            where l is
Eltseq(Sg(s[15],s[13],s[15]^Pi(s[12],s[13])));
  end if;

  return s;

end function;


/*------------------------------------------------------------
-----------
Return the product of two symmetrically represented elements of
J1
*/
Prod := function( x, u, y, v)
   t := Canon(Unify(x, u, y, v));
   return L11!t[1..11], t[12..#t];
```

```
end function;

/*-------------------------------------------------------------
-----------
Return the inverse of a symmetrically represented element of J1
*/
Invert := func< x, u | x^-1, [u[#u-i+1]^(x^-1) : i in [1..#u]] >;
```

Part III

Non-involutory symmetric generators

6

The (non-involutory) progenitor

6.1   Monomial automorphisms

Up until now, our progenitors have all been split extensions of groups in which the normal subgroup is a free product of cyclic groups of order 2. The complement acts on this free product by permuting a set of involutory generators of the cyclic subgroups. However, if the cyclic subgroups had order greater than 2, the normal subgroups would have additional automorphisms which, say, fix all but one of the symmetric generators and raise the remaining generator to a power of itself coprime to its order. If the symmetric generators have order m, so that the normal subgroup has shape $m^{\star n}$, then there are $\phi(m)$ such automorphisms for each generator, where $\phi(m)$ denotes the number of positive integers less than m and coprime to it. Automorphisms which permute the symmetric generators and raise them to such coprime powers are called *monomial automorphisms*; the reason for this nomenclature will become apparent in the following sections. It is clear that M, the *group of all monomial automorphisms* of the group $m^{\star n}$, has order $\phi(m)^n n!$, where $\phi(m)$ denotes the number of positive integers less than m and coprime to it. In practice, our generalized progenitors will have shape

$$m^{\star n} :_m N,$$

where $N \leq M$ acts transitively on the n cyclic subgroups and the subscript m on the colon indicates that the action is genuinely monomial and does not simply permute a set of generators.

As an example, let us consider a set of seven 3-cycles defined on the set $\Lambda = \{0, 1, \ldots, 6\}$ given by

$$\mathcal{T} = \frac{(1\,2\,4)\ (0\,1\,3)\ (6\,0\,2)\ (5\,6\,1)\ (4\,5\,0)\ (3\,4\,6)\ (2\,3\,5)}{t_0 \qquad t_1 \qquad t_2 \qquad t_3 \qquad t_4 \qquad t_5 \qquad t_6}.$$

The seven cyclic groups of order 3 generated by these elements are clearly in one-to-one correspondence with the lines of the Fano plane as labelled in Figure 5.15 (p. 163), and so the group $L_3(2)$ permutes them by conjugation. Indeed we see that the action is given by

$$a := (0\ 1\ 2\ 3\ 4\ 5\ 6) : (t_6\ t_5\ t_4\ t_3\ t_2\ t_1\ t_0)$$

and

$$b := (2\ 6)(4\ 5) : (t_0\ t_3^{-1})(t_1)(t_2\ t_2^{-1})(t_4\ t_4^{-1})(t_5\ t_6).$$

We may conveniently display these automorphisms as matrices in which the columns correspond to the symmetric generators (with t_0 in the seventh position). We place a 1 in the ijth position if the automorphism maps t_i to t_j and a -1 in the ijth position if it maps t_i to t_j^{-1}. Clearly there will be just one non-zero entry in each row and one in each column. Thus,

$$a \sim \begin{bmatrix} \cdot & \cdot & \cdot & \cdot & \cdot & 1 \\ 1 & \cdot & \cdot & \cdot & \cdot & \cdot \\ \cdot & 1 & \cdot & \cdot & \cdot & \cdot \\ \cdot & \cdot & 1 & \cdot & \cdot & \cdot \\ \cdot & \cdot & \cdot & 1 & \cdot & \cdot \\ \cdot & \cdot & \cdot & \cdot & 1 & \cdot \\ \cdot & \cdot & \cdot & \cdot & \cdot & 1 & \cdot \end{bmatrix}, \quad b \sim \begin{bmatrix} 1 & \cdot & \cdot & \cdot & \cdot & \cdot \\ \cdot & -1 & \cdot & \cdot & \cdot & \cdot \\ \cdot & \cdot & \cdot & \cdot & \cdot & -1 \\ \cdot & \cdot & \cdot & -1 & \cdot & \cdot \\ \cdot & \cdot & \cdot & \cdot & 1 & \cdot \\ \cdot & \cdot & \cdot & \cdot & 1 & \cdot \\ \cdot & -1 & \cdot & \cdot & \cdot & \cdot \end{bmatrix}.$$

A matrix in which there is precisely one non-zero term in each row and one in each column is said to be *monomial*. Thus a monomial matrix is a product of a permutation matrix and a non-singular diagonal matrix. Of course these seven 3-cycles do not generate a free product of cyclic groups but are readily seen to generate the alternating group A_7. Even so, the action defined by the above matrices may be interpreted as automorphisms of the free product $3^{\star 7}$, and so we see that there is a homomorphism

$$3^{\star 7} :_m L_3(2) \to A_7.$$

We obtained this monomial action of the control subgroup $L_3(2)$ by considering what happens in a homomorphic image of the progenitor. In Section 6.2 we shall see how to go directly from the control subgroup to its possible monomial actions.

6.2 Monomial representations

A *monomial representation* of a group G is a homomorphism from G into $GL_n(F)$, the group of non-singular $n \times n$ matrices over the field F, in which the image of every element of G is a monomial matrix over F. Thus the action of the image of a monomial representation on the underlying vector space is to permute the vectors of a basis while multiplying them by scalars. Every monomial representation of G in which G acts transitively on the 1-dimensional subspaces generated by the basis vectors is obtained by inducing a linear representation of a subgroup H up to G. If this linear representation is trivial, we obtain the permutation representation of G acting on the cosets of H. Otherwise we obtain a proper monomial representation. Now, an ordinary linear representation of H is a homomorphism of H onto C_m, say, a cyclic multiplicative subgroup of the complex numbers $\mathbb{C}$, and the resulting monomial matrices will involve complex mth roots of unity. But we can similarly define a linear representation into any finite field F which possesses mth roots of unity.

EXAMPLE **6.1** A 3-modular monomial representation of $2^{\cdot}S_4^+$, one of the double covers of the symmetric group S_4.

Consider the double cover of S_4, the symmetric group on four letters, in which the transpositions lift to involutions. This group G, which is isomorphic to the general linear group $GL_2(3)$ and whose ordinary character table is displayed in Table 6.1, is often denoted by $2^{\cdot}S_4^+$. The isoclinic group, see the ATLAS [25], p. xxiii, in which transpositions lift to elements of order 4, is denoted by $2^{\cdot}S_4^-$. Now, a subgroup H of index 4 in G is isomorphic to $2 \times S_3$. One of the three non-trivial linear representations of H maps the involution in its centre to 1, and so the resulting monomial representation is not faithful. The other two map this involution to -1 and give rise to equivalent monomial representations of degree 4, whose character is given by the bottom line of Table 6.1.[1]

---

[1] The character tables given in Tables 6.1 and 6.2 are displayed in ATLAS format. If a group G possesses an involution, z say, in its centre, then we may factor out the subgroup $\langle z \rangle$ to obtain $G/\langle z \rangle \cong H$. We write this as $G \cong 2.H$, where the lower 'dot' indicates that we are not specifying whether this is a split or non-split extension. Any element of H has two pre-images in G, with x and xz mapping to the same element of H for x any element of G. The elements x and xz may or may not be conjugate in G; or, to put it another way, a conjugacy class of H may be the image of one conjugacy class of G or of two. Of course, if one of x and xz has odd order, then the other has even order and they cannot be conjugate. Otherwise x and xz have the same (even) order, but they may still be non-conjugate. Table 6.1 displays the character table of H and indicates in the central double row whether the class corresponding to a column has one or two pre-images. The integers displayed give the orders of pre-images of elements in that class of H and, if a single integer is displayed in a particular column, then it means that all pre-images of elements in that class of H are conjugate. Thus the character tables of $2^{\cdot}S_4^+$ and $2^{\cdot}S_4^-$ (Table 6.1) both exhibit the 5×5 character table of S_4, and both groups have eight conjugacy classes. Moreover, in both cases the pre-images of elements in class $2A$ have order 4 and form one class in the double cover. However, in the first case pre-images of elements in class $2B$ have order 2 and in the second case they have order 4; in both cases they are all conjugate to one another. In both groups, if x is a pre-image of an element of S_4 of order 4, then x has order 8 and is not conjugate to xz. The faithful characters are given below the central double row. The character value given for each of these is the value on the upper class in the case when there are two classes of pre-images, with the value on the other class being the negative of the value indicated. Thus the first faithful character of degree 2 of $2^{\cdot}S_4^+$ written in full has the following form:

48	48	8	6	6	4	8	8
1X	2X	4X	3X	6X	2Y	8X	8Y
2	-2	0	-1	1	0	$i\sqrt{2}$	$-i\sqrt{2}$

Table 6.1. Character tables of $2\dot{}S_4^+$ and $2\dot{}S_4^-$

	$2\dot{}S_4^+$						$2\dot{}S_4^-$				
	24	8	3	4	4		24	8	3	4	4
	1A	2A	3A	2B	4A		1A	2A	3A	2B	4A
+	1	1	1	1	1	+	1	1	1	1	1
+	1	1	1	−1	−1	+	1	1	1	−1	−1
+	2	2	−1	0	0	+	2	2	−1	0	0
+	3	−1	0	1	−1	+	3	−1	0	1	−1
+	3	−1	0	−1	1	+	3	−1	0	−1	1
	1	4	3	2	8		1	4	3	4	8
	2		6		8		2		6		8
o	2	0	−1	0	$i\sqrt{2}$	−	2	0	−1	0	$-\sqrt{2}$
o	2	0	−1	0	$-i\sqrt{2}$	−	2	0	−1	0	$-\sqrt{2}$
+	4	0	1	0	0	−	4	0	1	0	0

Now, G may be shown to have the following presentation:

$$\langle x, y \mid y^3 = (xy)^4, x^2 = 1\rangle,$$

and the aforementioned representation is generated by

$$x = \begin{bmatrix} \cdot & 1 & \cdot & \cdot \\ 1 & \cdot & \cdot & \cdot \\ \cdot & \cdot & 1 & \cdot \\ \cdot & \cdot & \cdot & -1 \end{bmatrix}, \qquad y = \begin{bmatrix} -1 & \cdot & \cdot & \cdot \\ \cdot & \cdot & -1 & \cdot \\ \cdot & \cdot & \cdot & -1 \\ \cdot & -1 & \cdot & \cdot \end{bmatrix}.$$

Clearly these matrices will give a faithful representation over any field of characteristic other than 2; and so, in particular, they give a 3-modular monomial representation of the group.

EXAMPLE **6.2** A 5-modular monomial representation of $2\dot{}S_4^-$.

The unique subgroup of index 4 in $2\dot{}S_4^-$ has shape $2\dot{}S_3 \cong 3 : 4$; so it is generated by an element of order 3 and an element of order 4 inverting it. This group factored by its derived group is thus isomorphic to C_4 and so has a linear representation onto the subgroup of $\mathbb{C}$ generated by i, the square root of -1. Inducing up this linear representation produces an irreducible 4-dimensional monomial representation of $2\dot{}S_4^-$ whose character is given in the bottom row of the right hand side of Table 6.1. Now, a presentation of $2\dot{}S_4^-$ is given by

$$\langle x, y \mid x^4 = y^3 = (xy)^2\rangle,$$

and generating matrices over $\mathbb{C}$ are given by

$$
x := \begin{bmatrix} \cdot & 1 & \cdot & \cdot \\ \cdot & \cdot & 1 & \cdot \\ \cdot & \cdot & \cdot & 1 \\ -1 & \cdot & \cdot & \cdot \end{bmatrix}, \qquad y := \begin{bmatrix} \cdot & \cdot & -1 & \cdot \\ i & \cdot & \cdot & \cdot \\ \cdot & -i & \cdot & \cdot \\ \cdot & \cdot & \cdot & -1 \end{bmatrix}.
$$

But, of course, this representation may be written over any field which possesses fourth roots of unity, such as $\mathbb{Z}_5$, the field of integers modulo 5. So we replace i and $-$i in the above matrices by 2 and 3, respectively (or, indeed, by 3 and 2, respectively[2]), to obtain a 5-modular monomial representation of $2 \dot{} S_4^-$.

EXAMPLE **6.3** A 5-modular monomial representation of $SL_2(5)$.
  Consider now $G \cong 2\dot{}A_5 \cong SL_2(5)$, the double cover of the alternating group on five letters whose character table is displayed in Table 6.2.[3] A subgroup of index 5 has shape $2\dot{}A_4$ and its derived group contains its centre. Thus any linear representation would have the centre in its kernel, and the resulting monomial representation of G would not be faithful. A subgroup H of index 6, however, has $H/H' \cong C_4$ and $H' \cong C_5$. To obtain a faithful monomial representation, we must induce up a linear representation of H which maps a generator of this cyclic group of order 4 onto a primitive fourth root of unity in our field. A presentation for this group is given by

$$
\langle x, y \mid x^5 = y^4 = (xy)^3 = 1 = [x, y^2] \rangle,
$$

and over $\mathbb{C}$ the representation is given by

$$
x = \begin{bmatrix} 1 & \cdot & \cdot & \cdot & \cdot & \cdot \\ \cdot & \cdot & 1 & \cdot & \cdot & \cdot \\ \cdot & \cdot & \cdot & 1 & \cdot & \cdot \\ \cdot & \cdot & \cdot & \cdot & 1 & \cdot \\ \cdot & \cdot & \cdot & \cdot & \cdot & 1 \\ \cdot & 1 & \cdot & \cdot & \cdot & \cdot \end{bmatrix}, \qquad y = \begin{bmatrix} \cdot & 1 & \cdot & \cdot & \cdot & \cdot \\ -1 & \cdot & \cdot & \cdot & \cdot & \cdot \\ \cdot & \cdot & \cdot & \cdot & \cdot & -1 \\ \cdot & \cdot & \cdot & \cdot & i & \cdot \\ \cdot & \cdot & \cdot & -i & \cdot & \cdot \\ \cdot & \cdot & 1 & \cdot & \cdot & \cdot \end{bmatrix}.
$$

This complex representation is irreducible, and its character is given by the bottom line of Table 6.2. Now, the smallest finite field which contains primitive fourth roots of unity is $GF_5 \cong \mathbb{Z}_5$, the integers modulo 5.

---

[2] As will become apparent in the rest of this chapter, these two possibilities can lead to very different results.
  [3] The irrationality in Table 6.2 is given in ATLAS notation, where $b5$ denotes $(-1+\sqrt{5})/2$ and $b5\star$ denotes $(-1-\sqrt{5})/2$.

Table 6.2. Character table of $\mathrm{SL}_2(5)$

	$\mathrm{SL}_2(5)$				
	60	4	3	5	5
	1A	2A	3A	5A	B⋆
+	1	1	1	1	1
+	3	−1	0	−b5	⋆
+	3	−1	0	⋆	−b5
+	4	0	1	−1	−1
+	5	1	−1	0	0
	1	4	3	5	5
	2		6	10	10
−	2	0	−1	b5	⋆
−	2	0	−1	⋆	b5
−	4	0	1	−1	−1
−	6	0	0	1	1

Thus, a faithful 5-modular representation of G is obtained by replacing i by 2 and −i by 3 in the matrix representing y. (Or, indeed, i by 3 and −i by 2.)

EXAMPLE **6.4** A 7-modular monomial representation of the triple cover of the alternating group A_7.

As in our last example of a monomial representation, we take $G \cong 3^{\cdot}\mathrm{A}_7$, the triple cover of the alternating group on seven letters whose character table is displayed in Table 6.3. We seek a *faithful* monomial representation of this group. Now, a subgoup of index 7 in G is isomorphic to $3^{\cdot}\mathrm{A}_6$, which, being perfect, has the centre of G in the kernel of its unique linear representation; thus G has a unique 7-dimensional monomial representation, which is the permutation representation lifted from A_7. However, G possesses two classes of subgroups of index 15 which are isomorphic to $3 \times \mathrm{L}_2(7)$ and which are fused by the outer automorphism. Such a subgroup H possesses linear representations onto C_3, the group of complex cube roots of unity. Inducing such representations up to G, we obtain two faithful monomial representations of G of degree 15, whose characters are the second pair of faithful irreducible characters of degree 15 in Table 6.3.

In order to construct this representation explicitly, we take the following presentation of G:

$$\langle x, y \mid x^7 = y^4 = (xy^2)^3 = (x^3y)^3 = (yx)^5(xyx^2y^{-1})^2 = 1\rangle \cong 3^{\cdot}\mathrm{A}_7.$$

Coset enumeration over $\langle x \rangle$ verifies the order, the group is visibly perfect (abelianizing reduces it to the trivial group), and the homomorphism

Table 6.3. Character table of $3\,^{\cdot}A_7$

	3˙A₇								
	2520	24	36	9	4	5	12	7	7
	1A	2A	3A	3B	4A	5A	6A	7A	B⋆⋆
+	1	1	1	1	1	1	1	1	1
+	6	2	3	0	0	1	−1	−1	−1
o	10	−2	1	1	0	0	1	b7	⋆⋆
o	10	−2	1	1	0	0	1	⋆⋆	b7
+	14	2	2	−1	0	−1	2	0	0
+	14	2	−1	2	0	−1	−1	0	0
+	15	−1	3	0	−1	0	−1	1	1
+	21	1	−3	0	−1	1	1	0	0
+	35	−1	−1	−1	1	0	−1	0	0
	1	2	3	3	4	5	6	7	7
	3	6			12	15	6	21	21
	3	6			12	15	6	21	21
o2	6	2	0	0	0	1	2	−1	−1
o2	15	−1	0	0	−1	0	2	1	1
o2	15	3	0	0	1	0	0	1	1
o2	21	1	0	0	−1	1	−2	0	0
o2	21	−3	0	0	1	1	0	0	0
o2	24	0	0	0	0	−1	0	b7	⋆⋆
o2	24	0	0	0	0	−1	0	⋆⋆	b7

The format for the character table of a triple cover follows the same principle as for
a double cover, only now each element of the image (in this case A_7) has three pre-
images and a conjugacy class has either one or three pre-images. When a class has
three pre-images, a faithful character takes the value given on the class corresponding
to the topmost of the three classes indicated in the central triple row. If this value is
α, then our faithful character takes the value $\alpha\omega$ on the middle class and $\alpha\bar{\omega}$ on the
bottom class, where ω is a complex cube root of unity. Finally note that the symbol
o2 indicates that each row below the central triple row stands for two characters,
the one just described and the one obtained by interchanging ω and $\bar{\omega}$. Thus our
table for $3\,^{\cdot}A_7$ has $2+3\times 7 = 23$ classes and $9+2\times 7 = 23$ characters. Note that $b7$
denotes $(-1+\sqrt{7}i)/2$ and $b7\star\star$ denotes its complex conjugate. (In general, bp denotes
$(-1+\sqrt{\pm p})/2$ depending on whether $p \equiv \pm 1$ modulo 4.)

$x \mapsto (1\,2\,3\,4\,5\,6\,7)$, $y \mapsto (1\,7\,2\,4)(5\,6)$ demonstrates that it has A_7 as an
image. It is particularly suitable for our purposes as $\langle x, y^2 \rangle \cong L_2(7)$, the subgroup in which we are interested. Our elements x and y may be represented
by the following 15×15 complex matrices:

Of course, such a matrix may, as mentioned previously, be written as a
non-singular diagonal matrix followed by a permutation matrix:

$$x = (I_{15}, (1, 2, 3, 4, 5, 6, 7)(8, 9, 10, 11, 12, 13, 14)(15)),$$
$$y = (\text{diag}(\omega, \omega, 1, \bar{\omega}, \bar{\omega}, 1, \bar{\omega}, \bar{\omega}, \omega, \bar{\omega}, 1, 1, \omega, 1, \omega),$$
$$(1, 11, 8, 12)(2, 6, 13, 9)(4, 10, 5, 3)(7, 15)(14)).$$

For our purposes, we require a finite field with cube roots of unity. If we insist on a prime field then we may take $\mathbb{Z}_7$, the integers modulo 7, and replace ω and $\bar{\omega}$ by 2 and 4, respectively (or by 4 and 2); or we may simply interpret ω and $\bar{\omega}$ as elements in the Galois field of order 4 in the usual way.

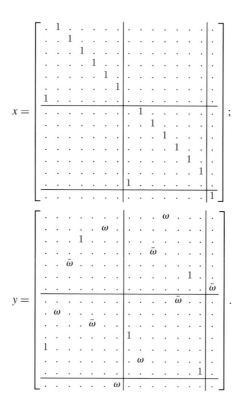

6.3   Monomial action of a control subgroup

We have seen that monomial automorphisms of a free product of cyclic groups can be exhibited as monomial matrices. Equally, though, a monomial matrix of degree n over a prime field $\mathbb{Z}_p$, say, defines a monomial automorphism of the free product $p^{\star n}$. For, if the unique non-zero entry in the ith row is a_{ij}, then we interpret this as saying that the corresponding automorphism maps t_i to $t_j^{a_{ij}}$, where

$$p^{\star n} = \langle t_1 \rangle \star \langle t_2 \rangle \cdots \langle t_n \rangle.$$

So an n-dimensional p-modular monomial representation of the group N tells us how N can act as control subgroup in a progenitor of shape

$$p^{\star n} :_m N.$$

EXAMPLE **6.5** A progenitor with symmetric generators of order 3.

Our monomial representation over $\mathbb{Z}_3$ in Example 6.1 shows how the group $2^{\cdot}S_4^+$ can act faithfully as control subgroup in the progenitor

$$P \cong 3^{*4} :_m 2^{\cdot}S_4^+.$$

To obtain a presentation for this group we must adjoin an element of order 3, t_1 say, to the control subgroup $N \cong 2^{\cdot}S_4^+$ such that $T_1 = \langle t_1 \rangle$ has just four images under its action. Specifically we require

$$x \sim (t_1 \ t_2)(t_3)(t_4 \ t_4^{-1}), \quad y \sim (t_1 \ t_1^{-1})(t_2 \ t_3^{-1} \ t_4 \ t_2^{-1} \ t_3 \ t_4^{-1}),$$

under conjugation, or, more concisely,

$$x \sim (1 \ 2)(3)(4 \ \bar{4}), \quad y \sim (1 \ \bar{1})(2 \ \bar{3} \ 4 \ \bar{2} \ 3 \ \bar{4}),$$

where in this abbreviated notation i is standing for t_i and $\bar{i}$ is standing for t_i^{-1}. This is easily seen to be achieved by

$$\langle x, y, t \mid y^3 = (xy)^4, x^2 = 1 = t^3 = t^y t = [t^{xy}, x] \rangle \cong 3^{*4} :_m 2^{\cdot}S_4^+,$$

since $N_N(\langle t_0 \rangle) = \langle y, x^{y^{-1}x} \rangle \cong 2 \times S_3$.

EXAMPLE **6.6** The group $2^{\cdot}S_4^-$ and a progenitor with symmetric generators of order 5.

The 5-modular representation of $2^{\cdot}S_4^-$, obtained from the matrices in Example 6.2 by replacing i by 2 and $-$i by 3, defines the progenitor

$$P \cong 5^{*4} :_m 2^{\cdot}S_4^-,$$

in which we denote the four symmetric generators by $\{t_1, t_2, t_3, t_4\}$. The elements x and y act on them by conjugation according to

$$x \sim (t_1 \ t_2 \ t_3 \ t_4 \ t_1^{-1} \ t_2^{-1} \ t_3^{-1} \ t_4^{-1}),$$
$$y \sim (t_1 \ t_3^{-1} \ t_2^2 \ t_1^{-1} \ t_3 \ t_2^{-2})(t_4 \ t_4^{-1}),$$

which we often abbreviate as follows:

$$x \sim (1 \ 2 \ 3 \ 4 \ 1^{-1} \ 2^{-1} \ 3^{-1} \ 4^{-1}), \quad y \sim (1 \ 3^{-1} \ 2^2 \ 1^{-1} \ 3 \ 2^{-2})(4 \ 4^{-1}).$$

Note that we have only given the action on half the powers of the symmetric generators, as the rest follow. The subgroup of $N \cong 2^{\cdot}S_4^-$ which normalizes $\langle t_1 \rangle$ is given by

$$N_N(\langle t_1 \rangle) = \langle xy \sim (1 \ 1^2 \ 1^{-1} \ 1^{-2})(2 \ 2^{-2} \ 2^{-1} \ 2^2)(3 \ 4^{-1} \ 3^{-1} \ 4),$$
$$y^x \sim (1 \ 1^{-1})(2 \ 4^{-1} \ 3^2 \ 2^{-1} \ 4 \ 3^{-2}) \rangle$$
$$\cong 3 : 4,$$

and so a presentation for the progenitor is given by

$$P = \langle x, y, t \mid x^4 = y^3 = (xy)^2, t^5 = t^{xy} t^{-2} = t^{y^x} t = 1 \rangle,$$

where t stands for the symmetric generator t_1.

Table 6.4. Character table of $SL_2(7)$

			SL$_2$(7)			
	168	8	3	4	7	7
	1A	2A	3A	4A	7A	B★★
+	1	1	1	1	1	1
o	3	−1	0	1	b7	★★
o	3	−1	0	1	★★	b7
+	6	2	0	0	−1	−1
+	7	−1	1	−1	0	0
+	8	0	−1	0	1	1
	1	4	3	8	7	7
	2		6	8	14	14
o	4	0	1	0	−b7	★★
o	4	0	1	0	★★	−b7
−	6	0	0	r2	−1	−1
−	6	0	0	−r2	−1	−1
−	8	0	−1	0	1	1

EXAMPLE 6.7 $SL_2(7)$ and a progenitor with symmetric generators of order 3.

The group $N \cong SL_2(7)$ contains a subgroup of shape $2 \times 7{:}3$ to index 8. We could map this subgroup onto a cyclic group of order 6 and thus obtain an 8-dimensional monomial representation of the group involving sixth roots of unity. However, we prefer to map the subgroup onto the cyclic group of order 2 and so obtain an 8-dimensional monomial representation of N whose non-zero entries are ± 1. This time we do not obtain an irreducible representation, but the sum of the two simple characters of degree 4 in Table 6.4.

Not surprisingly we choose to read this representation module 3, thus allowing ourselves to define a progenitor of shape

$$P \cong 3^{*8} :_m SL_2(7).$$

Rather than constructing matrices generating this representation of N, we prefer to go straight to the action of our control subgroup on the eight symmetric generators. In order to do this we first recall that the simple group $L_2(7)$ is generated by the two permutations $x \sim (\infty)(0\ 1\ 2\ 3\ 4\ 5\ 6)$ and $y \sim (\infty\ 0)(1\ 6)(2\ 3)(4\ 5)$, which satisfy the following presentation:

$$L_2(7) = \langle x, y \mid x^7 = y^2 = (xy)^3 = [x, y]^4 = 1 \rangle.$$

Involutions in $L_2(7)$ lift to elements of order 4 in $SL_2(7)$, and so a pre-image of y must square to the element of the control subgroup which inverts all the symmetric generators. Indeed, relaxing the above presentation so that each of the words mentioned is equal to the same central element, we have

$$\mathrm{SL}_2(7) = \langle x, y \mid x^7 = y^2 = (xy)^3 = [x, y]^4 \rangle. \tag{6.1}$$

By replacing a symmetric generator by its inverse if necessary, we may assume that a pre-image of x, $\hat{x}$ say, acts on the symmetric generators by conjugation as $\hat{x} \sim (t_\infty)(t_0\ t_1\ t_2\ t_3\ t_4\ t_5\ t_6)$ and that a pre-image of y, $\hat{y}$ say, acts as

$$\hat{y} \sim (t_\infty\ t_0^{-1}\ t_\infty^{-1}\ t_0)(t_1\ t_6^a\ t_1^{-1}\ t_6^{-a})(t_2\ t_3^b\ t_2^{-1}\ t_3^{-b})(t_4\ t_5^c\ t_4^{-1}\ t_6^{-c}),$$

where $a, b, c = \pm 1$. But, replacing $\hat{y}$ by its inverse if necessary, we may assume that $\hat{x}\hat{y}$ has order 3; this condition determines that $a = b = c = -1$, and the progenitor P is uniquely defined. In abbreviated notation, where i stands for t_i and $\bar{i}$ stands for t_i^{-1}, we thus have

$$\hat{x} \sim (0\ 1\ 2\ 3\ 4\ 5\ 6)(\infty), \quad \hat{y} \sim (\infty\ \bar{0}\ \bar\infty\ 0)(1\ \bar{6}\ \bar{1}\ 6)(2\ \bar{3}\ \bar{2}\ 3)(4\ \bar{5}\ \bar{4}\ 5).$$

If the central element of order 2 which inverts all the symmetric generators is denoted by $\hat{z}$, then it is readily checked that $\hat{x}\hat{z}$ and $\hat{y}$ satisfy the presentation in Equation (6.1).

EXAMPLE 6.8 $\mathrm{SL}_2(5)$ and a progenitor with symmetric generators of order 5.

The 5-modular representation of $\mathrm{SL}_2(5)$, obtained from the matrices in Example 6.3 by replacing i by 2 and $-$i by 3, defines the progenitor

$$P \cong 5^{*6} :_m \mathrm{SL}_2(5),$$

in which

$$x \sim (t_\infty)(t_0\ t_1\ t_2\ t_3\ t_4),$$
$$y \sim (t_\infty\ t_0\ t_\infty^{-1}\ t_0^{-1})(t_1\ t_4^{-1}\ t_1^{-1}\ t_4)$$
$$(t_2\ t_2^2\ t_2^4\ t_2^3)(t_3\ t_3^3\ t_3^4\ t_3^2).$$

We prefer to extend this to a progenitor:

$$P \cong 5^{*(6+6)} :_m (\mathrm{SL}_2(5) : 2) \cong 5^{*(6+6)} :_m 2^{\cdot}\mathrm{S}_5$$

by adjoining the outer automorphism of the control subgroup. The associated monomial representation has degree $(6+6) = 12$, with the two 6-dimensional subspaces invariant under $2^{\cdot}\mathrm{A}_5$ being interchanged by the outer automorphism. As presentation for the extended control subgroup, we take

$$2^{\cdot}\mathrm{S}_5 \cong \langle x, y \mid x^5 = y^4 = (xy^2)^3, (x^3y)^2 = 1 \rangle.$$

To obtain the progenitor from this we must adjoin an element r of order 5 which is normalized by a subgroup of the control subgroup of order 20, as described in Example 6.3:

$$P = 5^{\star\{6+6\}} :_m 2\text{`}S_5 \cong \langle x, y, r \mid x^5 = y^4 = (xy^2)^3, (x^3y)^2 = 1 = r^5 = r^x r,$$
$$r^{y^2x^2y^2} = (r^{y^2x^2})^3 \rangle.$$

This follows from the fact that $\langle x, (y^2x^2)y^2(y^2x^2)^{-1} \rangle$ is isomorphic to such a subgroup of order 20. If we let the two sets of six symmetric generators be $\mathcal{R} = \{r_\infty, r_0, \ldots, r_4\}$ and $\mathcal{S} = \{s_\infty, s_0, \ldots, s_4\}$, then the action on these satisfying the above presentation may be given by

$$x^2 = (r_\infty)(r_0\ r_1\ \cdots\ r_4)(s_\infty)(s_0\ s_1\ \cdots\ s_4),$$

$$y = (r_\infty\ s_2\ r_4\ s_1^{-1}\ \cdots)(r_0\ s_0^2\ r_0^{-2}\ s_0\ \cdots)$$
$$(r_1\ s_\infty^{-2}\ r_2\ s_4^{-2}\ \cdots)(r_3\ s_3\ r_3^2\ s_3^2\ \cdots),$$

where $x^5 = y^4$ inverts all the symmetric generators, and r corresponds to the symmetric generator r_∞. It is clear that x inverts $r = r_\infty$; moreover, $r_\infty^{y^2x^2} = r_0$ and $r_0^{y^2} = r_0^3$, confirming that a subgroup of order 20 normalises $\langle r \rangle$. It should be noted that $[r_\infty, x] = r_\infty^{-1}r_\infty^{-1} = r_\infty^{-2}$, and so the derived group of the progenitor contains all the symmetric generators. It also contains the derived group of the control subgroup and so has shape $5^{\star(6+6)} :_m \text{SL}_2(5)$. By a similar argument this group itself is perfect, and so any homomorphic image of P contains a perfect subgroup to index at most 2.

EXAMPLE 6.9 A progenitor with symmetric generators of order 7.

The monomial modular representation obtained in Example 6.4 defines the progenitor

$$P \cong 7^{\star 15} :_m 3\text{`}A_7$$

in the same way. As in Example 6.8, however, we prefer to extend the control subgroup to $3\text{`}S_7$. Note that this group does not possess a subgroup of index 15 as the two classes of subgroups isomorphic to $L_3(2)$ are fused by the outer automorphism; see p. 3 of the ATLAS [25]. Recall, moreover, that the outer automorphism of $3\text{`}A_7$ inverts the central elements of order 3, and so in the monomial representation obtained by inducing a non-trivial linear representation of $3 \times L_3(2)$ up to $3\text{`}S_7$ such a 'central' element will be represented by $\text{diag}(\omega^{15}, \bar{\omega}^{15})$ or its conjugate. Translating this into the language of progenitors, in

$$7^{\star(15+15)} :_m (3\text{`}S_7)$$

a 'central' element of order 3 will, by conjugation, square one set of 15 symmetric generators while fourth-powering the other. In keeping with the theme of this book, we choose to take the symmetric presentation for the control subgroup $3\text{`}S_7$, which was given in Theorem 4.3:

$$G \cong \frac{2^{\star7} : L_3(2)}{[(0,1,2,3,4,5,6)t_0]^6} \cong 3\text{`}S_7. \tag{6.2}$$

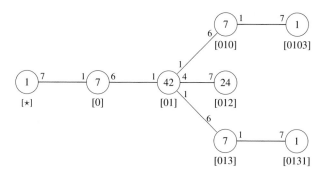

Figure 6.1. The Cayley diagram of $3\dot{}S_7$ over $L_3(2)$.

From Figure 6.1 we see that every element of N can be expressed, not necessarily uniquely, as a permutation of $L_3(2)$ acting on seven letters, followed by a word of length less than or equal to 4 in the seven symmetric generators. Now, when the normal subgroup of order 3 is factored out, the three single point orbits in Figure 6.1 are fused to give suborbits of lengths $1+7+14+8$, and so we can deduce that the 'central' elements of order 3 lie in the double cosets $[0103]$ $(= Nt_0t_1t_0t_3N)$ and $[0131]$. The images of the t_i in S_7 are given by $t_0 = (1\ 3)(2\ 6)(4\ 5)$, as above, and $t_i = t_0^{x^i}$, and so we find that $t_0t_1t_0t_3 \equiv (2\ 6)(4\ 5)$ modulo $Z(N')$. Thus, with $z = (2\ 6)(4\ 5)t_0t_1t_0t_3$, we have

$$Z(N') = \langle z \rangle = \langle x^3yx^{-2}t^{xt}t^{x^3} \rangle.$$

To obtain a presentation for the progenitor

$$7^{\star(15+15)} :_m (3\dot{}S_7),$$

we must adjoin an element s of order 7, which is centralized by our $N \cong L_3(2)$ and mapped to its square under conjugation by a central element of N'. This is achieved by adjoining s with

$$s^7 = [s, x] = [s, y] = s^{tx^2txtx^{-1}t}s^3 = 1,$$

where the monomial action of x, y and t is given by

$x = (I_{30}, (1\ 2\ 3\ 4\ 5\ 6\ 7)(8\ 9\ 10\ 11\ 12\ 13\ 14)(15)$
$\quad (16\ 17\ 18\ 19\ 20\ 21\ 22)(23\ 24\ 25\ 26\ 27\ 28\ 29)(30));$

$y = (\mathrm{diag}(1, 1, 1, 1, 1, 1, 1, 4, 2, 2, 4, 1, 1, 1, 1, 2, 4, 4, 1, 4, 2, 4, 4, 2, 2, 1, 2, 4, 2, 1),$
$\quad (1\ 6\ 5)(2\ 4\ 3)(7)(8)(9\ 14\ 11)(10)(12\ 13\ 15)$
$\quad (16\ 29\ 27)(17\ 26\ 25)(18\ 24\ 19)(20\ 23\ 22)(21)(28)(30));$

$t = (\mathrm{diag}(1, 4, 1, 2, 4, 2, 1, 1, 4, 2, 2, 2, 2, 4, 1, 1, 2, 4, 4, 4, 4, 2, 1, 1, 2, 4, 2, 1, 4, 1),$
$\quad (1\ 16)(2\ 22)(3\ 23)(4\ 20)(5\ 27)(6\ 29)(7\ 30)(8\ 28)$
$\quad (9\ 25)(10\ 21)(11\ 26)(12\ 19)(13\ 18)(14\ 17)(15\ 24)).$

If, as above, we let the two sets of symmetric generators be denoted by $\mathcal{R} = \{r_1, r_2, \ldots, r_{15}\}$ and $\mathcal{S} = \{s_1, s_2, \ldots, s_{15}\}$, this becomes:

$$x: (r_1\ r_2\ r_3\ r_4\ r_5\ r_6\ r_7)(r_8\ r_9\ r_{10}\ r_{11}\ r_{12}\ r_{13}\ r_{14})(r_{15})$$
$$(s_1\ s_2\ s_3\ s_4\ s_5\ s_6\ s_7)(s_8\ s_9\ s_{10}\ s_{11}\ s_{12}\ s_{13}\ s_{14})(s_{15});$$

$$y: (r_1\ r_6\ r_5)(r_2\ r_4\ r_3)(r_7)(r_8\ r_8^4\ r_8^2)(r_9\ r_{14}^2\ r_{11}^2)(r_{10}\ r_{10}^2\ r_{10}^4)(r_{12}\ r_{13}\ r_{15})$$
$$(s_1\ s_{14}^2\ s_{12}^4)(s_2\ s_{11}^4\ s_{10}^4)(s_3\ s_9^4\ s_4)(s_5\ s_8^4\ s_7^2)(s_6\ s_6^2\ s_6^4)(s_{13}\ s_{13}^4\ s_{13}^2)(s_{15});$$

$$t: (r_1\ s_1)(r_2\ s_7^4)(r_3\ s_8)(r_4\ s_5^2)(r_5\ s_{12}^4)(r_6\ s_{14}^2)(r_7\ s_{15})(r_8\ s_{13})$$
$$(r_9\ s_{10}^4)(r_{10}\ s_6^2)(r_{11}\ s_{11}^2)(r_{12}\ s_4^2)(r_{13}\ s_3^2)(r_{14}\ s_2^4)(r_{15}\ s_9);$$

where the generator s used in the presentation corresponds to s_{15}. It is worth stressing that our progenitor P is a uniquely defined group whose structure depends only on the control subgroup $3^{\cdot}S_7$. We may observe further that since, for instance, $[r_8, y] = r_8^{-1}r_8^4 = r_8^3$, the derived group of P contains the symmetric generators. It also contains the derived group of the control subgroup, and so is isomorphic to

$$7^{\star(15+15)} :_m 3^{\cdot}A_7$$

of index 2 in P. By a similar argument, the derived group of P is perfect. Moreover, the derived group of the control subgroup, which is isomorphic to $3^{\cdot}A_7$, acts non-identically on $\bar{\mathcal{R}} = \{\langle r_i \rangle \mid 1 \leq i \leq 15\}$ and $\bar{\mathcal{S}} = \{\langle s_i \rangle \mid 1 \leq i \leq 15\}$. Indeed, the subgroup $\langle x, y \rangle \cong L_2(7)$ mentioned above as commuting with $s = s_{15}$, has orbits of lengths $(1+14) + (7+8)$ on $\bar{\mathcal{R}} \cup \bar{\mathcal{S}}$, explicitly

$$\{r_1, r_2, \ldots, r_7\} \cup \{r_8, r_9, \ldots, r_{15}\} \cup \{s_1, s_2, \ldots, s_{14}\} \cup \{s_{15}\}.$$

In seeking interesting homomorphic images of P, therefore, we should consider what the images of $\langle r_1, s_{15} \rangle$, $\langle r_{15}, s_{15} \rangle$ and $\langle s_1, s_{15} \rangle$ might be.

7

Images of the progenitors in Chapter 6

As in the case of involutory generators, there are now two ways to proceed in order to find interesting homomorphic images of our progenitors. As before, every element in a progenitor can be written, essentially uniquely, in the form πw, where π is a monomial permutation and w is a reduced word in the symmetric generators. We can simply factor out the normal subgroup generated by a set of relators $\{\pi_1 w_1, \pi_2 w_2, \dots\}$, where Lemma 3.3 and other considerations may help in providing suitable relators. Alternatively, we can attempt to find a true transitive image of the progenitor in a symmetric group of designated degree. (Recall that a *true* image is one which contains an isomorphic copy of the control subgoup and in which the images of the symmetric generators are distinct and have the same orders as the original t_i.) Thus we are attempting to find true transitive permutation representations of the progenitor. Note that the word *true* has replaced *faithful* since we are seeking finite images of an infinite group.

7.1 The Mathieu group M_{11}

Let us first attempt the alternative approach of embedding the progenitor

$$3^{*4} :_m 2 \cdot S_4^+$$

in a symmetric group of low degree.

7.1.1 Preliminary images

In order to obtain a true image, we must include a faithful permutation representation of our control subgroup $N \cong 2 \cdot S_4^+$. So we must consider the action of N on the cosets of a subgroup of itself which does not contain the centre. But pre-images of non-trivial elements in the Klein fourgroup of S_4 have order 4, squaring to the central involution (the pre-image of the Klein fourgroup itself is a quaternion group), and so the subgroup we choose can only contain elements of order 3 and pre-images of transpositions. This leads us to subgroups isomorphic to S_3, and the minimal degree of a faithful permutation representation of N is 8. Care must be taken here, however, as the subgroup $2 \times S_3$ of N contains two classes of subgroups isomorphic to S_3 and they are *not* conjugate in N. The two permutation representations of degree 8 *are* conjugate to one another in S_8 and so there is essentially only one such representation to consider. However, the action on the symmetric generators will be slightly different in the two cases: we can let $y \sim (1\ \bar{1})(2\ 3\ 4\ \bar{2}\ 3\ \bar{4})$ in both cases; and we let $x \sim (1\ 2)(4\ \bar{4})$ in Case I and $x \sim (1\ 2)(3\ \bar{3})$ in Case II. The normalizer of t_1 in the control subgroup N is given by $\langle y, x^{y^{-1}x} \rangle$, where of course y acts in the same manner in both cases, but $x^{y^{-1}x}$ acts as $(2\ \bar{4})(3\ \bar{3})$ in Case I but as $(1\ \bar{1})(2\ 4)$ in Case II. So $x^{y^{-1}x}$ commutes with t_1 in Case I, but inverts it in Case II. Recall that when we write $x \sim (1\ 2)(3\ \bar{3})$ we mean that x acts by conjugation on the four symmetric generators as $(t_1\ t_2)(t_3\ t_3^{-1})$.

If we choose the eight letters to be $\{a, b, c, d, A, B, C, D\}$, then we may take the generators to be

$$x = (a\ b)(A\ B)(d\ D) \text{ and } y = (a\ A)(b\ C\ d\ B\ c\ D),$$

so that the central element y^3 is given by $(a\ A)(b\ B)(c\ C)(d\ D)$. If S_8 contains a true image of our progenitors, we must have an element of order 3 in S_8 which is inverted by y and which commutes with

$$x^{y^{-1}x} = (B\ d)(b\ D)(c\ C)$$

in Case I, and is inverted by it in Case II. Since $\langle y, x^{y^{-1}x} \rangle \cong 2 \times S_3$ has orbits of lengths 2 and 6, such a 3-element must have cycle shape $1^2.3^2$; we soon see that Case II cannot occur and the only possibility in Case I is $t_0 = (b\ c\ d)(B\ D\ C)$ or its inverse. This yields

$$
\begin{aligned}
t_0 &= (b\ c\ d)(B\ D\ C), \\
t_1 &= (a\ c\ D)(A\ d\ C), \\
t_2 &= (A\ b\ D)(a\ d\ B), \\
t_3 &= (a\ b\ C)(A\ c\ B),
\end{aligned}
$$

and we readily check that x and y conjugate these elements in the manner prescribed by the monomial representation over $\mathbb{Z}_3$. But $t_0 y^2 = (B\ C\ D)$ and x is an odd permutation, so it is easily seen that the image of our progenitor is the whole of S_8. For the sake of completeness, we ask what additional relation(s) are required to define the image, and we find that

$$\frac{3^{*4} :_m 2\,{}^{\cdot}S_4^+}{(0\ \bar{0})(1\ \bar{2}\ 3\ \bar{1}\ 2\ \bar{3}) = t_2 t_3^{-1} t_0 t_3 t_2 t_1 t_0^{-1}} \cong S_8.$$

The control subgroup $2 \dot{} S_4^+$ has (non-faithful) actions on one, two and three letters, and so it makes sense to map the progenitor into S_9, S_{10} or S_{11}. We shall consider briefly the case of degree 9 as it yields a rather interesting group; we shall then concentrate on degree 11, in which the control subgroup has orbits of lengths 8 and 3. Other low degrees are left to the interested reader.

We adjoin a letter, $\star$ say, and seek a permutation t_0 on the nine letters $\{\star, a, b, c, d, A, B, C, D\}$, which has order 3, moves $\star$, is inverted by y and commutes with or is inverted by $x^{y^{-1}x}$, where x and y act as before (and fix $\star$). Since t_0 must move $\star$ and must be inverted by y, we see that, without loss of generality, it must involve the 3-cycle $(\star\, a\, A)$. Indeed, this 3-cycle itself fulfils all the conditions, but rather obviously will lead to the symmetric group S_9. Otherwise there are just two possibilities:

$$t_0 = (\star\, a\, A)(b\, c\, d)(D\, C\, B) \text{ or } (\star\, a\, A)(d\, c\, b)(B\, C\, D).$$

The second of these can again be shown to generate the symmetric group S_9 (together with N), but the first leads to

$$
\begin{aligned}
t_0 &= (\star\, a\, A))(b\, c\, d)(B\, D\, C),\\
t_1 &= (\star\, b\, B)(a\, c\, D)(A\, d\, C),\\
t_2 &= (\star\, c\, C)(A\, b\, D)(a\, d\, B),\\
t_3 &= (\star\, d\, D)(a\, b\, C)(A\, c\, B).
\end{aligned}
$$

We note that these elements commute with one another, that $t_0 t_1 = t_2$, and that they generate an elementary abelian group of order 3^2, which must be normalized by the control subgroup. This leads to a group of shape $3^2 : 2 \dot{} S_4^+$, the *Hessian group*[1] consisting of all automorphisms of the unique Steiner system $S(2, 3, 9)$ (or *affine plane* of order 3) whose triples are the rows, columns, diagonals and generalized diagonals of the 'noughts and crosses' or 'tic-tac-toe' board. These triples may conveniently be thought of as the subsets of GF_9, the field of order 9, which contain three distinct elements summing to zero. The correspondence with our nine symbols is shown in Figure 7.1. The reader familiar with the Mathieu groups will recognize this as the stabilizer of a triple in M_{12}, which is often written $M_9 : S_3$.

Not surprisingly, one short additional relation is enough to define the Hessian group, and we have

$$\frac{3^{\star 4} \cdot_m 2 \dot{} S_4^+}{[(1\, 2\, 3)t_1]^3} \cong 3^2 : 2 \dot{} S_4^+ \cong M_9 : S_3.$$

---

[1] It was mentioned in Part I that the points of inflexion of a homogeneous curve in $\mathbb{C}^3$ are those points where the curve intersects its *Hessian*, which is the determinant of second partial derivatives. Now, a second partial derivative of a homogeneous cubic expression is linear, and so the Hessian is itself a cubic in the case when the original curve is cubic. In the general case, two cubics over $\mathbb{C}$ will intersect in nine distinct points, and it turns out that the line joining any pair of these points of inflexion will pass through a third. In this way, the nine points of inflexion form a Steiner system $S(2, 3, 9)$, which has the Hessian group as its group of automorphisms.

d	b	c		$-1+i$	i	$1+i$
A	★	a	~	-1	0	1
C	B	D		$-1-i$	$-i$	$1-i$

Figure 7.1. Affine plane of order 3.

With the generators x and y, and with $t = t_0$, this is described by the following presentation:

$$\langle x, y, t \mid y^3 = (xy)^4, x^2 = t^3 = t^y t = [t, x^{y^{-1}x}] = (y^2 t^x)^3 = 1 \rangle \cong M_9 : S_3.$$

7.1.2 The Mathieu group M_{11}

As our final attempt to embed this progenitor as a transitive subgroup of a symmetric group, we choose degree 11 and suppose the control subgroup acts with orbits $3 + 8$. Thus,

$$x = (v\ w)(a\ b)(A\ B)(d\ D),$$
$$y = (u\ v\ w)(a\ A)(b\ C\ d\ B\ c\ D),$$
$$x^{y^{-1}x} = (u\ w)(B\ d)(b\ D)(c\ C).$$

The subgroup $\langle y, x^{y^{-1}x} \rangle$ now has orbits $3 + 2 + 6$ and, since t_0 must move the 3-orbit, we see that it must have cycle shape $1^2.3^3$. This time Case I is impossible and both y and $x^{y^{-1}x}$ must invert t_0. There is only one possibility, and we obtain

$$t_0 = (u\ b\ B))(v\ c\ C)(w\ d\ D),$$
$$t_1 = (u\ a\ A)(v\ D\ d)(w\ c\ C),$$
$$t_2 = (u\ d\ D)(v\ a\ A)(w\ B\ b),$$
$$t_3 = (u\ C\ c)(v\ b\ B)(w\ a\ A).$$

These four elements of order 3 generate a 4-transitive subgroup of S_{11} which has order 7920 and preserves a Steiner system $S(4, 5, 11)$ whose special pentads are the 66 images of the subset $\{u, v, w, a, A\}$. It is, of course, the smallest sporadic simple group, the Mathieu group M_{11}. It can be verified mechanically that the additional relation $[(0\ 1)(2\ \bar{2})t_0]^5$ is sufficient to define the group, and so we have Theorem 7.1 as follows.

THEOREM 7.1

$$G = \frac{3^{\star 4} :_m 2 \cdot S_4^+}{[(0\ 1)(2\ \bar{2})t_0]^5} \cong M_{11}.$$

A corresponding presentation is given by

$$\langle x, y, t \mid y^3 = (xy)^4, x^2 = t^3 = t^y t = [t^{xy}, x] = (xt)^5 = 1 \rangle \cong M_{11}.$$

Labelling the 11 letters with the integers modulo 11 in the more familiar manner, we obtain the following:

$$
\begin{aligned}
x &= (1\ \mathrm{X})(2\ 6)(3\ 8)(4\ 5) \sim [0\ 1][2\ \bar{2}], \\
y &= (0\ \mathrm{X}\ 1)(2\ 7\ 4\ 8\ 9\ 5)(3\ 6) \sim [0\ \bar{0}][1\ \bar{2}\ 3\ \bar{1}\ 2\ \bar{3}], \\
t_0 &= (1\ 5\ 4)(2\ 0\ 8)(7\ 9\ \mathrm{X}), \\
t_1 &= (\mathrm{X}\ 4\ 5)(6\ 0\ 3)(7\ 9\ 1), \\
t_2 &= (1\ 2\ 8\ )(\mathrm{X}\ 3\ 6)(0\ 5\ 4), \\
t_3 &= (7\ 0\ 9)(3\ 6\ 1)(8\ 2\ \mathrm{X}),
\end{aligned}
$$

where X stands for 10 and the action on the symmetric generators is written in square brackets. Invariably these highly symmetric generating sets carry a great deal of information about the structure of the group. In this case we have

$$\langle t_i, t_j \rangle \cong A_5, \ \langle t_i, t_j, t_k \rangle \cong L_2(11),$$

for i, j, k distinct, and

$$\langle t_0 t_1 t_0, yx \rangle \cong M_{10} \cong A_6 \cdot 2.$$

It is left as an exercise for the reader to obtain M_{11} acting transitively on 12 letters in an analogous manner by letting the control subgroup act with orbits 4+8 on the set $\{t, u, v, w\} \cup \{a, b, c, d, A, B, C, D\}$; so we take

$$
\begin{aligned}
x &= (t\ u)(a\ b)(A\ B)(d\ D), \\
y &= (u\ v\ w)(a\ A)(b\ C\ d\ B\ c\ D), \\
x^{y^{-1}x} &= (u\ w)(B\ d)(b\ D)(c\ C).
\end{aligned}
$$

7.2   The Mathieu group M₂₃

We now turn our attention to the progenitor[2] of shape

$$P \cong 5^{*4} :_m 2\dot{\ }S_4^-,$$

as described in Example 6.2. So the action by conjugation on the four symmetric generators of the elements x and y is given by the monomial matrices to be

$$x \sim (1\ 2\ 3\ 4\ 1^{-1}\ 2^{-1}\ 3^{-1}\ 4^{-1}), \quad y \sim (1\ 3^{-1}\ 2^2\ 1^{-1}\ 3\ 2^{-2})(4\ 4^{-1}).$$

Note that we have only given the action on half the powers of the symmetric generators, as the rest follow. The subgroup of $N \cong 2\dot{\ }S_4^-$ which normalizes $\langle t_1 \rangle$ is given by

$$
\begin{aligned}
N_N(\langle t_1 \rangle) &= \langle xy \sim (1\ 1^2\ 1^{-1}\ 1^{-2})(2\ 2^{-2}\ 2^{-1}\ 2^2)(3\ 4^{-1}\ 3^{-1}\ 4), \\
&\qquad y^x \sim (1\ 1^{-1})(2\ 4^{-1}\ 3^2\ 2^{-1}\ 4\ 3^{-2}) \rangle \\
&\cong 3:4,
\end{aligned}
$$

---

[2] This progenitor was investigated by Stephen Stanley in his Ph.D. thesis (ref. [82], p. 121), along with many other monomial progenitors with a small number of symmetric generators.

This was, of course, obtained by writing the monomial matrices of Example 6.2 over $\mathbb{Z}_5$ by replacing i and $-$i by 2 and 3, respectively; recall that we could equally well have let i be replaced by 3 and $-$i by 2. As usual, we attempt to embed P as a transitive subgroup of a symmetric group of low degree. Firstly we ask: what is the lowest degree of a faithful representation of $N \cong 2{\cdot}S_4^-$? We seek the largest subgroup of N which does not contain the centre; but every involution in S_4 lifts to elements of order 4 in $2{\cdot}S_4^-$, and so the largest subgroup of N which does not contain the centre has order 3. Thus the lowest degree of a faithful permutation representation of N is 16, and Sylow's theorem ensures that this representation is unique. There are one or two transitive images in S_n for $16 < n < 23$ which the reader may like to investigate, but it is degree 23 which particularly interests us. It can be shown that the Mathieu group M_{23} contains two non-conjugate copies of $2{\cdot}S_4^-$, N_1 and N_2 say, and these have orbits of lengths $(1+6+16)$ and $(3+4+16)$, respectively. In Figure 7.2 we have $N_1 = \langle x_1, y \rangle$ and $N_2 = \langle x_2, y \rangle$. In each case there is just one contender for the cyclic subgroup $\langle t_1 \rangle$, and we may be led to believe that these give rise to the only two ways in which M_{23} is an image of P. However, if we take the alternative monomial representation over $\mathbb{Z}_5$ in which i is replaced by 3 rather than 2, we obtain a further possibility for the control subgroup N_1. The three possibilities for t_1 are given in Figure 7.3, and we have

$$\langle N_1, t_1^a \rangle = \langle N_1, t_1^b \rangle = \langle N_2, s_1 \rangle \cong M_{23}.$$

These three sets of four symmetric generators of order 5 are completely different from one another. Indeed, we have

$$\langle t_i^a, t_j^a \rangle \cong A_6, \quad \langle t_i^a, t_j^a, t_k^a \rangle \cong 2^4 : A_8, \quad \langle t_1^a, t_2^a, t_3^a, t_4^a \rangle \cong M_{23},$$

$$\langle t_i^b, t_j^b \rangle \cong M_{22}, \quad \langle t_i^b, t_j^b, t_k^b \rangle \cong M_{23}, \quad \langle t_1^b, t_2^b, t_3^b, t_4^b \rangle \cong M_{23},$$

$$\langle s_i, s_j \rangle \cong L_3(4), \quad \langle s_i, s_j, s_k \rangle \cong M_{22}, \quad \langle s_1, s_2, s_3, s_4 \rangle \cong M_{23},$$

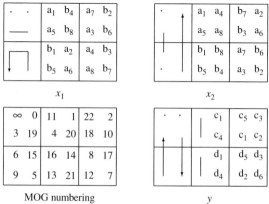

.	.	a_1	b_4	a_7	b_2
		a_5	b_8	a_3	b_6
		b_1	a_2	a_4	b_3
		b_5	a_6	a_8	b_7

x_1

.		a_1	a_4	b_7	a_2
		a_5	a_8	b_3	a_6
		b_1	b_8	a_7	b_6
.		b_5	b_4	a_3	b_2

x_2

∞	0	11	1	22	2
3	19	4	20	18	10
6	15	16	14	8	17
9	5	13	21	12	7

MOG numbering

.	.		c_1	c_5	c_3
			c_4	c_1	c_2
			d_1	d_5	d_3
			d_4	d_2	d_6

y

Figure 7.2. Generators of $2{\cdot}S_4^-$ of degree 23.

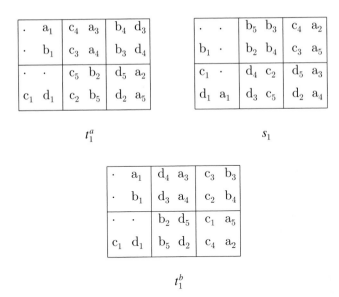

Figure 7.3. Symmetric generators for M_{23}.

for i, j, k distinct. In the following MAGMA printout, generators x_1, x_2 and y for the control subgroup shown diagrammatically in Figure 7.2 are given as permutations of the set $\{1, 2, \ldots, 23\}$, so the symbol 0 has been replaced by 23. The three sequences of four symmetric generators are given by tsa $= \{t_1^a, t_2^a, t_3^a, t_4^a\}$, tsb $= \{t_1^b, t_2^b, t_3^b, t_4^b\}$ and ss $= \{s_1, s_2, s_3, s_4\}$, where t_1^a, t_1^b and s_1 are displayed in figure 7.3. The procedure 'subs' returns the orders of the subgroups generated by two generators, three generators and four generators, when applied to a sequence of generators, and confirms the orders of the subgroups given above. The procedure 'act4' when applied as `act4(tt,uu)` returns the way the permutation uu acts on the sequence tt. So the output achieved by `act4(tsa,x1)` at the end of the following computer printout tells us that x_1 acts by conjugation as $(t_1^a\, t_2^a\, t_3^a\, t_4^a\, (t_1^a)^{-1} \cdots)$, and `act4(tsa,y)` tells us that y acts as $(t_1^a\, (t_3^a)^{-1}\, (t_2^a)^2\, (t_1^a)^{-1} \cdots)(t_4^a\, (t_4^a)^{-1})$, as required:

```
> s23:=Sym(23);
> x1:=s23!(1,8,22,16,20,12,18,13)(2,21,4,17,10,14,11,7)
> (3,19)(5,15,6,9);
> x2:=s23!(3,6)(5,15,19,23)(11,2,12,1,4,10,8,20)
> (16,7,18,21,13,17,22,14);
> y:=s23!(9,6,3)(19,15,5)(11,4)(16,13)(2,20,22,10,1,18)
> (14,12,17,21,8,7);
>
> nn1:=sub<s23|x1,y>;
> nn2:=sub<s23|x2,y>;
>
> t1a:=s23!(1,20,7,23,17)(2,10,8,5,12)(4,11,16,9,13)
```

```
  (14,18,22,21,19);
> t1b:=s23!(1,20,17,23,7)(2,10,13,19,16)(4,11,14,5,21)
  (8,18,22,12,9);
> s1:=s23!(5,2,17,7,10)(3,4,1,20,11)(6,14,18,22,21)
  (9,12,13,16,8);
>
> m23:=sub<s23|x1,y,t1a>;
>
> tsa:=[t1a,t1a^x1,t1a^(x1^2),t1a^(x1^3)];
> tsb:=[t1b,t1b^x1,t1b^(x1^2),t1b^(x1^3)];
> ss:=[s1,s1^x2,s1^(x2^2),s1^(x2^3)];
>
> procedure subs(tt);
procedure> Order(sub<m23|tt[1],tt[2]>),
procedure> Order(sub<m23|tt[1],tt[2],tt[3]>),
procedure> Order(sub<m23|tt[1],tt[2],tt[3],tt[4]>);
procedure> end procedure;
>
> procedure act4(ts,uu);
procedure> for i,j,k in [1..4] do
procedure|for|for|for> if ts[i]^uu eq ts[j]^k then
procedure|for|for|for|if> print i,j,k;
procedure|for|for|for|if> end if;
procedure|for|for|for> end for;end for;end for;
procedure> end procedure;
>
> subs(tsa);
360 40320 10200960
> subs(tsb);
443520 10200960 10200960
> subs(ss);
20160 443520 10200960
>
> act4(tsa,x1);
1 2 1
2 3 1
3 4 1
4 1 4
> act4(tsa,y);
1 3 4
2 1 2
3 2 3
4 4 4
```

As is always the case, every element of the progenitor may be written πw, where π is an element of the control subgroup and w is a word in the symmetric generators. Thus, in order to find additional relations which will define the group M_{23} in the three cases, we seek short words in the symmetric generators which lie in N. We obtain Theorem 7.2.

THEOREM 7.2

(i) Symmetric presentation

$$\frac{5^{\star 4} :_m 2^{\cdot}S_4^-}{\left(\begin{smallmatrix} \cdot & 2 & \cdot & \cdot \\ \cdot & \cdot & 3 & \cdot \\ 1 & \cdot & \cdot & \cdot \\ \cdot & \cdot & \cdot & 1 \end{smallmatrix}\right)\, t_1 t_2 t_1 t_3^{-1} t_2^{-2} t_1^{-2} t_3^2 t_2^{-1},\, \left[\left(\begin{smallmatrix} 2 & \cdot & \cdot & \cdot \\ \cdot & 3 & \cdot & \cdot \\ \cdot & \cdot & \cdot & 4 \\ \cdot & \cdot & 1 & \cdot \end{smallmatrix}\right) t_3\right]^5} \cong M_{23},$$

which is realized by the following presentation:

$$\langle x, y, t \mid \quad x^4 = y^3 = (xy)^2,\, t^5 = t^{xy} t^{-2} = t^{y^x} t$$
$$= (xyt^{x^3})^5 = y^2 tt^x t (t^{x^2})^{-1} (t^x)^{-2} t^{-2} (t^{x^2})^2 (t^x)^{-1} = 1 \rangle \cong M_{23}.$$

(ii) Symmetric presentation

$$\frac{5^{\star 4} :_m 2^{\cdot}S_4^-}{\left(\begin{smallmatrix} \cdot & \cdot & 2 & \cdot \\ \cdot & 2 & \cdot & \cdot \\ 2 & \cdot & \cdot & \cdot \\ \cdot & \cdot & \cdot & 3 \end{smallmatrix}\right)\, t_1 t_2 t_4^{-2} t_1 t_3^{-1} t_2} \cong M_{23},$$

with presentation

$$\langle x, y, t \mid \quad x^4 = y^3 = (xy)^2,\, t^5 = t^{xy} t^2 = t^{y^x} t = 1,$$
$$tt^x (t^{x^{-1}})^2 tt^y t^x = (yx)^{xy} \rangle \cong M_{23}.$$

(iii) Symmetric presentation

$$\frac{5^{\star 4} :_m 2^{\cdot}S_4^-}{\left(\begin{smallmatrix} \cdot & \cdot & 4 & \cdot \\ 2 & \cdot & \cdot & \cdot \\ \cdot & 3 & \cdot & \cdot \\ \cdot & \cdot & \cdot & 4 \end{smallmatrix}\right)\, t_1 t_2 t_1^2 t_3^{-1} t_1^{-1} t_2^2 t_1^2 t_3^2,\, \left[\left(\begin{smallmatrix} \cdot & 1 & \cdot & \cdot \\ \cdot & \cdot & 1 & \cdot \\ \cdot & \cdot & \cdot & 1 \\ 4 & \cdot & \cdot & \cdot \end{smallmatrix}\right) t_1\right]^{11}} \cong M_{23},$$

with presentation

$$\langle x, y, t \mid \quad x^4 = y^3 = (xy)^2,\, t^5 = t^{xy} t^{-2} = t^{y^x} t =,$$
$$ytt^x t^2 t^y t^{-1} t^{y^2} t^2 t^{xy^{-1}} = (xt)^{11} = 1 \rangle \cong M_{23}.$$

7.3 The Mathieu group M_{24}

Perhaps the main observation which led to our concept of symmetric generation was the fact that the group M_{24} is a homomorphic image of the progenitor

$$P = 2^{\star 7} : L_2(7).$$

Indeed, if one attempts to embed this progenitor in the symmetric group S_{24} with the control subgroup $L_2(7)$ acting transitively on the 24 points, then one immediately writes down generators for M_{24}.

This chapter, however, is devoted to genuinely monomial control subgroups, and so we start with the progenitor

$$3^{*8} :_m SL_2(7),$$

which was constructed in Example 6.7 and in which the control subgroup is generated by

$$x \sim (\infty)(0\ 1\ 2\ 3\ 4\ 5\ 6) \quad \text{and} \quad y \sim (\infty\ \bar{0}\ \bar{\infty}\ 0)(1\ \bar{6}\ \bar{1}\ 6)(2\ \bar{3}\ \bar{2}\ 3)(4\ \bar{5}\ \bar{4}\ 5),$$

in the usual abbreviated notation. This progenitor was investigated thoroughly by Stanley [82], and additional relators which define M_{24} were obtained. We shall restrict ourselves to seeking low degree permutation representations of the progenitor and, in this manner, will obtain generators for M_{24}. Firstly, note that a faithful transitive permutation representation of the control subgroup $SL_2(7)$ must be over a subgroup which contains no elements of even order (since the involutions of $L_2(7)$ lift in $SL_2(7)$ to elements of order 4 squaring to the central involution). The largest subgroup of odd order is the Frobenius group 7:3, giving a permutation action on 16 points in which the central involution acts fixed-point-free. Stanley showed that the only true images of this progenitor of degree less than 24 are the alternating groups A_{17} and A_{23}, so we concentrate on embedding the progenitor in S_{24}. The action of $SL_2(7)$ on 16 points is unique up to relabelling, and so can be taken to be the action on the eight symmetric generators and their inverses. We then let $SL_2(7)$ act on seven letters $\{a_0, a_1, \ldots, a_6\}$ (note that this can be done in two different ways), and adjoin a fixed point $\star$. So the control subgroup has orbits of lengths $1 + 7 + 16$. We must now find an element of order 3 which commutes with a Frobenius group of shape 7:3 and is inverted by the central involution. In addition, this symmetric generator must fuse the three orbits and so must have cycle shape 3^8. For convenience we display the 24 points with these labels in Figure 7.5.

In Figure 7.4 the permutations x and y generate $SL_2(7)$ and the final element of order 3, which fixes the top row and rotates the three remaining rows downwards, normalizes x to give a Frobenius group of shape 7:3. The central involution of our $SL_2(7)$ is of course y^2, which bodily interchanges the second and third bricks of the array. It turns out there are just two possible symmetric generators satisfying the above requirements: the element t_∞ shown in Figure 7.4 and an element derived from this one by multiplying by $(\star\ \infty\ \bar{\infty})$. Adjunction of the first gives the group M_{24} and the second gives the alternating group A_{24}.

The resulting set of symmetric generators $\{t_\infty, t_0, \ldots, t_6\}$ for M_{24} is of further interest in that $\langle t_\infty, t_0, t_1, t_5 \rangle \cong M_{12}$, a fact that was exploited by Stanley in obtaining defining relations for the larger Mathieu group.

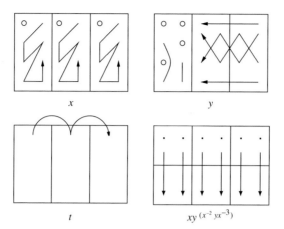

Figure 7.4. Generators for $\mathrm{SL}_2(7)$ and a symmetric generator for M_{24}.

$\star$	a_0	∞	0	$\bar{\infty}$	$\bar{0}$
a_1	a_3	1	3	$\bar{1}$	$\bar{3}$
a_2	a_6	2	6	$\bar{2}$	$\bar{6}$
a_4	a_5	4	5	$\bar{4}$	$\bar{5}$

Figure 7.5. Labelling of the 24 points showing the action of $N \cong \mathrm{SL}_2(7)$.

7.4 Factoring out a 'classical' relator

As a classical example, we let p be a prime and consider

$$\mathcal{F} = \langle t_1 \rangle \star \langle t_2 \rangle \cong \mathrm{C}_p \star \mathrm{C}_p = p^{\star 2}.$$

Then if λ is a generator of the cyclic group $\mathbb{Z}_p^{\times}$, the group of monomial automorphisms of $\mathcal{F}$ is isomorphic to $\mathrm{C}_{p-1} \wr 2$ and is generated by

$$\pi : t_1 \mapsto t_1^{\lambda}, t_2 \mapsto t_2,$$

and

$$\sigma : t_1 \leftrightarrow t_2.$$

We abbreviate these monomial actions as follows:

$$\pi \sim \begin{pmatrix} \lambda & \cdot \\ \cdot & 1 \end{pmatrix} \quad \text{and} \quad \sigma \sim \begin{pmatrix} \cdot & 1 \\ 1 & \cdot \end{pmatrix},$$

with the obvious meaning. It turns out that the projective general linear group $\mathrm{PGL}_2(p)$ is an image of

$$P \cong p^{\star 2} :_m \mathrm{D}_{2(p-1)},$$

where

$$N = \langle \pi\pi^{-\sigma}, \sigma \rangle = \left\langle \begin{pmatrix} \lambda & \cdot \\ \cdot & \lambda^{-1} \end{pmatrix}, \begin{pmatrix} \cdot & 1 \\ 1 & \cdot \end{pmatrix} \right\rangle \cong D_{2(p-1)}.$$

Indeed, the classical presentations of Coxeter and Moser [27], p. 95, (see also Stanley [82], p. 130), in our language take the following form:

$$\frac{p^{\star 2} :_m D_{2(p-1)}}{\left[\begin{pmatrix} \cdot & 1 \\ 1 & \cdot \end{pmatrix} t_1 \right]^3} \cong \mathrm{PGL}_2(p), \tag{7.1}$$

where the left-hand side denotes the progenitor $p^{\star 2} :_m D_{2(p-1)}$ factored by the relator

$$\left[\begin{pmatrix} \cdot & 1 \\ 1 & \cdot \end{pmatrix} t_1 \right]^3,$$

which may be rewritten as the relation

$$\begin{pmatrix} \cdot & 1 \\ 1 & \cdot \end{pmatrix} = t_1 t_2 t_1.$$

We realize the image $\mathrm{PGL}_2(p)$ regarded as the group of linear fractional transformations of $\mathrm{PG}_1(p) = \mathbb{Z}_p \cup \{\infty\}$ as follows:

$$\pi\pi^{-\sigma} = \begin{pmatrix} \lambda & \cdot \\ \cdot & \lambda^{-1} \end{pmatrix} \sim \eta \mapsto \lambda\eta \quad \text{and} \quad \sigma = \begin{pmatrix} \cdot & 1 \\ 1 & \cdot \end{pmatrix} \sim \eta \mapsto -\frac{1}{\eta},$$

and

$$t_1 \sim \eta \mapsto \eta + 1 \quad \text{and} \quad t_2 \sim \eta \mapsto \frac{\eta}{1-\eta}.$$

For p odd, in order to obtain the simple group, we let

$$N = \langle (\pi\pi^{-\sigma})^2, \sigma \rangle = \left\langle \begin{pmatrix} \mu & \cdot \\ \cdot & \mu^{-1} \end{pmatrix}, \begin{pmatrix} \cdot & 1 \\ 1 & \cdot \end{pmatrix} \right\rangle \cong D_{p-1},$$

where μ (which we take, without loss of generality, to be λ^2) is a generator for the quadratic residues of $\mathbb{Z}_p^\times$. We then obtain the following:

$$\frac{p^{\star 2} :_m D_{p-1}}{\left[\begin{pmatrix} \cdot & 1 \\ 1 & \cdot \end{pmatrix} t_1 \right]^3} \cong \begin{cases} \mathrm{L}_2(p) \times 2 & \text{if } p \equiv 1 (\mathrm{mod}\ 4), \\ \mathrm{L}_2(p) & \text{if } p \equiv 3 (\mathrm{mod}\ 4). \end{cases} \tag{7.2}$$

To quotient out the central involution in the case when $p \equiv 1(\mathrm{mod}\ 4)$, we add the additional relator

$$\left[\begin{pmatrix} \cdot & \lambda^{-2} \\ \lambda^2 & \cdot \end{pmatrix} t_1^\lambda \right]^3.$$

Note that in all of the above presentations, whether the group we have presented is $\mathrm{PGL}_2(p)$, $\mathrm{L}_2(p) \times 2$ or $\mathrm{L}_2(p)$, we have $\langle t_1, t_2 \rangle \cong \mathrm{L}_2(p)$. Thus, if the two symmetric generators of order p are denoted by t_1 and t_2, the control subgroup is generated by automorphisms of $p^{\star 2} = \langle t_1 \rangle \star \langle t_2 \rangle$. In Presentation (7.1) the element $\pi\pi^{-\sigma}$ conjugates t_1 to t_1^{λ} and t_2 to $t_2^{\lambda^{-1}}$ and so acts as

$$(t_1, t_1^{\lambda}, t_1^{\lambda^2}, \ldots, t_1^{\lambda^{-2}}, t_1^{\lambda^{-1}})(t_2, t_2^{\lambda^{-1}}, t_2^{\lambda^{-2}}, \ldots, t_2^{\lambda^2}, t_2^{\lambda})$$

on the non-trivial powers of t_1 and t_2; the involution σ interchanges t_1 and t_2 by conjugation, so we write that it has action (t_1, t_2). Note that we can determine the action of the given automorphisms on t_i^j for all i and j; in particular, (t_1, t_2) is an abbreviation for $(t_1, t_2)(t_1^2, t_2^2) \cdots (t_1^{p-1}, t_2^{p-1})$. The subscript '$m$' on the colons in the progenitors above conveys the fact that the action is properly monomial.

Note that given an $n \times n$ monomial matrix A over $\mathbb{Z}_m$, where the non-zero entries of A are units, we can define the (group) action of A on $F \cong m^{\star n}$ by $t_i^A = t_j^{a_{ij}}$, where a_{ij} is the unique non-zero entry in the ith row of A.

7.4.1   The unitary group $U_3(5)$

It turns out that the progenitor of shape

$$P = 5^{\star(6+6)} : 2^{\cdot}S_5,$$

which was described fully in Example 6.8, has the unitary group $U_3(5)$ as an image, a fact which was exploited in Curtis [35] to obtain a presentation of the group $HS : 2$. In that paper, the control subgroup was extended by an element a of order 4 which squares to the central involution, which inverts all the symmetric generators, and is inverted by outer elements in the group $2^{\cdot}S_5$, thus mapping onto $U_3(5) : 2$ rather than $U_3(5)$. It acts on the symmetric generators by $r_i^a = r^3$, $s_i^a = s^2$, for all i. A presentation for this larger control subgroup is given by

$$M \cong \langle x, y, a; x^5 = y^4 = (xy^2)^3 = a^2, (x^3y)^2 = 1 = [a, x] = ay^{-1}ay \rangle,$$

but we shall consider only the smaller group $\langle x, y \rangle \cong 2^{\cdot}S_5$ here. For convenience, we reproduce the action of our generators as given in Example 6.8, noting that x is an element of order 10 fifth-powering to the central involution:

$$x^2 : (r_\infty)(r_0, r_1, r_2, r_3, r_4)(s_\infty)(s_0, s_1, s_2, s_3, s_4),$$

$$y : (r_\infty, s_2, r_4, s_1^{-1}, \ldots)(r_0, s_0^2, r_0^3, s_0, \ldots)(r_1, s_\infty^{-2}, r_2, s_4^3, \ldots)(r_3, s_3, r_3^2, s_3^2, \ldots).$$

As stated in ref. [35], the relator (x^3yr_∞) ensures that $\langle r_i, s_j \rangle \cong A_5$ for all $i \neq j$, and maps the progenitor onto the triple cover $3^{\cdot}U_3(5)$. Factoring by the further relator $(y^3r_\infty)^7$ yields the simple group.

Since we have given presentations which correspond to our symmetric presentation, we can confirm claims made above by conventional single coset enumeration. However, it is of interest to see how the double coset enumerator copes with symmetric generators of order greater than 2. In

r	1	2	3	4	0	∞
1	1	2	3	4	5	6
2	7	8	9	10	11	12
3	13	14	15	16	17	18
4	19	20	21	22	23	24

s	1	2	3	4	0	∞
1	25	26	27	28	29	30
2	31	32	33	34	35	36
3	37	38	39	40	41	42
4	43	44	45	46	47	48

Figure 7.6. Labelling of the powers of r and s.

this case we have 12 symmetric generators of order 5, and so have a total of $12 \times 4 = 48$ elements of order 5 being permuted (imprimitively with blocks of size 4) by the control subgroup. In Figure 7.6 we label the various powers of our symmetric generators with the integers $1, \ldots, 48$; thus, for example, r_3^2 is labelled 14 and s_∞^4 is labelled 48. We now input x and y as permutations in S_{48}. The additional relations then take the following form:

$$(x^3 y)^3 = r_\infty^4 s_2 r_\infty^4 \sim [24, 26, 24]$$

and

$$y^5 = r_\infty^4 s_1 r_4 s_2^4 r_\infty^4 s_1^4 r_4^4 \sim [24, 25, 4, 44, 6, 43, 22].$$

In addition we need to tell the program that the generators have order 5, so we input the relator $\langle [6, 6, 6, 6, 6], \mathrm{Id}(nn) \rangle$, and we must input a permutation which inverts every symmetric generator; this need not be an element of our control subgroup, but here we have inv := y^4:

```
>s48:=Sym(48)
>xx:=s48!(1, 22, 2, 23, 3, 19, 4, 20, 5, 21)(6, 24)
     (7, 16, 8, 17, 9, 13, 10, 14, 11,15)(12, 18)
     (25, 46, 26, 47, 27, 43, 28, 44, 29, 45)(30, 48)
     (31, 40, 32, 41, 33, 37, 34, 38, 35, 39)(36, 42);

> yy:=s48!(1, 42, 2, 40, 19, 36, 20, 34)(3, 27, 9, 33, 21, 45,15, 39)
      (4, 43, 24, 44, 22,25, 6, 26)(5, 35, 17, 29, 23, 41, 11, 47)
      (7, 30, 8, 28, 13, 48, 14, 46)(10,37, 18, 38, 16, 31, 12, 32);

>nn:=sub<s48|xx,yy>;

> RRb:=[<[6,6,6,6,6],Id(nn)>, <[24,26,24],(xx^3*yy)^3>,
> <[24,25,4,44,6,43,22],yy^5>];

> CTb:=DCEnum(nn,RRb,[nn]:Print:=5,Grain:=100,Inv:=yy^4);

Index: 525 === Rank: 7 === Edges: 75  === Time: 0.406

> CTb[4];
[
    [],
    [ 1 ],
    [ 1, 3 ],
```

```
    [ 1, 3, 3 ],
    [ 1, 29, 3, 38, 16 ],
    [ 1, 1, 26, 26, 26 ],
    [ 1, 28, 17, 32 ]
]
> CTb[7];
[ 1, 48, 120, 120, 96, 20, 120 ]
>
```

So we see that, for instance, the fourth double coset found $Nr_1 r_3 r_3 N$ contains 120 single cosets.

7.4.2 The Held group

More ambitiously, we now aim to find a finite image of the progenitor

$$P = 7^{\star(15+15)} : (3 \dot{} S_7),$$

in which, of course, a 'central' element of order 3 will, by conjugation, square one set of 15 symmetric generators while fourth-powering the other. Prompted by the comment at the very end of Example 6.9, we consider what the images of $\langle r_1, s_{15} \rangle$, $\langle r_{15}, s_{15} \rangle$ and $\langle s_1, s_{15} \rangle$ might be. As usual, in order to find a non-trivial finite homomorphic image of P we apply Lemma 3.3. Now,

$$C_N(\langle s_{15}, r_7 \rangle) = \langle y, (xy)^{x^2} \rangle \cong S_4,$$

and the centralizer in N of this subgroup is just $\langle z, t \rangle \cong S_3$, where $\langle z \rangle = Z(N')$; and so we seek a group generated by two elements of order 7, normalized by an element of order 3 which squares one of them and fourth-powers the other, and an involution which interchanges them. Lemma 3.3 asserts that both these automorphisms may be inner, i.e. that we can have

$\langle z, t \rangle \leq \langle s_{15}, r_7 \rangle$. We seek a minimal example in which this holds. In the language of progenitors we are thus seeking a homomorphic image of

$$7^{*2} : K, \text{ where } S_3 \cong K = \left\langle \begin{pmatrix} 2 & \cdot \\ \cdot & 4 \end{pmatrix}, \begin{pmatrix} \cdot & 1 \\ 1 & \cdot \end{pmatrix} \right\rangle,$$

which is generated by the images of the two symmetric generators of order 7. Such a group is clearly perfect,[3] and, since we are looking for a minimal example, we may assume it is simple. A presentation of this progenitor is given by

$$\langle u, a, b \mid u^7 = a^3 = b^2 = (ab)^2 = u^a u^{-2} = 1 \rangle.$$

Consideration of small perfect groups soon tells us that the smallest example satisfying all these conditions is the linear group $L_2(7)$, and, if we require the element ub to have order 3, we obtain a presentation for the group. A corresponding relator by which to factor our progenitor P is $(ts)^3$, and we obtain Theorem 7.3.

THEOREM 7.3

$$G = \frac{7^{\cdot(15+15)} : 3 \cdot S_7}{(ts)^3} \cong \text{He},$$

where $s = s_{15}$ and t is an involution with the required property.

As a presentation we have

$$\begin{aligned}
\text{He} = \langle x, y, t, s \mid x^7 &= y^3 = (xy)^2 = [x, y]^4 = t^2 \\
&= [t, y] = [t, (xy)^{x^2}] = (xt)^6 = s^7 \\
&= [s, x] = [s, y] = s^{tx^2 txtx^{-1} t} s^3 = (st)^3 = 1 \rangle.
\end{aligned}$$

Using the coset enumerator in the MAGMA package [19], we find that the subgroup $\langle x, y, t \rangle \cong 3 \cdot S_7$ has index 266 560 and so He has order 4 030 387 200. The group is easily seen to be the Held sporadic simple group (see refs. [42], [43] and [25], p. 104). This presentation can be extended to the automorphism group of He by adjoining a further involutory generator a which commutes with x, y and t and inverts s; thus a inverts all the symmetric generators.

Verification using the double coset enumerator

As a second example of how the double coset enumerator can handle non-involutory symmetric generators, we verify that this symmetric presentation does define the Held group. This time we have 30 symmetric generators of order 7 and so we must input our control subgroup as permutations on $30 \times (7 - 1) = 180$ elements, with blocks of imprimitivity of size 6 corresponding to the non-trivial powers of a particular element. Our monomial

---

[3] Since its derived subgroup contains the two elements of order 7, which we are told generate.

representation was written over cube roots of unity (not sixth roots) and so N has two orbits of length 90 on these 180 elements, consisting of a set of symmetric generators, their squares and their fourth powers. We label these $1, \dots, 90$, and we let the inverse of i be denoted by $i + 90$:

```
> n<x,y,t>:=Group<x,y,t|x^7=y^3=(x*y)^2=(x,y)^4=t^2=
  (t,y)=
>(t,(x*y)^(x^2))=(x*t)^6=1>;
> h:=sub<n|x,y>;
> Index(n,h);
90
> f,np,k:=CosetAction(n,h);
> s180:=Sym(180);

> x180:=s180!([i^f(x):i in [1..90]] cat [i^f(x)+90:
  i in [1..90]]);
> y180:=s180!([i^f(y):i in [1..90]] cat [i^f(y)+90:
  i in [1..90]]);
> t180:=s180!([i^f(t):i in [1..90]] cat [i^f(t)+90:
  i in [1..90]]);
> inv:=s180!([i+90:i in [1..90]] cat [i:i in [1..90]]);

> n180b:=sub<s180|x180,y180,t180,inv>;
> #n180b;
30240
> 1^x180;1^y180;1^t180;
1
1
2
> Fix(sub<n180|x180,y180>);
{ 1, 89, 90, 91, 179, 180 }
> 1^(t180*x180^2*t180*x180*t180*x180^-1*t180);
89
```

The blocks of imprimitivity of size 6 corresponding to powers of a symmetric generator are not now labelled consecutively, as was the case for $U_3(5)$. However, s_{15} is clearly labelled 1, and the above fixed point set shows us that its powers are labelled $\{1, 89, 90, 179, 180\}$. Of course, we do know that s_{15}^{-1} is labelled 91, and the following command tells us (by comparison with the presentation) that s_{15}^4 is labelled 89. We can feed in this information as $s_{15}^4 s_{15}^4 s_{15}^{-1} \sim [89, 89, 91] = \mathrm{Id}(N)$. We of course also include the relation which says that $s_{15}^7 = \mathrm{Id}(N)$, and finally we need the additional relation which says that $s_{15}^{-1} r_7^{-1} s_{15}^{-1} \sim [91, 92, 91] = t180$:

```
> RR:=[<[1,1,1,1,1,1,1],Id(n180b)>,<[89,89,91],
  Id(n180b)>,
> <[91,92,91],t180>];
```

```
> CT:=DCEnum(n180b,RR,[n180b]:Print:=0,Grain:=100,
  Inv:=inv);
> CT[1];
266560

> CT[2];
41
> CT[7];
[ 1, 180, 7560, 3780, 630, 15120, 15120, 7560, 15120, 15120,
15120, 3780, 126,2160, 630, 30240, 15120, 15120, 7560, 3780,
3780, 315, 7560, 15120, 7560, 2520,7560, 7560, 2160, 15120,
1890, 3780, 7560, 2520, 1890, 105, 1260, 1260, 1890,
1260, 63 ]
>
```

The printout CT[1] shows us that there are 266 560 cosets of N in G; CT[2] tells us that there are 41 double cosets; and CT[7] tells us how many single cosets there are in each of the double cosets. Of course we are also given representatives for each of these 41 double cosets, but we have chosen not to reproduce them.

Subgroups of He generated by subsets of the symmetric generators

As was mentioned in the previous examples, the symmetric generators often contain a great deal of information about the subgroup structure of the group they generate. In order to investigate the subgroups generated by subsets of our 30 generators of order 7, we need to be conversant with the action of S_7 on 30 letters. Now,

$$A_7 \leq A_8 \cong L_4(2),$$

and the 30 letters may be taken to be the 15 1-dimensional subspaces (the *points*) and the 15 3-dimensional subspaces (the *hyperplanes*) of a 4-dimensional vector space over $\mathbb{Z}_2$. These actions are seen clearly in the Mathieu group M_{24} (see Figure 5.2): the subgroup fixing an octad, a point in it and a point outside it is isomorphic to A_7, acting simultaneously on the seven remaining points of the fixed octad and the 15 remaining *points* of the complementary 16-ad. The *hyperplanes* then correspond to the 15 octads disjoint from the fixed octad and containing the fixed point of the 16-ad; thus points are contained in seven hyperplanes. The outer elements of S_7 correspond to polarities which interchange points and hyperplanes. One may readily read off the action of elements of A_7 on points and hyperplanes from the Miracle Octad Generator of ref. [29], and the correspondence with our generators $\mathcal{R} \cup \mathcal{S}$ is given in Figure 7.7, together with the action of the polarity t.

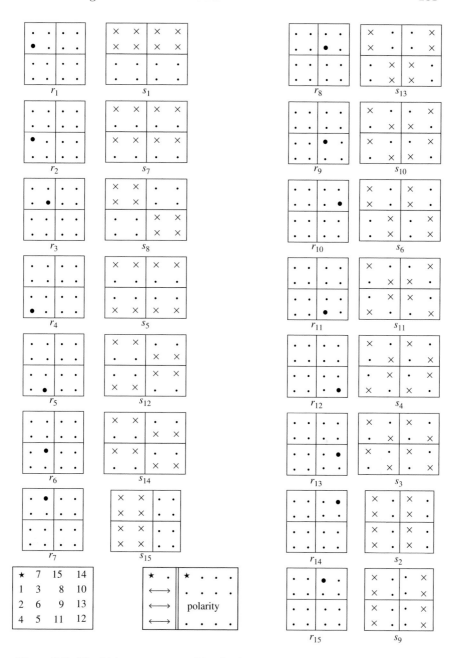

Figure 7.7. The 30 letters permuted by S_7 showing the polarity t.

As seen above, the stabilizer in S_7 of the hyperplane s_{15}, say, is isomorphic to $L_3(2)$ with orbits of lengths 7 and 8 on the points. Fixing a point in the 7-orbit reduces this to a subgroup isomorphic to S_4. In our construction, with s_{15} as the hyperplane and r_7 as the point, we imposed a relation which forced $\langle r_7, s_{15} \rangle \cong L_3(2)$, and thus

$$\langle s, t, y, (xy)^{x^2} \rangle \cong L_3(2) \times S_4,$$

a maximal subgroup of He. If instead we fix a point in the 8-orbit, we are left with a Frobenius group of shape $7{:}3$, which normalizes the point but does not centralize it. Thus

$$C_N(\langle r_{15}, s_{15} \rangle) = \langle x \rangle \cong C_7.$$

In fact, $\langle r_{15}, s_{15} \rangle \cong 7^{1+2}$, an extra-special group, and

$$\begin{aligned}
N_{\text{He}}(\langle r_{15}, s_{15} \rangle) &= \langle r_{15}, s_{15} \rangle : N_N(\langle r_{15}, s_{15} \rangle) = \langle s, y^{x^2 y^{-1}}, (x^{yx^{-2}} r^{x^3})^3 \rangle \\
&= \langle s_{15}, (0,4,5)(1,6,2), (0,3,2,6,4,1,5)t_1 t_6 t_3 \rangle \\
&\cong 7^{1+2} : (S_3 \times 3),
\end{aligned}$$

a further maximal subgroup. In addition, the control subgroup N itself is maximal, as is the normalizer of a symmetric generator:

$$\begin{aligned}
N_G(\langle s \rangle) &= \langle s, x, y, z \rangle \\
&= \langle s_{15}, (0,1,2,3,4,5,6), (6,5,1)(4,3,2), (6,2)(4,5)t_0 t_1 t_0 t_3 \rangle \\
&\cong 7 : 3 \times L_3(2).
\end{aligned}$$

This leaves us with possibly the most interesting case: the 4-dimensional space containing the 15 points and 15 hyperplanes also contains 35 2-dimensional subspaces, each of which contains three points and is contained in three hyperplanes. A set of six such symmetric generators is thus normalized by a subgroup of N of order $(3 \times 7!)/35 = 2^4 3^3$, of which a subgroup isomorphic to V_4 centralizes all six generators. We can see from Table 7.1 that $\mathcal{D} = \{r_1, r_3, r_7, s_1, s_8, s_{15}\}$ is such a set. So $D = \langle \mathcal{D} \rangle$ is a group generated by six elements of order 7 normalized by N_D, a group of monomial automorphisms of order 432, of which a subgroup isomorphic to V_4 acts trivially. Indeed we can see that

$$\begin{aligned}
t &: (r_1\ s_1)(r_3\ s_8)(r_7\ s_{15}), \\
yx &: (r_1\ r_7)(r_3)(s_1\ s_8^2)(s_{15}), \\
y^{x^2} &: (r_1\ r_7\ r_3)(s_1\ s_1^2\ s_1^4)(s_8\ s_8^4\ s_8^2)(s_{15}),
\end{aligned}$$

and a coset enumeration over $\langle s, t, yx, y^{x^2} \rangle$ yields index 8330. We can, of course, investigate the structure of this subgroup through the resulting permutation action but, in the spirit of the present work, we note that it is a homomorphic image of

$$7^{\star(3+3)} : (2^2 \cdot 3^{1+2} : 2^2),$$

where the monomial automorphisms in N_D are generated by

$$
\left[\begin{array}{ccc|ccc}
. & . & 4 & . & . & . \\
1 & . & . & . & . & . \\
. & 2 & . & . & . & . \\
\hline
. & . & . & 1 & . & . \\
. & . & . & . & 2 & . \\
. & . & . & . & . & 4
\end{array}\right]
;
\left[\begin{array}{ccc|ccc}
. & . & . & 1 & . & . \\
. & . & . & . & 1 & . \\
. & . & . & . & . & 1 \\
\hline
1 & . & . & . & . & . \\
. & 1 & . & . & . & . \\
. & . & 1 & . & . & .
\end{array}\right]
;
\left[\begin{array}{ccc|ccc}
. & . & 4 & . & . & . \\
. & 1 & . & . & . & . \\
2 & . & . & . & . & . \\
\hline
. & . & . & . & . & 4 \\
. & . & . & . & 1 & . \\
. & . & . & 2 & . & .
\end{array}\right].
$$

$$(u) \qquad\qquad (v) \qquad\qquad (w)$$

This automorphism group is defined (modulo the central V_4) by

$$\langle u, v, w \mid u^3 = v^2 = w^2 = (vw)^2 = 1, u^v u^w = [u, v]\rangle \cong 3^{1+2} : 2^2.$$

If we let q stand for the symmetric generator s_8^2 in the above, we see that $q^u = q^2$ and $[q, w] = 1$. The additional relator by which we factored P becomes $(vq)^3$, and we find

$$\langle u, v, w, q \mid u^3 = v^2 = w^2 = (vw)^2 = q^7 = q^u q^{-2}$$
$$= [q, w] = (vq)^3 = 1, u^v u^w = [u, v]\rangle \cong L_3(4) : S_3,$$

which can be shown by explicitly exhibiting permutations of $L_3(4) : S_3$ which satisfy the presentation, together with a coset enumeration over, for instance, $\langle u, v, w\rangle$. The full subgroup of index 8330 has shape $2^2 \cdot L_3(4) : S_3$, and the outer automorphism of He which commutes with N and inverts the symmetric generators completes this to $2^2 \cdot L_3(4) : (S_3 \times 2)$.

From the ATLAS (ref. [25], p. 104), we see that representatives of five of the conjugacy classes of maximal subgroups of He can be described naturally in terms of the control subgroup and symmetric generators. In Table 7.1 we exhibit this information by giving, in the last four cases, a subgroup of N, its orbits on the symmetric generators, and the normalizer of the subgroup generated by the symmetric generators it fixes. Each of these is maximal.

Other maximal subgroups of He

Although the permutation representation of He on 8330 letters emerges naturally in this approach, the minimal action on 2058 letters with point-stabilizer isomorphic to $S_4(4) : 2$ (whose order is not divisible by seven) seems less promising. In fact, we can obtain copies of this maximal subgroup as follows. Firstly, note that the pre-image in $3 \cdot S_7$ of the centralizer of a transposition in S_7 is a group isomorphic to $S_3 \times S_5$. As an example of this we take

$$N_T = \langle(6, 5, 1)(4, 3, 2)t_2, (2, 6)(4, 5), t_0\rangle = \langle yt^{x^2}, (xy)^{x^{-2}}, t\rangle \cong S_3 \times S_5.$$

It turns out that $\langle N_T, r_7 s_{15}^{-2} = s^t s^{-2}\rangle \cong S_4(4) : 2$, and so we may readily obtain the permutation representation of minimal degree. Unfortunately it is not quite as easy as usual to extend this to a maximal subgroup

Table 7.1. Normalizers of subgroups generated by symmetric generators

Subgroup of $N \cong 3\dot{} S_7$	Orbits on $(15+15)$ symmetric generators	Corresponding maximal subgroup
$3\dot{} S_7$	transitive	$3\dot{} S_7$
$L_3(2)$	$(1+14)+(7+8)$	$7:3 \times L_3(2)$
C_7	$(1+7+7)+(1+7+7)$	$7^{1+2}:(S_3 \times 3)$
S_4	$(1+6+8)+(1+6+8)$	$S_4 \times L_3(2)$
V_4	$(1^3+4^3)+(1^3+4^3)$	$2^2\dot{} L_3(4):S_3$

of Aut He $\cong$ He : 2, as the outer half of $S_4(4):4$ contains no involutions. However, for completeness we mention that

$$\langle (6,5,1)(4,3,2)t_2, as_{15}r_1^3 r_3^{-1}\rangle = \langle yt^{x^2}, as(s^{tx})^3 (s^{tx^3})^{-1}\rangle \cong S_4(4):4.$$

Now, a subgroup of N isomorphic to $3\dot{} S_6$ is given by $\langle y, xtx^4\rangle$. If we adjoin the element $s_{15}^2 r_7^2 s_{15}^{-1}$ to this, we obtain the maximal subgroup

$$\langle y, xtx^4, s^2(s^t)^2 s^{-1}\rangle \cong 2^6 : 3\dot{} S_6.$$

Of course, the two conjugacy classes of subgroups of this shape are interchanged by our outer automorphism a.

Our element t is a representative of the conjugacy class of involutions denoted by $2B$ in the ATLAS. Its centralizer is given by

$$\langle y, r_1^2 s_1^{-1} r_1^4, r_1^2 s_1^3 r_1^3\rangle \cong 2^{1+6}:L_3(2),$$

where $r_1 = s^{tx}$ and $s_1 = s^{txt}$. This maximal subgroup is normalized by the outer automorphism a. The maximal subgroups of He which were worked out by Butler [18] are listed on p. 104 of the ATLAS. We have given generators for a representative of each class except those of shape $7^2:2\dot{} L_2(7)$ and $5^2:4\dot{} A_4$. In fact the former of these is simply $N_{\mathrm{He}}(\langle x, s_{15}t_{15}^{-1}\rangle)$ and the latter is $N_{\mathrm{He}}(\mathrm{Syl}_5(\mathrm{He}))$, and so representatives can be readily obtained using MAGMA. In addition, Aut He has two classes of maximal subgroups known as novelties whose intersections with He are not maximal in the simple group. These subgroups, which have shapes $(S_5 \times S_5):2$ and $2^{4+4}.(S_3 \times S_3).2$, were found by Wilson [87].

7.4.3   The Harada–Norton group[4]

In Section 7.4.2 we took $N \cong 3\dot{} S_7$ as our control subgroup and used a $(15+15)$-dimensional faithful monomial representation of N to construct a progenitor of shape

$$7^{\star(15+15)}:3\dot{} S_7.$$

---

[4]For further detailed information about this construction, the reader is referred to Bray and Curtis [16].

The subgroup of N fixing one of the symmetric generators, which is isomorphic to $L_2(7)$, acts with orbits $(1+14)+(7+8)$ on the 30 cyclic subgroups of order 7. Fixing a further symmetric generator in the 7-orbit is a subgroup isomorphic to S_4 which acts with orbits $(1+6+8)+(1+6+8)$. Normalizing the subgroup generated by these two symmetric generators of order 7, r_0 and s_0 say, we have the 'central' element of order 3, which may be taken to square the r_i and fourth-power the s_i, and an involution, commuting with the aforementioned S_4, interchanging them (replacing s_0 by a power if necessary). The subgroup of N isomorphic to S_4 mentioned above centralizes r_0 and s_0. Thus, in our usual notation these two automorphisms of 7^{*2} are denoted by

$$\begin{pmatrix} 2 & \cdot \\ \cdot & 4 \end{pmatrix} \quad \text{and} \quad \begin{pmatrix} \cdot & 1 \\ 1 & \cdot \end{pmatrix}.$$

Therefore, factoring by the relator

$$\left[\begin{pmatrix} \cdot & 1 \\ 1 & \cdot \end{pmatrix} r_0 \right]^3$$

ensures that $\langle r_0, s_0 \rangle \cong L_2(7)$, or an image thereof. Factoring one of the two progenitors of shape $7^{*(15+15)} :_m 3{\cdot}S_7$ by a relator corresponding to this results in the Held group He, a sporadic simple group of order $4\,030\,387\,200$; with the other progenitor we obtain the trivial group. The outer automorphism of He is obtained by adjoining an element of order 2 which commutes with the control subgroup and inverts all the symmetric generators.

It turns out that the Harada–Norton group, HN, can be obtained in a remarkably analogous manner, using a progenitor with symmetric generators of order 5. We take as our control subgroup the group N of shape $2{\cdot}\mathrm{HS}{:}2$ in which the outer involutions lift to elements of order 2. (The isoclinic variant of this group, namely $2{\cdot}\mathrm{HS}{\cdot}2$, has no outer involutions.) This group contains a subgroup $H \cong (2 \times U_3(5)){\cdot}2 \cong U_3(5){:}4$, which is generated by $U_3(5)$ together with an element of order 4 acting on it as an outer automorphism and squaring to the central involution. Thus, $H/H' \cong C_4$. In the usual way, we map a generator of H/H' onto a primitive fourth root of unity in an appropriate field. We induce the corresponding linear representation of H up to N to obtain a faithful monomial $(176+176)$-dimensional representation of N. Over the complex numbers $\mathbb{C}$, this gives an irreducible representation whose restriction to $2{\cdot}\mathrm{HS}$ has character which is the sum of the two 176-dimensional characters given in Table 7.2. Of course, the field with fourth roots of unity which interests us is $\mathbb{Z}_5$, which enables us to define a progenitor of shape

$$P = 5^{*(176+176)} :_m 2{\cdot}\mathrm{HS}{:}2.$$

Table 7.2. Some characters of $2{\cdot}\mathrm{HS}$ (taken from the ATLAS [25])

44 352 000	7680	2880	360	3840	256	64	500	300	25	36	24	7	16	16	16	20	20	11	11	12	15	20	20
p power	A	A	A	A	A	A	A	A	A	AB	AA	A	B	C	C	AA	BB	A	A	BA	BA	AA	AA
p' part	A	A	A	A	A	A	A	A	A	AB	AA	A	A	A	A	AA	BB	A	A	AA	BA	AA	AA
ind 1A	2A	2B	3A	4A	4B	4C	5A	5B	5C	6A	6B	7A	8A	8B	8C	10A	10B	11A	B**	12A	15A	20A	B**

ind	1A	2A	2B	3A	4A	4B	4C	5A	5B	5C	6A	6B	7A	8A	8B	8C	10A	10B	11A	B**	12A	15A	20A	B**
+	1	1	1	1	1	1	1	1	1	1	1	1	1	1	1	1	1	1	1	1	1	1	1	1
+	22	6	−2	4	−6	2	2	−3	2	2	−2	0	1	0	0	0	1	−2	0	0	0	−1	−1	−1
+	77	13	1	5	5	5	1	2	−3	2	1	1	0	1	−1	−1	−2	1	0	0	−1	0	0	0
+	154	10	10	1	−2	6	−2	4	4	−1	1	1	0	0	0	0	0	0	0	0	1	1	−2	−2
+	154	10	−10	1	−10	−2	2	4	4	−1	−1	1	0	0	2	−2	0	0	0	0	−1	1	0	0
+	154	10	−10	1	−10	−2	2	4	4	−1	−1	1	0	0	−2	2	0	0	0	0	−1	1	0	0
+	175	15	11	4	15	−1	3	0	5	0	2	0	0	−1	1	1	−1	−1	−1	−1	0	−1	0	0
+	231	7	−9	6	15	−1	−1	6	1	1	0	−2	0	−1	−1	−1	2	1	0	0	0	1	0	0

ind	1	2		3	4		4	5	5	5	12	6	7	8	8	8	10	20	11	11	12	15	20	20
	2	2		6	4		4	10	10	10	12	6	14	8	8	8	10		22	22	12	30	20	20

ind	1A	2A	2B	3A	4A	4B	4C	5A	5B	5C	6A	6B	7A	8A	8B	8C	10A	10B	11A	B**	12A	15A	20A	B**
+	56	8	0	2	0	0	0	6	−4	1	0	2	0	0	0	0	−2	0	1	1	0	2	0	0
o	176	16	0	5	16i	0	0	1	6	1	0	1	1	0	0	0	1	0	0	0	−i	0	i	−i
o	176	16	0	5	−16i	0	0	1	6	1	0	1	1	0	0	0	1	0	0	0	i	0	−i	i

As in the Held case, there are two non-isomorphic progenitors of this shape, depending on whether we choose 2 or 3 as our primitive fourth root of unity.

Let $\mathcal{T} = \mathcal{R} \cup \mathcal{S} = \{r_0, r_1, \ldots, r_{175}, s_0, s_1, \ldots, s_{175}\}$ be our set of symmetric generators, and let $\tilde{\mathcal{T}} = \{\langle r_i \rangle, \langle s_i \rangle : i = 0, 1, \ldots, 175\}$ be the set of cyclic subgroups they generate, arranged so that $N' \cong 2{\cdot}\mathrm{HS}$ has orbits $\bar{\mathcal{R}} = \{\langle r_0 \rangle, \langle r_1 \rangle, \ldots, \langle r_{175} \rangle\}$ and $\bar{\mathcal{S}} = \{\langle s_0 \rangle, \langle s_1 \rangle, \ldots, \langle s_{175} \rangle\}$ on $\tilde{\mathcal{T}}$. Then H may be chosen to normalize $\langle r_0 \rangle$, whence H' commutes with $\langle r_0 \rangle$, and both H and H' have orbits $(1 + 175) + (50 + 126)$ on $\tilde{\mathcal{T}} = \bar{\mathcal{R}} \cup \bar{\mathcal{S}}$. We now choose $\langle s_0 \rangle$ to lie in the 50-orbit of H (which is in $\bar{\mathcal{S}}$). Then $\langle r_0, s_0 \rangle$ is centralized (in N) by a subgroup K of H' isomorphic to A_7, which has orbits $(1 + 7 + 42 + 126) + (1 + 7 + 42 + 126)$ on $\tilde{\mathcal{T}}$. The subgroup $\langle r_0, s_0 \rangle$ is normalized in N by an element of order 4, which squares the r_i and cubes the s_i, and an involution interchanging r_0 and s_0. Together, these two elements extend K to $\mathrm{N}_N(K) \cong (2^2 \times A_7){:}2$. Thus, the subgroup $\langle r_0, s_0 \rangle \cong 5^{*2} \cong \langle r_0 \rangle \star \langle s_0 \rangle$ is normalized in N by a subgroup isomorphic to D_8, which, in the usual notation, acts as

$$\begin{pmatrix} 2 & \cdot \\ \cdot & 3 \end{pmatrix} \quad \text{and} \quad \begin{pmatrix} \cdot & 1 \\ 1 & \cdot \end{pmatrix}.$$

Therefore, factoring by the relator

$$\left[\begin{pmatrix} \cdot & 1 \\ 1 & \cdot \end{pmatrix} r_0 \right]^3,$$

where the matrix is one of the non-central involutions of

$$\mathrm{N}_N(K) \cong (2^2 \times A_7){:}2$$

that commutes with $K \cong A_7$, ensures that $\langle r_0, s_0 \rangle \cong L_2(5)$, or an image thereof. Factoring one of the two possible progenitors by such a relation yields Theorem 7.4.

THEOREM 7.4

$$G = \frac{5^{*(176+176)} :_m 2{\cdot}\mathrm{HS}{:}2}{(ts)^3} \cong \mathrm{HN},$$

a sporadic simple group of order $273\,030\,912\,000\,000$.

As before, factoring the other progenitor $5^{*(176+176)} :_m 2{\cdot}\mathrm{HS}{:}2$ by the corresponding relation gives rise to the trivial group. The outer automorphism of HN is obtained by adjoining an element of order 4 which commutes with N' and squares all the symmetric generators (and is thus inverted by the elements in $N \setminus N'$).[5]

---

[5] For further details, the reader is referred to Bray and Curtis [16].

7.5   The Suzuki chain and the Conway group

In Section 5.4 we repeatedly applied Lemma 3.3 to progenitors with involutory symmetric generators in which the centralizer in the control subgroup N of a 2-point stabilizer N_{ij} was cyclic of order 2, generated by an involution π_{ij} interchanging i and j. We factored by an additional relation $\pi_{ij} = t_i t_j t_i$ and obtained in turn $U_3(3){:}2$, $J_2{:}2$, $G_2(4){:}2$ and finally $3{\cdot}\text{Suz}{:}2$. At this point, the procedure appeared to terminate, but we mentioned that in this section we would modify the method slightly and hence extend the chain.

We have seen that the complex monomial representations we make use of to define our progenitors with non-involutory symmetric generators are written over some particular roots of unity, mth roots say. We seek a finite field $\mathbb{F}_q$ which contains mth roots of unity and, if q is a prime, interpret our monomial matrix as acting on a free product of n copies of the cyclic group C_q, where n is the degree of the representation. Now, q need not, of course, be prime, but we may have $q = p^r$, where p is prime and $r > 1$. In this case, our symmetric generators are elementary abelian subgroups rather than cyclic groups; we refer to them as *symmetrically generating subgroups* or simply *sg-subgroups*.

Till now when we have needed cube roots of unity we have chosen $\mathbb{Z}_7$, the integers modulo 7, but we could just as easily have chosen $\mathbb{F}_4$. For example, if we took $N \cong A_5$ as our control subgroup and restricted to a subgroup of index 5 isomorphic to A_4, then we could take the linear representation of A_4 onto the complex cube roots of unity and induce this representtaion up to N to obtain a 5-dimensional monomial representation of A_5, all of whose non-zero entries are cube roots of unity. We could then interpret these roots of unity as lying either in $\mathbb{Z}_7$ or in $\mathbb{F}_4$, and so obtain the following progenitors:

$$P_1 = 7^{\star 5} : A_5 \text{ and } P_2 = (2^2)^{\star 5} : A_5.$$

In the second case, the notation 2^2 stands for an elementary abelian group of order 4, in this case a Klein fourgroup V_4, so that the control subgroup is permuting five copies of V_4 which are unrelated to one another (except that they have the same identity).

It is our intention to take a somewhat grander control subgroup, namely $N \cong 3{\cdot}\text{Suz}:2$. We restrict to a subgroup of index 1782, which, since the group $G_2(4)$ possesses no triple cover and the elements in the outer half on N invert generators of its normal subgroup of order 3, has shape $(3 \times G_2(4)) : 2$. We map this group onto the symmetric group S_3 and interpret this group as acting on a copy of V_4 as its full group of automorphisms, generated by multiplication by an element of order 3 and the field automorphism. So if our copy of V_4 has elements $\{1, v_1, v_\omega, v_{\bar\omega}\}$ then multiplication by ω acts as $(v_1, v_\omega, v_{\bar\omega})$ and the field automorphism acts as $(v_1)(v_\omega, v_{\bar\omega})$. This further subtlety was necessary, of course, because we have included the

Table 7.3. Labelling of the elements of two copies of V_4 embedded in S_5

	First V_4		Second V_4		
Generator	Label	Element in S_5	Generator	Label	Element in S_5
u_1	1	(b c)(d e)	v_1	4	(a c)(d e)
u_ω	2	(b d)(e c)	v_ω	5	(a d)(e c)
$u_{\bar\omega}$	3	(b e)(c d)	$v_{\bar\omega}$	6	(a e)(c d)

outer automorphism of 3˙Suz. Note that it will result in the image group G possessing subgroups of shape $(A_4 \times G_2(4)) : 2$.

We are, in effect, inducing up a semi-linear representation of S_3 (multiplication by ω, together with complex conjugation). Since this is a novel idea, we firstly consider the smallest possible case, in which we have just two copies of V_4. We shall embed these two copies of V_4 in S_5 acting on the set $\Lambda = \{a, b, c, d, e\}$. We have labelled the six non-trivial elements of the two fourgroups as integers $1, \ldots, 6$ as we shall need to write the additional relations in the standard sequence-permutation form (see Table 7.3).

Multiplication by ω corresponds to the permutation (1 2 3)(4 5 6) and can be realized within S_5 by conjugation by $(c\ d\ e)$. The field automorphism $\sigma : \omega \leftrightarrow \bar\omega$ corresponds to the permutation (2 3)(5 6) and is realized within S_5 by $(d\ e)$. Finally note that the two copies of V_4 are interchanged by (1 4)(2 5)(3 6), which is realized in S_5 by $(a\ b)$. Together these three elements generate $2 \times S_3 \cong D_{12}$. We can represent these actions on the free product $(2^2)^{\ast 2}$ in the usual matrix manner as

$$\begin{pmatrix} \omega & \\ & \omega \end{pmatrix}, \begin{pmatrix} \sigma & \\ & \sigma \end{pmatrix}, \begin{pmatrix} & 1 \\ 1 & \end{pmatrix},$$

respectively. Observe further that (1 4)(2 6)(3 5), which corresponds in S_5 to $(a\ b)(d\ e)$, would be represented by the matrix

$$\begin{pmatrix} & \sigma \\ \sigma & \end{pmatrix},$$

and its product with u_1 has order 3. So we see that a homomorphism exists:

$$\frac{(2^2)^{\ast 2} : D_{12}}{\left[\begin{pmatrix} & \sigma \\ \sigma & \end{pmatrix} u_1\right]^3} \mapsto S_5.$$

In fact this is an isomorphism as we shall demonstrate using the double coset enumerator, at the same time showing how the input must be modified to cope with the slightly changed circumstances. If the progenitor is $(2^2)^{\ast n} : N$, then we input N as permutations on the $3n$ non-trivial elements in the sg-subgroups. We must now identify a triple whose product is the identity, say $\{i, j, k\}$, and factor by the relation $< [i, j, k], Id(N) >$. So we have:

```
> nnd12:=sub<Sym(6)|(1,5,3,4,2,6),(1,4)(2,6)(3,5)>;
> RRs5:=[<[1,2,3],Id(nnd12)>,<[1,4,1],nnd12.2>];
> ENs5:=DCEnum(nnd12,RRs5,[nnd12]:Print:=5,Grain:=100);

Index: 10 === Rank: 3 === Edges: 7 === Time: 0.016
> ENs5[4];
[
    [],
    [ 1 ],
    [ 1, 5 ]
]
> ENs5[7];
[ 1, 6, 3 ]
```

So, the image group is a pre-image of S_5 of order $12 \times 10 = 120 = 5!$, and so the map is an isomorphism. This example, which could readily be proved by hand to yield S_5, gives us a clue as to how to proceed with the larger progenitor

$$P = (2^2)^{\star 1782} : (3 \dot{} \mathrm{Suz} : 2).$$

A non-trivial element z in the centre of N' may be taken to map $z : (u_1 \; u_\omega \; u_{\bar\omega})$ for each of the 1782 fourgroups; it is inverted by any element in $N \setminus N'$. Now the normalizer in N of one of the sg-subgroups, U say, is isomorphic to $(3 \times G_2(4)) : 2$, which contains a subgroup isomorphic to $G_2(4) : 2$ fixing an involution in U (but interchanging the other two). This group has orbits of lengths $1 + 416 + 1365$ on the sg-subgroups, and fixing a fourgroup V in the 416 orbit is a subgroup isomorphic to $J_2 : 2$, which must fix one of the three involutions in each of the two fixed sg-subgroups and interchange the other two as it must invert z. But this last subgroup can be seen from the ATLAS [25], p. 131, to commute with an involution in $N \setminus N'$ which interchanges the two fourgroups, and so, if U and V have non-trivial elements $\{u_1, u_\omega, u_{\bar\omega}\}$ and $\{v_1, v_\omega, v_{\bar\omega}\}$, respectively, then this element, π say, will have action $(u_1 \; v_1)(u_\omega \; v_{\bar\omega})(u_{\bar\omega} \; v_\omega)$, where u_1 and v_1 are the two fixed elements. In order to ensure that, as above, $\langle U, V \rangle \cong A_5$ (or the trivial group), we must factor by the relator $(\pi u_1)^3$. The situation is slightly more complicated than has been described here because of the triple cover. There are of course three pre-images of π in the triple cover, but only one of them is fixed by the copy of J_2 we are considering. We now claim the following.

THEOREM 7.5

$$G = \frac{(2^2)^{\star 1782} : 3 \dot{} \mathrm{Suz} : 2}{(\pi u_1)^3} \cong \mathrm{Co}_1,$$

Conway's largest simple group.

Verification In order to use the enumerator to verify this claim, we first need to obtain the control subgroup as permutations on $3 \times 1782 = 5346$ letters. We do this by taking Soicher's Coxeter-style presentation from the ATLAS [25], p. 131, and note that the subgroup generated by $\{a, b, c, d, e, f, g\}$ is isomorphic to $G_2(4) : 2$ with the required index.

```
> G<a,b,c,d,e,f,g,h>:=Group<a,b,c,d,e,f,g,h|a^2=b^2=
  c^2=d^2=e^2=
> f^2=g^2=h^2=(a*b)^3=(a*c)^2=(a*d)^2=(a*e)^2=(a*f)^2=
  (a*g)^2=
> (a*h)^2=(b*c)^3=(b*d)^2=(b*e)^2=(b*f)^2=(b*g)^2=
  (b*h)^2=
> (c*d)^8=(c*e)^2=(c*f)^2=(c*g)^2=(c*h)^2=
> (d*e)^3=(d*f)^2=(d*g)^2=(d*h)^2=
> (e*f)^3=(e*g)^2=(e*h)^2=(f*g)^3=(f*h)^2=(g*h)^3=
> (c*d)^4*a=(b*c*d*e)^8=1>;
> H:=sub<G|a,b,c,d,e,f,g>;
> Index(G,H);
5346
> CT:=CosetTable(G,H);
> nn:=CosetTableToPermutationGroup(G,CT);
```

So *nn* is the required permutation group of degree 3×1782; we must now identify the three involutions in one of the sg-subgroups, so that we can set their product equal to the identity. These triples are of course simply the orbits of the normal subgroup of order 3. The second relation is equally easy to find in our notation, for we have $\langle a, b, c, d, e, f \rangle \cong J_2 : 2$, whose centralizer is $\langle h \rangle$ – the eighth generator of *nn*. We require the product of this element h and u_1 to have order 3, and record this as the relation $< [1, 1^h, 1], h >$, noting that the machine records the image of h as a permutation on 5346 letters as *nn*.8:

```
> znn:=Centre(DerivedGroup(nn));
> #znn;
3
> 1^znn;
GSet{ 1, 5345, 5346 }
> RR:=[<[1,5345,5346],Id(nn)>,<[1,1^nn.8,1],nn.8>];
> EN:=DCEnum(nn,RR,[nn]:Print:=5,Grain:=100);

Index: 1545600 === Rank: 5 === Edges: 19 === Time: 16.828
> EN[4];
[
    [],
    [ 1 ],
    [ 1, 2291 ],
    [ 1, 31 ],
    [ 1, 2291, 8 ]
```

```
]
> EN[7];
[ 1, 5346, 1111968, 405405, 22880 ]
>
```

Thus we confirm that the index is as required, and so if we verify that Co_1 is an image of G, then the theorem is verified.

7.6 Systematic approach

7.6.1 The progenitors

In this section we perform a similar systematic search to that conducted in Section 3.11, but in the case when the symmetric generators have order 3. The results are taken from Bray and Curtis [15]. Thus we seek homomorphic images of a progenitor of the form

$$3^{*n} : N,$$

where N, a transitive permutation group on n letters, may act either monomially or as permutations on the symmetric generators. We shall limit ourselves to the permutation progenitors $3^{*3} : S_3$, $3^{*4} : S_4$, $3^{*5} : A_5$, $3^{*5} : S_5$, $3^{*6} : L_2(5)$ and $3^{*7} : L_3(2)$, and the monomial progenitors $3^{*2} :_m D_8$, $3^{*3} :_m S_3$, $3^{*4} :_m S_4$, $3^{*3} :_m 2^2 S_3$, $3^{*4} :_m 2 \cdot S_4{}^+$, $3^{*5} :_m S_5$, $3^{*6} :_m L_2(5)$ and $3^{*7} :_m L_3(2)$. Here, when regarded as permutation groups on the $T_i = \langle t_i \rangle$, A_n and S_n denote (respectively) the alternating and symmetric groups in their natural actions on n points, and $L_n(q)$ denotes the projective special linear group in n dimensions over $\mathbb{F}_q$ acting on 1-dimensional subspaces (i.e. points).

For all these monomial progenitors except $3^{*3} :_m 2^2 S_3$ and $3^{*4} :_m 2 \cdot S_4{}^+$, there is a unique subgroup of index 2 in N_0, so these progenitors are uniquely determined (up to isomorphism). For the case $N \cong 2 \cdot S_4{}^+$, we saw in Example 6.1 that N_0 had three subgroups of index 2, but up to automorphisms of N we obtained just one faithful monomial representation of N on 3^{*4}, thus $3^{*4} :_m 2 \cdot S_4{}^+$ is uniquely determined. (Also, $2 \wr S_4 \cong 2^4 : S_4$, the full monomial automorphism group of 3^{*4}, has just one conjugacy class of subgroups isomorphic to $2 \cdot S_4{}^+$.) Now, the full monomial automorphism group of 3^{*3} is $2 \wr S_3 \cong 2^3 : S_3 \cong S_4 \times 2$. This contains three subgroups of index 2, two of which have shape $2^2 : S_3 \cong S_4$. This gives rise to two isomorphism classes for a progenitor of shape $3^{*3} :_m 2^2 S_3$. The line-stabilizer in $3^{*7} :_m L_3(2)$, i.e. $N_N(\langle t_0, t_1, t_3 \rangle)$ in the notation of Table 7.4, is a progenitor of shape $3^{*3} :_m 2^2 S_3$. This progenitor, which we shall henceforth write as $3^{*3} :_m S_4$, is the one of shape $3^{*3} :_m 2^2 S_3$, which we shall study here.

Presentations for all the progenitors we shall study here are given in Tables 7.5 and 7.4. Table 7.5 lists presentations for the permutation progenitors and Table 7.4 deals with the proper monomial progenitors. Our

Table 7.4. Presentations of the progenitors $3^{\star n} :_m N$ that we are considering

(vii)	$\langle x, y, t \mid x^4 = y^2 = (yx)^2 = 1 = t^3 = [t, y] = (x^2 t)^2 \rangle \cong 3^{\star 2} :_m D_8$
	$x \sim (0, 1, \bar{0}, \bar{1}), \; y \sim (1, \bar{1}), \; t \sim t_0$
(viii)	$\langle x, y, t \mid x^3 = y^2 = (yx)^2 = 1 = t^3 = (yt)^2 \rangle \cong 3^{\star 3} :_m S_3$
	$x \sim (0, 1, 2), \; y \sim (0, \bar{0})(1, \bar{2}), \; t \sim t_0$
(ix)	$\langle x, y, t \mid x^4 = y^2 = (yx)^3 = 1 = t^3 = (yt)^2 = (yt^x)^2 \rangle \cong 3^{\star 4} :_m S_4$
	$x \sim (0, \bar{1}, 2, \bar{3}), \; y \sim (0, \bar{0})(1, \bar{1})(2, \bar{3}), \; t \sim t_0$
(x)	$\langle x, y, t \mid x^4 = y^2 = (yx)^3 = 1 = t^3 = [t, y] = (x^2 t)^2 \rangle \cong 3^{\star 3} :_m S_4$
	$x \sim (0, 1, \bar{0}, \bar{1})(2, \bar{2}), \; y \sim (1, \bar{2}), \; t \sim t_0$
(xi)	$\langle x, y, t \mid x^4 = (yx)^3, y^2 = 1 = t^3 = [t, y] = (yt^x)^2 \rangle \cong 3^{\star 4} :_m 2\dot{}S_4{}^+$
	$x \sim (0, \bar{1}, \bar{2}, 3, \bar{0}, 1, 2, \bar{3}), \; y \sim (1, \bar{1})(2, 3), \; t \sim t_0$
(xii)	$\langle x, y, t \mid x^5 = y^3 = (xy)^2 = 1 = t^3 = [t, x] = (t^{yx^2} xy)^2 \rangle \cong 3^{\star 6} :_m L_2(5)$
	$x \sim (0, 1, 2, 3, 4), \; y \sim (\infty, 0, 1)(2, 4, \bar{3}), \; t \sim t_\infty$
(xiii)	$\langle x, y, t \mid x^5 = y^2 = (yx)^4 = [x, y]^3 = 1 = t^3 = (yt)^2 = (yt^x)^2 = (yt^{x^2})^2 \rangle$
	$\cong 3^{\star 5} :_m S_5$
	$x \sim (0, 1, 2, 3, 4), \; y \sim (0, \bar{0})(1, \bar{1})(2, \bar{2})(3, 4), \; t \sim t_0$
(xiv)	$\langle x, y, t \mid x^7 = y^2 = (xy)^3 = [x, y]^4 = 1 = t^3 = [t^{x^4}, xy] = (ty^{x^4})^2 (= [t, y]) \rangle$
	$\cong 3^{\star 7} :_m L_3(2)$
	$x \sim (0, 1, 2, 3, 4, 5, 6), \; y \sim (1, \bar{1})(3, \bar{3})(2, \bar{6})(4, 5), \; t \sim t_0$

presentations for the progenitors are readily obtained from classical presentations of the control subgroups (see Coxeter and Moser [27]) extended by a generator of order 3 which is centralized (or normalized) by a subgroup of the required index. For instance, in Table 7.5(vi), our symmetric generator $t = t_0$ is centralized by $(xy)^{x^3}$ and y^{x^4}, which together generate a subgroup of $\langle x, y \rangle \cong L_3(2)$ isomorphic to S_4. In fact, this subgroup contains y, so that the relation $[t, y] = 1$ also holds in our group.

7.6.2   Additional relations

Unfortunately, Lemma 3.3 is not as useful at determining which elements can lie in $\langle t_0, t_1 \rangle \cap N$ as it was in the permutation case. Indeed, for eight of our progenitors, namely, $3^{\star 3} : S_3$, $3^{\star 2} :_m D_8$, $3^{\star 3} :_m S_3$, $3^{\star 4} :_m S_4$, $3^{\star 3} :_m S_4$, $3^{\star 4} :_m 2\dot{}S_4{}^+$, $3^{\star 6} :_m L_2(5)$ and $3^{\star 7} :_m L_3(2)$, this criterion places no restriction on $\langle t_0, t_1 \rangle \cap N$ whatsoever. In these cases, inspection of the following tables shows that we can have $\langle t_0, t_1 \rangle \cap N = N$; indeed, we can even have $\langle t_0, t_1 \rangle = G$.

Table 7.5. Presentations of the progenitors $3^{\star n}:N$ that we are considering

(i)	$\langle x, y, t \mid x^3 = y^2 = (yx)^2 = 1 = t^3 = [t, y] \rangle \cong 3^{\star 3}:S_3$
	$x \sim (0, 1, 2),\ y \sim (1, 2),\ t \sim t_0$
(ii)	$\langle x, y, t \mid x^4 = y^2 = (yx)^3 = 1 = t^3 = [t, y] = [t^x, y] \rangle \cong 3^{\star 4}:S_4$
	$x \sim (0, 1, 2, 3),\ y \sim (2, 3),\ t \sim t_0$
(iii)	$\langle x, y, t \mid x^5 = y^3 = (xy)^2 = 1 = t^3 = [t, y] = [t, y^{x^2}] \rangle \cong 3^{\star 5}:A_5$
	$x \sim (0, 1, 2, 3, 4),\ y \sim (4, 2, 1),\ t \sim t_0$
(iv)	$\langle x, y, t \mid x^5 = y^2 = (yx)^4 = [x, y]^3 = 1 = t^3 = [t, y] = [t^x, y] = [t^{x^2}, y] \rangle \cong 3^{\star 5}:S_5$
	$x \sim (0, 1, 2, 3, 4),\ y \sim (3, 4),\ t \sim t_0$
(v)	$\langle x, y, t \mid x^5 = y^3 = (xy)^2 = 1 = t^3 = [t, x] = [t^{yx^2}, xy] \rangle \cong 3^{\star 6}:L_2(5)$
	$x \sim (0, 1, 2, 3, 4),\ y \sim (\infty, 0, 1)(2, 4, 3),\ t \sim t_\infty$
(vi)	$\langle x, y, t \mid x^7 = y^2 = (xy)^3 = [x, y]^4 = 1 = t^3 = [t^{x^4}, xy] = [t, y^{x^4}] \rangle \cong 3^{\star 7}:L_3(2)$
	$x \sim (0, 1, 2, 3, 4, 5, 6),\ y \sim (2, 6)(4, 5),\ t \sim t_0$

Other useful relations by which to factor the progenitors are those of the form $(\pi w)^s = 1$, where $\pi \in N$ and w is a short word in the symmetric generators. In this context we define the orders of elements of the form πt_i^a to be the *first order parameters* of the symmetric presentation; similarly, the orders of elements of the form $\pi t_i^a t_j^b$ are *second order parameters*, and so on. In Tables 7.6 and 7.7 we give, for each progenitor, a list of elements whose orders form a complete set of first order parameters, together with other useful parameters.

7.6.3   Low degree representations of progenitors

We shall often adopt the approach which by-passes a search for suitable additional relations and, instead of attempting to construct a Cayley diagram, seeks a low degree permutation or matrix representation of the progenitor. In Curtis [31] it is shown, in effect, that representing the progenitors $3^{\star 5}:A_5$ and $2^{\star 7}:L_3(2)$ on 12 and 24 letters, respectively, with the control subgroup acting transitively in each case, results immediately in the Mathieu groups M_{12} and M_{24}. Indeed, curiously, the smallest true permutation representation of an 'interesting' progenitor often turns out to produce a sporadic simple group.

We shall illustrate this approach here by considering permutation representations of $3^{\star 4}:_m 2{}^{\cdot}S_4{}^+$ of degree less than or equal to 11. Modulo (outer) automorphisms, $N \cong 2{}^{\cdot}S_4{}^+$ possesses just one faithful permutation representation of degree 8, and this is the only faithful (transitive) representation of degree less than 12. The control subgroup N has a unique

Table 7.6. Extra relations for the progenitors $3^{\star n}:N$
(permutation action)

$3^{\star 3}:S_3$	$3^{\star 5}:A_5$
$[(0,1,2)t_0]^a = 1$	$[(0,1,2,3,4)t_0]^a = 1$
$[(0,1)t_0]^b = 1$	$[(0,2,4,1,3)t_0]^b = 1$
$[t_0 t_1^{-1}]^c = 1$	$[(0,1,2)t_0]^c = 1$
$[(0,1,2)t_0 t_1^{-1}]^d = 1$	$[t_0 t_1]^{d/2} = 1$
$[(0,1)t_0 t_2]^e = 1$	$[t_0^{-1} t_1]^e = 1$
$[(0,1)t_0 t_2^{-1}]^f = 1$	$[(0,1,2)t_0^{-1} t_1]^f = 1$
	$(4,2,1) = [t_0 t_3]^{s_1}$
	$(4,2,1) = [t_0^{-1} t_3]^{s_2}$

$3^{\star 4}:S_4$	$3^{\star 5}:S_5$
$[(0,1,2,3)t_0]^a = 1$	$[(0,1,2)(3,4)t_0]^a = 1$
$[(0,1,2)t_0]^b = 1$	$[(0,1,2,3,4)t_0]^b = 1$
$[(0,1)(2,3)t_0]^c = 1$	$[(0,1,2,3)t_0]^c = 1$
$[(0,1)t_0]^d = 1$	$[(0,1,2)t_0]^d = 1$
$[t_0^{-1} t_1]^e = 1$	$[(0,1)t_0]^e = 1$
$[(0,1,2)t_0 t_1^{-1}]^f = 1$	$[t_0^{-1} t_1]^f = 1$
$[(0,1)t_0 t_2]^p = 1$	$[(0,1,2)t_0^{-1} t_1]^p = 1$
$[(0,1)t_0 t_2^{-1}]^q = 1$	$[(0,1)t_0 t_2]^q = 1$
$[t_0 t_1]^{s_1} = (2,3)$	$[(0,1)t_0 t_2^{-1}]^r = 1$
$[t_0^{-1} t_1]^{s_2} = (2,3)$	

$3^{\star 6}:L_2(5)$	$3^{\star 7}:L_3(2)$
$[(0,1,2,3,4)t_0]^a = 1$	$[(0,1,2,3,4,5,6)t_0]^a = 1$
$[(0,2,4,1,3)t_0]^b = 1$	$[(0,6,5,4,3,2,1)t_0]^b = 1$
$[(\infty,0,1)(2,4,3)t_\infty]^c = 1$	$[t_0 t_1 t_3]^{c/3} = 1$
$[(\infty,0,1)(2,4,3)t_2]^d = 1$	$[(0,1,6)(2,3,5)t_0]^d = 1$
$[(\infty,0)(1,4)t_\infty]^e = 1$	$[t_0 t_1]^{e/2} = 1$
$[(\infty,0)(1,4)t_1]^f = 1$	$[(2,6)(4,5)t_1 t_3]^{f/2} = 1$
$[t_\infty^{-1} t_0]^p = 1$	$[(0,3)(2,6,4,5)t_2]^p = 1$
$[t_2 t_3]^{s_1} = (\infty,0)(1,4)$	$[t_1 t_3]^{s_1} = (2,6)(4,5)$
$[t_2 t_3^{-1}]^{s_2} = (\infty,0)(1,4)$	$[t_1 t_3^{-1}]^{s_2} = (2,6)(4,5)$

transitive representation for each of the degrees 1, 2 and 3. The imprimitive 8-point action N may be generated by $a = (0)(1,2,3)(\bar{0})(\bar{1},\bar{2},\bar{3})$ and $b = (0,1)(\bar{0},\bar{1})(2,\bar{2})(3)(\bar{3})$, whose actions in the one-, two- and three-letter representations are shown in Table 7.8. Our 3-element t_0 must commute with a and be inverted by the central element $(ab)^4 = (0,\bar{0})(1,\bar{1})(2,\bar{2})(3,\bar{3})$. Furthermore, one of $(2,3)(\bar{2},\bar{3})(0,\bar{0})$ and $(2,\bar{3})(\bar{2},3)(1,\bar{1})$ must commute with t_0, and the other invert it. It is left to the reader to verify that the five cases exhibited in Table 7.8 are the only possibilities. The group $3^2:2\,{}^{\cdot}S_4^+ \cong M_9:S_3$ of Table 7.8 is the famous *Hessian group*, which is the full automorphism group of the Steiner system, $S(2,3,9)$.

Table 7.7. Extra relations for the progenitors $3^{\star n} :_m N$ (monomial action)

$3^{\star 3} :_m D_8$	$3^{\star 3} :_m S_3$	$3^{\star 4} :_m S_4$
$[(0,1,\bar{0},\bar{1})t_0]^a = 1$	$[(0,1,2)t_0]^a = 1$	$[(0,\bar{1},2,\bar{3})t_0]^a = 1$
$[(0,1)t_0]^b = 1$	$[(0,1,2)t_0^{-1}]^b = 1$	$[(0,\bar{1},2,\bar{3})t_0^{-1}]^b = 1$
$[(1,\bar{1})t_0 t_1]^c = 1$	$[(0,\bar{1})(2,\bar{2})t_0]^c = 1$	$[(0,\bar{1})(2,\bar{2})(3,\bar{3})t_0]^c = 1$
$[(0,1)t_0 t_1 t_0^{-1}]^d = 1$	$[(0,1,2)t_0 t_1^{-1}]^d = 1$	$[(0,1,2)t_0]^d = 1$
$[(0,1,\bar{0},\bar{1})t_0 t_1 t_0]^e = 1$	$[t_0 t_1]^e = 1$	$[(0,1,2)t_0^{-1}]^e = 1$
	$[(0,\bar{1})(2,\bar{2})t_0 t_2]^f = 1$	$[(0,1)(2,3)t_0]^f = 1$
	$[(0,\bar{1})(2,\bar{2})t_2 t_0]^p = 1$	$[(0,1,2)t_0 t_1^{-1}]^p = 1$
	$[(0,1,2)t_0 t_2 t_1]^q = 1$	$[(0,\bar{1})(2,\bar{2})(3,\bar{3})t_0 t_2]^q = 1$
		$[(0,\bar{1})(2,\bar{2})(3,\bar{3})t_2 t_0]^r = 1$

$3^{\star 3} :_m S_4$	$3^{\star 4} :_m 2 \cdot S_4^+$	$3^{\star 6} :_m L_2(5)$
$[(0,1,\bar{0},\bar{1})(2,\bar{2})t_0]^a = 1$	$[(0,\bar{1},\bar{2},3,\bar{0},1,2,\bar{3})t_0]^a = 1$	$[(0,1,2,3,4)t_0]^a = 1$
$[(0,1,2)t_0]^b = 1$	$[(0,\bar{1},2,\bar{0},1,\bar{2})(3,\bar{3})t_0]^b = 1$	$[(0,1,2,3,4)t_0^{-1}]^b = 1$
$[(1,\bar{2})t_1]^c = 1$	$[(0,\bar{2},\bar{0},2)(1,\bar{3},\bar{1},3)t_0]^c = 1$	$[(0,2,4,1,3)t_0]^c = 1$
	$[(0,2,1)t_0]^d = 1$	$[(0,2,4,1,3)t_0^{-1}]^d = 1$
	$[(1,\bar{1})(2,3)t_2]^e = 1$	$[(\infty,0,1)(2,4,\bar{3})t_\infty]^e = 1$
		$[(\infty,0,1)(2,4,\bar{3})t_\infty^{-1}]^f = 1$
		$[(\infty,0,1)(2,4,\bar{3})t_2]^p = 1$
		$[(\infty,0,1)(2,4,\bar{3})t_2^{-1}]^q = 1$
		$[(\infty,0)(1,4)(2,\bar{2})(3,\bar{3})t_\infty]^r = 1$
		$[(\infty,0)(1,4)(2,\bar{2})(3,\bar{3})t_1]^s = 1$

$3^{\star 5} :_m S_5$	$3^{\star 7} :_m L_3(2)$
$[(0,\bar{1})(2,\bar{3},4,\bar{2},3,\bar{4})t_0]^a = 1$	$[(0,1,2,3,4,5,6)t_0]^a = 1$
$[(0,\bar{1})(2,\bar{3},4,\bar{2},3,\bar{4})t_0^{-1}]^b = 1$	$[(0,6,5,4,3,2,1)t_0]^b = 1$
$[(0,\bar{1})(2,\bar{3},4,\bar{2},3,\bar{4})t_2]^c = 1$	$[(1,\bar{1})(3,\bar{3})(2,\bar{6})(4,5)t_2]^c = 1$
$[(0,1,2,3,4)t_0]^d = 1$	$[(1,\bar{1})(0,3,\bar{0},\bar{3})(2,6,4,\bar{5})t_0]^d = 1$
$[(0,\bar{1},2,\bar{3})(4,\bar{4})t_0]^e = 1$	$[(1,\bar{1})(0,3,\bar{0},\bar{3})(2,6,4,\bar{5})t_2]^e = 1$
$[(0,\bar{1},2,\bar{3})(4,\bar{4})t_0^{-1}]^f = 1$	$[(1,\bar{1})(0,3,\bar{0},\bar{3})(2,6,4,\bar{5})t_2^{-1}]^f = 1$
$[(0,1,2)t_0]^p = 1$	$[(0,\bar{1},6)(2,\bar{3},\bar{5})t_0]^p = 1$
$[(0,1)(2,3)t_0]^q = 1$	$[(0,\bar{1},6)(2,\bar{3},\bar{5})t_0^{-1}]^q = 1$
$[(0,\bar{1})(2,\bar{2})(3,\bar{3})(4,\bar{4})t_0]^r = 1$	$[(0,\bar{1},6)(2,\bar{3},\bar{5})t_2]^r = 1$

In order to demonstrate representing progenitors as matrices, we consider a family of images of $3^{\star 6} :_m L_2(5)$. As is well known [40], the projective special linear group $L_2(q)$ contains subgroups isomorphic to A_5 if, and only if, $q \equiv \pm 1$ or $0 \pmod 5$, and modulo automorphisms of $L_2(q)$ there is just one class of such subgroups. Furthermore, A_5 contains a unique class of subgroups isomorphic to D_{10}. Now, an element of order 5 in such a D_{10} will commute with an element of order 3 in $L_2(q)$ just when $q \equiv \pm 1 \pmod{15}$, and moreover this element is unique up to inversion and is inverted by the involutions in the D_{10}. This shows that when $q \equiv \pm 1 \pmod{15}$, $L_2(q)$ contains a (true) homomorphic image of the progenitor $3^{\star 6} :_m L_2(5)$, and

Table 7.8.   Low degree permutation representations of $3^{*4} :_m 2 \cdot S_4{}^+$

Name	8 letters	1 letter	2 letters	3 letters
a	$(0)(1,2,3)(\bar{0})(\bar{1},\bar{2},\bar{3})$	(u)	$(v)(w)$	(x,y,z)
b	$(0,1)(\bar{0},\bar{1})(2,\bar{2})(3)(\bar{3})$	(u)	(v,w)	$(x)(y,z)$
$(ab)^4$	$(0,\bar{0})(1,\bar{1})(2,\bar{2})(3,\bar{3})$	(u)	$(v)(w)$	$(x)(y)(z)$
—	$(2,3)(\bar{2},\bar{3})(0,\bar{0})$	(u)	(v,w)	$(x)(y,z)$
—	$(2,\bar{3})(\bar{2},3)(1,\bar{1})$	(u)	(v,w)	$(x)(y,z)$

	Action on 8, 9 or 11 letters		Group generated	
$t_0^{(A)}$	$(1,2,3)(\bar{3},\bar{2},\bar{1})$		S_8	
$t_0^{(B)}$	$(u,0,\bar{0})$		S_9	
$t_0^{(C)}$	$(u,0,\bar{0})(3,2,1)(\bar{1},\bar{2},\bar{3})$		$3^2{:}2\cdot S_4{}^+$	
$t_0^{(D)}$	$(u,0,\bar{0})(1,2,3)(\bar{3},\bar{2},\bar{1})$		S_9	
$t_0^{(E)}$	$(x,1,\bar{1})(y,2,\bar{2})(z,3,\bar{3})$		M_{11}	

in an essentially unique way. From the classification of subgroups of $L_2(q)$ [40] and the fact that any image of $3^{*6} :_m L_2(5)$ is perfect (as can be seen by abelianizing the presentation of Table 7.4(xii)), we see that this image can only be a subfield subgroup $L_2(r)$, where q is a power of r. This image will be the whole group $L_2(q)$ for $q = p^m$ (p prime) if, and only if, q is the minimal power of p for which the above congruences hold, namely $p^m \equiv \pm 1$ (mod 15) and $p^k \not\equiv \pm 1$ (mod 15) whenever $1 \le k < m$. Thus, for example, $L_2(31^2)$ is not an image of this progenitor but $L_2(11^2)$, $L_2(16)$ and $L_2(31)$ are.

In order to construct these embeddings in a generic fashion, we give matrices over a finite extension of the rationals, which we can then reduce modulo any prime we wish (except 5). We comment that this is essentially the unique 3-dimensional representation of the progenitor $3^{*6} :_m L_2(5)$. The matrices x, y and t given below satisfy the presentation of Table 7.4(xii), so $N = \langle x, y \rangle \cong A_5$ and $\langle t \rangle$ has just six images under conjugation by N:

$$
x \sim \begin{bmatrix} 0 & 0 & 1 \\ -1 & 0 & -\beta \\ 0 & -1 & -\beta \end{bmatrix}, \quad
y \sim \begin{bmatrix} 0 & 1 & 0 \\ 0 & 0 & 1 \\ 1 & 0 & 0 \end{bmatrix}
$$

and

$$
t \sim \frac{1}{5} \begin{bmatrix}
(2\beta+1)\gamma - 2\beta - 1 & (\beta+3)\gamma - 5 & (\beta+3)\gamma - 3\beta + 1 \\
(\beta-2)\gamma - 2\beta - 1 & 3\beta - 1 & (-\beta+2)\gamma - 4\beta - 2 \\
(-\beta-3)\gamma + 5 & (-\beta-3)\gamma + 3\beta - 1 & (-2\beta-1)\gamma - \beta + 2
\end{bmatrix},
$$

where $\gamma = y_{15} = z_{15} + z_{15}^{-1}$ and $\beta = b_5 = \gamma^3 - 3\gamma = z_5 + z_5^{-1}$ ($z_n = e^{2\pi i/n}$ denotes a primitive nth root of unity), as in the ATLAS notation for irrationalities. This representation can be reduced modulo any prime $p \neq 5$ to give a

3-dimensional representation of $3^{*6} :_m L_2(5)$ over $\mathbb{F}_q = \mathbb{F}_p(y_{15})$. (Since all four algebraic conjugates of y_{15} are in $\mathbb{Z}[y_{15}]$, this q is independent of the p-modular analogue of y_{15} chosen.) If $p = 3$, then $t = I_3$; in all other cases the reduced t has order 3 and $\langle t \rangle$ has six conjugates under $N = \langle x, y \rangle$. Note that since xt^y has trace $\frac{1}{5}((2\beta + 1)\gamma - \beta - 3)$, this representation cannot be realized over a proper subfield. (If $F = \mathbb{F}_p$ or $\mathbb{Q}$, the only subfields of $F(y_{15})$ are F, $F(b_5)$ and $F(y_{15})$, these fields not necessarily being distinct.)

The group $\langle x, y, t \rangle$ preserves the quadratic form with matrix given by

$$\begin{bmatrix} 2 & -\beta & -\beta \\ -\beta & 2 & -\beta \\ -\beta & -\beta & 2 \end{bmatrix},$$

which has determinant $2\beta + 4$, so the form is non-singular modulo all primes but 2 (and 5, since the group is not defined for this prime). Now, x, y and t have determinant 1, so we have $\langle x, y, t \rangle \le SO_3(q) \cong PGL_2(q)$, and, since any image of $3^{*6} :_m L_2(5)$ is perfect, we actually have $\langle x, y, t \rangle \le PSL_2(q) = L_2(q)$. Equality follows from a discussion earlier in this section. Note that the 2-modular reduction of the above representation does give $L_2(16)$, even though the above quadratic form is singular. In the characteristic 0 case, the full outer automorphism group of the progenitor $3^{*6} :_m L_2(5)$ corresponds to a cyclic group of order 4 generated by a conjugate of a field automorphism (of order 4).

7.6.4   Automorphisms of the progenitors

Knowledge of the (outer) automorphism group of a progenitor is often useful in cutting down the number of sets of parameters we have to consider. For instance, the isomorphism class of the group

$$\frac{3^{*4} :_m S_4}{[(0, 1, 2, 3)t_0]^a, \ [(0, 1, 2, 3)t_0^{-1}]^b, \ [(0, 1)t_0]^c}$$

is independent of the order of a, b and c. Any automorphism (of a progenitor) is a combination of the following four types:

(i)   inner;

(ii) those that fix $\bar{\mathcal{T}}$ and lie in $N_{rmM(n,T)}(\bar{N})$ but not $\bar{N}$;

(iii) those that commute with N but move $\bar{\mathcal{T}}$;

(iv) those that swap N and $T_0 : N_0$,

where $M(n, T) \cong \text{Aut}(T) \wr S_n$ denotes the full group of monomial automorphisms of the free product T^{*n} and $\bar{N} = N/C_N(\mathcal{T})$. This follows from considering a progenitor $T^{*n} : N$ as a free product of N and $T_0 : N_0$ amalgamated over their intersection, N_0. Note that an assumption that $N \le M(n, T)$ (which holds in all the cases considered here) ensures that the progenitor has trivial

Table 7.9. Outer automorphism groups of our progenitors

Progenitor, P	Type (ii)	Type (iii)	Type (iv)	Out(P)
$3^{\star 3}:S_3$	2	1	1	2
$3^{\star 4}:S_4$	2	1	1	2
$3^{\star 5}:A_5$	2^2	1	1	2^2
$3^{\star 5}:S_5$	2	1	1	2
$3^{\star 6}:L_2(5)$	2^2	1	1	2^2
$3^{\star 7}:L_3(2)$	2	1	1	2
$3^{\star 3}:_m S_3$	2	1	2	2^2
$3^{\star 4}:_m S_4$	2	3	1	D_6
$3^{\star 5}:_m S_5$	2	1	1	2
$3^{\star 2}:_m D_8$	1	1	1	1
$3^{\star 3}:_m S_4$	2	1	1	2
$3^{\star 4}:_m 2{\cdot}S_4{}^+$	1	1	1	1
$3^{\star 6}:_m L_2(5)$	4	1	1	4
$3^{\star 7}:_m L_3(2)$	2	1	1	2

centre, and so the inner automorphisms of a progenitor may be identified in a natural manner with elements of the progenitor itself.

The automorphisms of types (i) and (ii) are relatively straightforward. As can be seen from Table 7.9, those of types (iii) and (iv) are rare. We have already encountered an automorphism of type (iii) when dealing with involutory generators in ref. [36] in relation to the progenitor $2^{\star 3}:S_3$. Indeed, this progenitor possesses two sets of symmetric generators, namely $\{t_0, t_1, t_2\}$ and $\{t_0(1,2), t_1(2,0), t_2(0,1)\}$, which are interchanged by an outer automorphism commuting with $N \cong S_3$. The only progenitor in the present work with an automorphism of type (iii) is $3^{\star 4}:_m S_4$, which possesses three sets of symmetric generators. These are $\mathcal{T} = \{t_i \mid i = 0,1,2,3\}$, $\mathcal{R} = \{r_i = t_i(j,k,l) \mid i = 0,1,2,3\}$ and $\mathcal{S} = \{s_i = t_i(j,l,k) \mid i = 0,1,2,3\}$, where i, j, k, l is an even permutation of $0, 1, 2, 3$. This progenitor admits an automorphism of order 3 which commutes with $N \cong S_4$ and acts as (r_i, s_i, t_i) on the sets of symmetric generators. Such additional sets of symmetric generators are said to be *auxiliary*. The reader will observe that both these examples are special cases of a family of progenitors of shape $p^{\star(p+1)}:_m \mathrm{PGL}_2(p)$ which possess a type (iii) automorphism of order p.

A presentation for the progenitor $3^{\star 3}:_m S_3$ is given in Table 7.4(viii), in terms of generators x, y and t. This group has an outer automorphism α of order 2 which interchanges x and t whilst fixing y. Thus, in the first case, the symmetric generators are $\{t, t^x, t^{x^2}\}$ with control subgroup $N = \langle x, y \rangle$, and in the second case the symmetric generators are $\{x, x^t, x^{t^2}\}$ with control subgroup $N^\alpha = \langle t, y \rangle$.

We tabulate the outer automorphism groups of the progenitors under consideration in Table 7.9. A fuller explanation of this classification of automorphisms is to be found in Bray [14].

7.6.5   Non-tabulated results

Certain of our progenitors possess too few known images to warrant separate tables, although the images they do have are of great interest. Indeed, it was the realization that the Mathieu group M_{12} could be readily constructed as an image of $3^{*5}:A_5$ which motivated the programme of work that led to this book. Thus we have

$$\frac{3^{*5}:A_5}{(2,3,4)=(t_0 t_1^{-1})^2} \cong M_{12} \times 3,$$

whose centre is generated by $[(0,1,2,3,4)t_0]^8$.

As described in the text, the progenitor $3^{*7}:_m L_3(2)$ maps onto the alternating group A_7, the image being realized by, for example, the single relator $[(1,\bar{1})(0,3,\bar{0},\bar{3})(2,6,4,\bar{5})t_2]^3$. The corresponding permutation progenitor gives rise to

$$\frac{3^{*7}:L_3(2)}{(t_0 t_1)^2 = 1} \cong A_9 \times L_3(2)$$

and

$$\frac{3^{*7}:L_3(2)}{(t_0 t_1)^2 = [(0,1,2,3,4,5,6)t_0]^9 = 1} \cong A_9.$$

More interestingly, however, the Hall–Janko group J_2 and its double cover are both images of

$$\frac{3^{*7}:L_3(2)}{(2,6)(4,5)=(t_1 t_3^{-1})^2}.$$

Explicitly, factoring by any one of the relators $[(0,1,2,3,4,5,6)t_0]^{10}$, $[(0,3)(2,6,4,5)t_2]^7$ or $[(0,1,6)(2,3,5)t_0]^8$ yields $2\dot{\,}J_2$, while factoring by $(t_0 t_1 t_3)^2$ yields $J_2 \times 3$. The simple group J_2 is obtained by factoring by $(t_0 t_1 t_3)^2$ and any one of the other three relators. It is worth noting that the subgroup generated by an oval of symmetric generators such as $\{t_0, t_3, t_5, t_6\}$ is isomorphic to the unitary group $U_3(3)$ of (minimal) index 100.

For the progenitor $3^{*6}:L_2(5)$, we have the remarkably simple result that

$$\frac{3^{*6}:L_2(5)}{(\infty,0)(1,4)=(t_2 t_3^{-1})^2} \cong 2\dot{\,}J_2 \times 3.$$

In this group $(\infty,0,1)(2,4,3)t_\infty$ has order 60, and factoring by $[(\infty,0,1)(2,4,3)t_\infty]^{30}$, $[(\infty,0,1)(2,4,3)t_\infty]^{20}$ or $[(\infty,0,1)(2,4,3)t_\infty]^{10}$ gives $J_2 \times 3$, $2\dot{\,}J_2$ or J_2, respectively. This progenitor also has an image $A_8 \times A_5$ obtained by factoring by $(t_\infty t_0)^2$, the image A_8 being obtained by factoring by the further relator $[(0,1,2,3,4)t_0]^7$. We also found that

$$\frac{3^{*6}:L_2(5)}{[(0, 1, 2, 3, 4)t_0]^{13} = (t_\infty t_0^{-1})^2 = 1} \cong 3^{\cdot}O_7(3),$$

the triple cover of the simple group $P\Omega_7(3)$, as in the notation of the ATLAS (and Artin). A faithful permutation representation for this group of degree 2268 may be obtained over the subgroup $\langle t_0, t_1, t_2, t_3, t_4 \rangle \cong L_4(3)$.

7.7 Tabulated results

The information given in Tables 7.6–7.19 is, for the most part, self-explanatory. Tables 7.6 and 7.7 describe the particular parameters used for each of our progenitors, while Tables 7.5 and 7.4 allow the reader to write down an explicit presentation for each of the groups given. Note that, in spite of our requirement that $\mathcal{T}$ generates G, we have allowed some interesting cases when this is not quite true!

The notation for group structures is essentially that of the ATLAS [25], the main difference being in the notation p^{m+n}, which we are reserving for an elementary abelian group of order p^{m+n}, with an indication that when it is regarded as an $\mathbb{F}_p$-module for the group acting on it there is just one non-trivial submodule, and this has dimension m. (Note, however, that the notation $p^n.K$ does *not* imply that the p^n is an irreducible $\mathbb{F}_p$-module for K.)

In the ATLAS , the notation p^{m+n} is used for a special group whose centre has order p^m; we shall use the notation $p^{m|n}$ instead, but we do keep the notation $p_\pm^{1+2n}$ for extraspecial groups. The notation $[2^3.3^2.A_6]$ for a group with composition factors C_2 (three times), C_3 (twice) and A_6 (once) extends the use of $[n]$ denoting a soluble group of order n.

More importantly, the parameters given in boldface type (see Tables 7.10–7.19) are those used to define the group; parameters given in small numerals are the values assumed in the resulting group G.

Verification that the images have the shapes claimed is usually achieved as follows. Firstly, we perform a coset enumeration over an identifiable subgroup using the algebra package MAGMA [19] to obtain the order. We then seek a faithful permutation action of small degree, and use it to provide information about the group such as its composition factors, derived series, and so on. This information is then used to determine the structure of the group.

The interested reader is referred to Bray [14], where various generic symmetric presentations for the alternating groups and a family of orthogonal groups are proved valid and others conjectured. Special cases of these presentations appear in Tables 7.10–7.19.

We have omitted most soluble images from the tables, and many of the images listed can be specified by alternative parameters. For a fuller listing, the reader is referred to Bray [14] or to the World Wide Web at http://www.mat.bham.ac.uk/spres/.

Table 7.10. Some finite images of the progenitor $3^{\cdot 3}:S_3$

Parameters						Order of G	Shape of $\langle t_0, t_1 \rangle$	Shape of $\langle \mathcal{T} \rangle$	Shape of G
a	b	c	d	e	f				
5	4	3	2	6	4	120	A_4	A_5	S_5
4	8	7	4	8	6	336	$L_2(7)$	$L_2(7)$	$PGL_2(7)$
15	12	3	2	6	4	360	$A_4 \times 3$	$A_5 \times 3$	$S_5 \times 3$
15	4	3	6	6	4	360	A_4	A_5	$(A_5 \times 3):2$
6	5	5	5	5	3	660	A_5	G	$L_2(11)$
7	13	7	13	3	7	1092	G	G	$L_2(13)$
5	8	3	4	6	8	1920	$4^2:3$	$2^4:A_5$	$2^4:S_5$
7	5	5	6	7	7	2520	A_5	G	A_7
21	39	7	13	3	7	3276	G	G	$L_2(13) \times 3$
5	8	6	4	6	8	3840	$[2^5 3]$	$2^{1+4} \cdot A_5$	$2^{1+4} \cdot S_5$
15	7	4	4	5	7	20 160	$L_3(2)$	G	$A_8 \cong L_4(2)$
15	7	4	4	5	14	40 320	$L_3(2)$	G	$2 \cdot A_8$
15	8	4	7	4	6	40 320	$L_3(2)$	A_8	S_8
7	6	4	7	8	6	40 320	$4^2:3$	$L_3(4)$	$L_3(4):2_3$
7	6	12	7	8	6	120 960	$3 \times 4^2:3$	$3 \cdot L_3(4)$	$3 \cdot L_3(4):2_3$
15	8	4	21	4	6	120 960	$L_3(2)$	A_8	$(A_8 \times 3):2$
5	10	15	5	10	10	124 800	$U_4(3)$	$U_4(3)$	$U_4(3):2$
14	5	5	6	14	7	645 120	A_5	G	$(2^{4_1} \times 2^{4_2}):A_7$
8	6	13	8	12	8	63 078 912	$13^2:3$	$L_3(3)^2$	$(L_3(3)^2):2$

Table 7.11. Some finite images of the progenitor $3^{\cdot 4}:S_4$

Parameters										Order of G	Shape of $\langle t_0, t_1 \rangle$	Shape of $\langle t_0, t_1, t_2 \rangle$	Shape of G
a	b	c	d	e	f	p	q	s_1	s_2				
6	5	4	4	3	2	6	4	–	–	720	A_4	A_5	S_6
6	15	12	12	3	2	6	4	–	–	2160	$A_4 \times 3$	$A_5 \times 3$	$S_6 \times 3$
13	13	8	8	6	6	13	13	2	3	5616	$SL_2(3)$	G	$L_3(3)$
7	8	12	12	4	4	8	7	3	2	6048	$SL_2(3)$	G	$U_3(3)$
12	15	4	4	3	6	6	4	–	–	8640	A_4	A_5	$(A_6 \times A_4):2_1$
12	5	8	8	3	4	6	8	–	–	11 520	$4^2:3$	$2^4:A_5$	$2^4 \cdot S_6$
21	24	12	12	4	4	24	7	3	2	18 144	$SL_2(3)$	G	$U_3(3) \times 3$
9	7	10	5	5	6	7	7	–	–	181 440	A_5	A_7	A_9
12	10	5	10	5	11	12	12	–	–	190 080	A_5	M_{12}	$M_{12}:2$
7	8	12	12	4	4	8	7	3	–	774 144	$[2^4 3]$	G	$2^{1+6}:U_3(3)$
8	20	5	10	5	5	10	6	–	–	777 600	A_5	$3 \cdot (A_6 \times A_6)$	$3 \cdot (A_6 \times A_6):2$
18	6	7	14	9	9	18	14	–	–	1 524 096	$L_2(8)$	$L_2(8)^2$	$L_2(8)^2:S_3$
8	60	5	10	5	15	10	6	–	–	2 332 800	A_5	$3 \cdot (A_6 \times A_6)$	$3 \cdot (A_6 \times A_6):S_3$
21	15	14	7	4	4	5	7	–	–	9 999 360	$L_3(2)$	A_8	$L_5(2)$
9	14	10	5	5	6	14	7	–	–	46 448 640	A_5	$2^{4_1 \oplus 4_2}:A_7$	$2^{8_2}:A_9$

Table 7.12. Some finite images of the progenitor $3^{\star 5}:S_5$

Parameters									Order of G	Shape of $\langle t_0, t_1, t_2 \rangle$	Shape of $\langle \mathcal{T} \setminus \{t_4\} \rangle$	Shape of G
a	b	c	d	e	f	p	q	r				
10	**7**	6	5	4	3	2	6	4	5040	A_5	A_6	S_7
10	7	6	**5**	4	3	2	6	4	5040	A_5	A_6	S_7
30	21	**6**	15	12	3	2	6	4	15 120	$A_5 \times 3$	$A_6 \times 3$	$S_7 \times 3$
30	21	6	15	12	3	**2**	6	4	15 120	$A_5 \times 3$	$A_6 \times 3$	$S_7 \times 3$
30	35	12	15	**4**	3	6	6	4	302 400	A_5	A_6	$(A_7 \times A_5):2$
6	15	12	**6**	6	4	6	6	8	368 640	$[2^6 3]$	$[2^8 3]$	$2^{2|8}:(3 \times S_5)$
12	**13**	18	12	6	**2**	3	6	4	12 130 560	$2^{1+2}:A_4$	$3^4:(2^{1+2}:A_4)$	$L_4(3):2_2$
14	11	9	**7**	5	5	6	7	7	19 958 400	A_7	A_9	A_{11}
30	63	21	15	**7**	**4**	4	**5**	7	20 158 709 760	$L_4(2)$	$L_5(2)$	$L_6(2)$

Table 7.13. Some finite images of the progenitor $3^{\star 2}:_m D_8$

Parameters					Order of G	Shape of $\langle t_0, t_1 \rangle$	Shape of G
a	b	c	d	e			
6	**5**	4	3	4	120	A_5	S_5
6	**5**	4	**3**	4	120	A_5	S_5
6	10	4	6	4	240	A_5	$S_5 \times 2$
7	8	8	6	3	336	$L_2(7)$	$PGL_2(7)$
7	8	8	6	**3**	336	$L_2(7)$	$PGL_2(7)$
12	10	**4**	6	4	480	A_5	$(A_5 \times 2^2):2$
10	12	**5**	10	12	1320	$L_2(11)$	$PGL_2(11)$
28	**8**	8	6	12	1344	$L_3(2)$	$(L_3(2) \times 4):2$
10	**12**	10	10	12	2640	$L_2(11)$	$PGL_2(11) \times 2$
9	16	18	16	9	4896	$L_2(17)$	$PGL_2(17)$
11	**11**	12	12	6	6072	$L_2(23)$	$L_2(23)$
20	**9**	20	10	18	6840	$L_2(19)$	$PGL_2(19)$
12	**10**	8	6	8	7680	$2^4.A_5$	$2^{4+1+1}.S_5$
10	**17**	10	17	10	8160	$L_2(16)$	$L_2(16):2$
24	60	24	12	**4**	8640	$[2.3^2 A_5]$	$[2^4 3^2 A_5]$
28	14	26	**4**	28	8736	$L_2(13)$	$(L_2(13) \times 2^2):2$
13	**12**	8	12	8	11 232	$L_3(3)$	$L_3(3):2$
10	**13**	10	12	6	15 600	$L_2(25)$	$P\Sigma L_2(25)$
21	42	40	**8**	**5**	68 880	$L_2(41)$	$PGL_2(41)$
15	**10**	10	10	10	124 800	$U_3(4)$	$U_3(4):2$

Table 7.14. Some finite images of the progenitor $3^{*3} :_m S_3$

a	b	c	d	e	f	p	q	Order of G	Shape of $\langle t_0, t_1 \rangle$	Shape of $\langle \mathcal{T} \rangle$	Shape of G
2	3	4	3	3	2	4	3	24	A_4	A_4	S_4
5	5	3	2	2	5	5	5	60	A_4	G	A_5
3	6	4	3	3	4	2	3	72	A_4	A_4	$3{:}S_4$
6	12	8	6	6	2	8	3	144	$SL_2(3)$	$SL_2(3)$	$3{:}GL_2(3)$
4	4	7	4	4	3	3	3	168	G	G	$L_2(7)$
4	4	14	4	4	6	6	3	336	$L_2(7)$	$L_2(7)$	$L_2(7) \times 2$
7	4	6	7	7	8	8	2	336	$7{:}3$	$L_2(7)$	$PGL_2(7)$
15	15	6	2	6	10	10	5	360	$A_4 \times 3$	$A_5 \times 3$	$A_5 \times D_6$
4	5	5	5	5	5	4	3	360	A_5	G	A_6
7	9	9	7	7	7	7	2	504	G	G	$L_2(8)$
6	6	5	5	5	11	11	5	660	A_5	G	$L_2(11)$
4	21	6	21	21	8	8	3	1008	$7{:}3 \times 3$	$L_2(7) \times 3$	$(L_2(7) \times 3){:}2$
21	12	6	7	21	8	8	2	1008	$7{:}3 \times 3$	$L_2(7) \times 3$	$(L_2(7) \times 3){:}2$
7	13	7	7	7	6	7	2	1092	G	G	$L_2(13)$
7	6	6	13	13	3	7	7	1092	$13{:}3$	G	$L_2(13)$
4	5	30	5	5	10	12	3	2160	$A_5 \times 3$	$3{\cdot}A_6$	$3{\cdot}A_6 \times 2$
5	12	10	15	15	4	10	12	2160	$A_5 \times 3$	$A_6 \times 3$	$A_6 \times D_6$
7	7	12	7	7	12	4	2	2184	$L_2(13)$	$L_2(13)$	$PGL_2(13)$
17	9	8	8	8	3	9	9	2448	G	G	$L_2(17)$
9	8	9	17	17	3	9	9	2448	G	G	$L_2(17)$
6	5	7	4	4	5	7	4	2520	$L_2(7)$	G	A_7
9	10	5	5	5	9	19	5	3420	A_5	G	$L_2(19)$
7	7	10	6	5	6	6	7	5040	A_5	A_7	S_7
11	12	11	11	11	3	4	4	6072	G	G	$L_2(23)$

Table 7.15. Some finite images of the progenitor $3^{*4} :_m S_4$

a	b	c	d	e	f	p	Order of G	Shape of $\langle t_0, t_1 \rangle$	Shape of $\langle t_0, t_1, t_2 \rangle$	Shape of G
4	4	4	3	6	6	3	288	A_4	A_4	$A_4{:}S_4$
3	5	5	5	4	5	5	360	A_5	G	A_6
5	5	3	5	5	4	2	360	A_4	A_5	A_6
6	10	10	5	4	5	5	720	A_5	A_6	$A_6 \times 2$
10	10	6	5	5	4	2	720	A_4	A_5	$A_6 \times 2$
10	10	6	15	15	12	2	2160	$A_4 \times 3$	$A_5 \times 3$	$A_6 \times D_6$
6	20	20	15	4	15	15	8640	$A_5 \times A_4$	$A_6 \times A_4$	$A_6 \times S_4$
20	20	6	15	15	4	6	8640	A_4	A_5	$A_6 \times S_4$
10	10	6	5	5	8	4	11 520	$4^2{:}3$	$2^4{:}A_5$	$2^{4+1}{:}A_6$
6	10	10	5	8	5	5	11 520	A_5	$2^4{:}A_6$	$2^{4+1}{:}A_6$
12	10	12	5	10	10	5	190 080	M_{12}	M_{12}	$M_{12}{:}2$
12	12	10	10	10	5	1	190 080	A_5	M_{12}	$M_{12}{:}2$
15	15	5	20	20	5	5	388 000	A_5	G	$3{\cdot}(A_6 \times A_6)$
5	15	15	20	5	20	10	388 000	$A_5 \times A_4$	$3{\cdot}A_6 \times A_5$	$3{\cdot}(A_6 \times A_6)$
6	6	8	14	14	8	4	3 753 792	$L_2(7)$	$L_3(7)$	$L_3(7) \times 2$
7	13	6	13	8	6	13	4 245 696	3^{1+2}_+	$L_3(3)$	$G_2(3)$
6	13	7	13	6	8	6	4 245 696	$L_2(7)$	G	$G_2(3)$
7	6	13	6	8	13	7	4 245 696	G	G	$G_2(3)$

Table 7.16. Some finite images of the progenitor $3^{*3} :_m S_4$

Parameters			Order of G	Shape of $\langle t_0, t_1 \rangle$	Shape of $\langle \mathcal{T} \rangle$	Shape of G
a	b	c				
6	7	5	2520	A_5	G	A_7
6	7	10	5040	A_5	A_7	$A_7 \times 2$
6	21	5	7560	A_5	G	$3 \cdot A_7$
6	13	10	15 600	A_5	$L_2(25)$	$P\Sigma L_2(25)$

Table 7.17. Some finite images of the progenitor $3^{*4} :_m 2\cdot S_4{}^+$

Parameters					Order of G	Shape of $\langle t_0, t_1 \rangle$	Shape of $\langle t_0, t_1, t_2 \rangle$	Shape of $\langle \mathcal{T} \rangle$	Shape of G
a	b	c	d	e					
6	24	13	39	12	33696	$L_3(3)$	$L_3(3)$	$L_3(3)$	$(L_3(3) \times 3){:}2$
6	8	13	13	12	11232	$L_3(3)$	$L_3(3)$	$L_3(3)$	$L_3(3){:}2$
22	5	6	6	10	15840	A_5	$L_2(11)$	M_{11}	$M_{11} \times 2$
8	6	4	3	6	432	3^2	3^2	3^2	$3^2{:}2\cdot S_4{}^+$
11	5	6	6	5	7920	A_5	$L_2(11)$	G	M_{11}

Table 7.18. Some finite images of the progenitor $3^{*6} :_m L_2(5)$

Parameters										Order of G	Shape of $\langle t_\infty, t_0 \rangle$	Shape of $\langle t_\infty, t_0, t_1 \rangle$	Shape of $\langle t_\infty, t_0, t_2 \rangle$	Shape of G
a	b	c	d	e	f	p	q	r	s					
5	7	29	15	14	15	7	15	29	14	12 180	G	G	G	$L_2(29)$
4	16	16	16	8	15	16	5	8	15	14 880	G	G	G	$L_2(31)$
19	19	15	15	6	6	11	11	10	5	175 560	A_5	$L_2(11)$	G	J_1
89	11	44	45	4	45	22	22	45	44	352 440	G	G	G	$L_2(89)$
9	9	12	12	12	12	15	15	6	4	1 451 520	A_4	$2^3{:}A_4$	A_5	$S_6(2)$
45	45	60	60	12	12	15	15	6	4	87 091 200	A_4	$2^3{:}A_4$	A_5	$S_6(2) \times A_5$

Table 7.19. Some finite images of the progenitor $3^{\star 5} :_m S_5$

Parameters									Order of G	Shape of $\langle t_0, t_1 \rangle$	Shape of $\langle t_0, t_1, t_2 \rangle$	Shape of $\langle \mathcal{T} \setminus \{t_4\} \rangle$	Shape of G
a	b	c	d	e	f	p	q	r					
3	3	4	7	5	5	5	4	3	2520	A_4	A_5	A_6	A_7
3	3	4	7	5	5	5	4	**3**	2520	A_4	A_5	A_6	A_7
6	6	**4**	21	10	10	15	12	6	15 120	$A_4 \times 3$	$A_5 \times 3$	$A_6 \times 3$	$A_7 \times S_3$
12	12	6	15	8	8	**6**	6	4	92 160	A_4	$2^4{:}3$	$2^6{:}3$	$2^8{:}(3{:}S_5)$
4	12	10	**8**	12	10	11	**4**	12	190 080	$A_4 \times 3$	M_{12}	M_{12}	$M_{12}{:}2$
6	6	12	35	20	20	15	4	6	302 400	A_4	A_5	A_6	$A_7 \times S_5$
4	12	10	24	12	10	33	6	12	570 240	$A_4 \times 3$	$M_{12} \times 3$	$M_{12} \times 3$	$(M_{12} \times 3){:}2$
12	12	6	**13**	8	8	12	6	4	12 130 560	A_4	$2^{1+2}{:}A_4$	$3^4{:}(2^{1+2}{:}A_4)$	$L_4(3) \times 2$

Table 7.20. Symmetric presentations for some sporadic simple groups or their decorations

Group	Symmetric presentation	Page
M_{11}	$\dfrac{3^{\star 4} :_m 2^{\cdot} S_4{}^+}{[(0\ 1)(2\ \bar{2})t_0]^5}$	266
$3 \times M_{12}$	$\dfrac{3^{\star 5} : A_5}{(2\ 3\ 4) = (t_0^{-1} t_1)^2}$	173
J_1	$\dfrac{2^{\star 11} : L_2(11)}{(\sigma_{018} t_0)^5}$	138
$2^{\cdot} M_{22} : 2$	$\dfrac{2^{\star 15} : A_7}{(\pi_{ijk} t_i)^4}$	129
$J_2 : 2$	$\dfrac{2^{\star 36} : (U_3(3) : 2)}{\pi_{ij} = t_i t_j t_i}$	170
M_{23}	$\dfrac{5^{\star 4} : 2^{\cdot} S_4{}^-}{\text{relations}}$	271
$HS : 2$	$\dfrac{2^{\star 50} : (U_3(5) : 2)}{\pi_{ijk} = t_i t_k t_i t_j}$	151
$J_3 : 2$	$\dfrac{2^{\star 120} : (L_2(16) : 4)}{(\pi t_i)^5}$	175
M_{24}	$\dfrac{2^{\star 7} : L_3(2)}{[t_0^{I_1}, t_3], (2\ 4)(5\ 6) = (t_0 t_3)^3, [(1\ 2\ 4)(3\ 6\ 5)t_3]^{11}}$	173
$McL : 2$	$\dfrac{2^{\star 672} : M_{22}}{\nu = (t_1 t_2)^2}$	222
He	$\dfrac{7^{\star(15+15)} : 3^{\cdot} S_7}{(\sigma t_0)^3}$	278
$3^{\cdot} Suz : 2$	$\dfrac{2^{\star 416} : (G_2(4) : 2)}{\pi_{ij} = t_i t_j t_i}$	172
$O'N : 2$	$\dfrac{2^{\star 12} : M_{11}}{(t_\infty t_0)^4, (\sigma^3 t_\infty t_3)^5, (\sigma(t_\infty t_0)^2)^5}$	235
Co_2	$\dfrac{2^{\star \binom{23}{3}} : M_{23}}{t_{ab} t_{ac} t_{ad} = \nu}$	219
HN	$\dfrac{5^{\star(176+176)} : (2^{\cdot} HS : 2)}{(\sigma t_0)^3}$	287
$2 \times Fi_{23}$	$\dfrac{2^{\star 5775} : S_{12}}{\left[\begin{smallmatrix} 1234 \\ 5678 \\ 90xy \end{smallmatrix}\right]\left[\begin{smallmatrix} 1256 \\ 3478 \\ 90xy \end{smallmatrix}\right]\left[\begin{smallmatrix} 1278 \\ 3456 \\ 90xy \end{smallmatrix}\right] = (1\ 2)(3\ 4)(5\ 6)(7\ 8)}$	233
$\cdot O$	$\dfrac{2^{\star \binom{24}{4}} : M_{24}}{t_{ab} t_{ac} t_{ad} = \nu}$	204
$2 \times J_4$	$\dfrac{2^{\star 3795} : M_{24}}{t_A t_B t_A t_D = \nu_2}$	199
$3^{\cdot} Fi_{24}$	$\dfrac{2^{\star \frac{1}{2}\binom{12}{6}} : (O_{10}^-(2) : 2)}{\left(\left[\begin{smallmatrix} 123456 \\ 7890xy \end{smallmatrix}\right] t\right)^3}$	233

7.8   Some sporadic groups

We conclude by listing in Table 7.20 symmetric presentations for 19 sporadic groups dealt with in this book, together with a reference to the page on which the presentation is described in more detail.

References

[1] A. E. Brouwer, A. M. Cohen and A. Neumaier. *Distance-Regular Graphs*. Ergebnisse der Mathematik und ihrer Grenzgebiete, vol. 18 (3) (Berlin: Springer, 1989).

[2] D. V. P. Alexander, A. Ivanov and S. V. Shpectorov. Non-abelian representations of some sporadic geometries. *J. Algebra*, **181** (1996), 523–557.

[3] M. Aschbacher. *Finite Group Theory* (Cambridge: Cambridge University Press, 1986).

[4] M. Aschbacher. The existence of J_3 and its embedding into E_6. *Geom. Dedicata*, **35** (1990), 143–154.

[5] M. Ashworth. On the Mathieu groups ma(12) and ma(24) and related Steiner systems. M.Sc. thesis, University of Birmingham, 1997.

[6] J. H. Conway. An algorithm for double coset enumeration. In M. D. Atkinson, ed., Computational Group Theory (New York: Academic Press, 1984), pp. 33–37.

[7] R. A. Parker. The computer calculation of modular characters (The MeatAxe). In M. D. Atkinson, ed., *Computational Group Theory* (New York: Academic Press, 1984), pp. 267–274.

[8] H. F. Baker. Note introductory to the study of Klein's group of order 168. *Proc. Cambridge, Phil. Soc.*, **31** (1935), 468–481.

[9] B. Baumeister. A computer-free construction of the third group of Janko. *J. Algebra*, **192** (1997), 780–809.

[10] D. Benson. The simple group J_4. Ph.D. thesis, University of Cambridge (1980).

[11] S. Bolt. Some applications of symmetric generation. Ph.D. thesis, University of Birmingham (2002).

[12] J. Bradley. Symmetric generation of some sporadic simple groups. Ph.D. thesis, University of Birmingham (2005).

[13] J. Bradley and R. T. Curtis. Symmetric generation of J_3:2, the automorphism group of the third Janko group. *J. Algebra*, **304** (2006), 256–270.

[14] J. N. Bray. Symmetric presentations of finite groups. Ph.D. thesis, University of Birmingham (1997).

[15] J. N. Bray and R. T. Curtis. A systematic approach to symmetric presentations II: Generators of order 3. *Math. Proc. Cambridge Phil. Soc.*, **128** (2000), 1–20.

[16] J. N. Bray and R. T. Curtis. Monomial modular representations and symmetric generation of the Harada–Norton group. *J. Algebra*, **268** (2003), 723–743.

[17] J. N. Bray and R. T. Curtis. Double coset enumeration of symmetrically generated groups. *J. Group Theory*, **7** (2004), 167–185.

[18] G. Butler. The maximal subgroups of the sporadic simple group of Held. *J. Algebra*, **58** (1981), 67–81.

[19] W. Bosma and J. Cannon. *Handbook of MAGMA Functions* (Sydney: University of Sydney, 1994).

[20] J. H. Conway. A perfect group of order 8, 315, 553, 613, 086, 720, 000 and the sporadic simple groups. *Proc. Natl Acad. Sci.*, **61** (1968), 398–400.

[21] J. H. Conway. A characterization of Leech's lattice. *Invent. Math.*, **7** (1969), 137–142.

[22] J. H. Conway. A group of order 8, 315, 553, 613, 086, 720, 000. *Bull. London Math. Soc.*, **1** (1969), 79–88.

[23] J. H. Conway. *Three Lectures on Exceptional Groups* (New York: Academic Press, 1971), pp. 215–247.

[24] J. H. Conway. *Hexacode and Tetracode – MOG and MINIMOG* (New York: Academic Press, 1984), pp. 359–365.

[25] J. H. Conway, R. T. Curtis, S. P. Norton, R. A. Parker and R. A. Wilson. *An ATLAS of finite groups* (Oxford: Oxford University Press, 1985).

[26] J. H. Conway and N. J. A. Sloane. *Sphere Packings, Lattices and Groups*. Grundlehren der Mathematischen Wissenschaften, vol. 290 (Berlin: Springer, 1988).

[27] H. S. M. Coxeter and W. O. J. Moser. *Generators and Relations for Discrete Groups*, 4th edn (New York: Springer, 1984).

[28] R. T. Curtis. On the Mathieu group M_{24} and related topics. Ph.D. thesis, University of Cambridge (1972).

[29] R. T. Curtis. A new combinatorial approach to M$_{24}$. *Math. Proc. Cambridge Phil. Soc.*, **79** (1976), 25–42.

[30] R. T. Curtis. Eight octads suffice. *J. Comb. Theory*, **36**: 1 (1984), 116–123.

[31] R. T. Curtis. Natural constructions of the Mathieu groups. *Math. Proc. Cambridge Phil. Soc.*, **106** (1989), 423–429.

[32] R. T. Curtis. Geometric interpretations of the 'natural' generators of the Mathieu groups. *Math. Proc. Cambridge Phil. Soc.*, **107** (1990), 19–26.

[33] R. T. Curtis. Symmetric presentations I: Introduction, with particular reference to the Mathieu groups M$_{12}$ and M$_{24}$. *J. London Math. Soc.*, **165** (1990), 380–396.

[34] R. T. Curtis. Symmetric presentation II: The Janko group J$_1$. *J. London Math. Soc.*, **47** (1993), 294–308.

[35] R. T. Curtis. Symmetric generation of the Higman-Sims group. *J. Algebra*, **171** (1995), 567–586.

[36] R. T. Curtis, A. M. Hammas and J. N. Bray. A systematic approach to symmetric presentation I: Involutory generators. *Math. Proc. Cambridge Phil. Soc.*, **119** (1996), 23–34.

[37] R. T. Curtis and Z. Hasan. Symmetric representation of elements of the Janko group J$_1$. *J. Symb. Comp.*, **22** (1996), 201–214.

[38] R. T. Curtis and R. A. Wilson, eds. *An Atlas of Sporadic Group Representations* (Cambridge: Cambridge University Press, 1990).

[39] R. T. Curtis and R. A. Wilson, eds. *Presentations of Fischer Groups* (Cambridge: Cambridge University Press, 1990).

[40] L. E. Dickson. *Linear Groups, With an Exposition of the Galois Field Theory* (Leipzig: Teubner, 1901; reprinted by Dover, 1958).

[41] GAP group. *Groups, algorithms and programming* (Aachen: GAP, 2006).

[42] A. Hammas. Symmetric presentations of some finite groups. Ph.D. thesis, University of Birmingham (1991).

[43] G. Havas. Coset enumeration strategies. Technical Report 200, University of Queensland (1991).

[44] D. Held. The groups related to M$_{24}$. *J. Algebra*, **13** (1969), 253–296.

[45] D. Held. *Some Groups Related to M$_{24}$* (New York: Benjamin, 1969), pp. 121–124.

[46] D. Higman and J. McKay. On Janko's simple group of order 50, 234, 960. *Bull. London Math. Soc.*, **1** (1969), 89–94.

[47] D. Higman and C. Sims. A simple group of order 44,352,000. *Math Z.*, **105** (1968), 110–113.

[48] G. Higman. On the simple group of D. G. Higman and C. C. Sims. *Illus. J. Math.*, **13** (1969), 74–80.

[49] H. Hilton. *Plane Algebraic Curves* (Oxford: Oxford University Press, 1920).

[50] A. J. Hoffman and R. Singleton. On Moore graphs with diameters 2 and 3. *IBM, J. Res. Dev.*, **4** (1960), 497–504.

[51] A. A. Ivanov. On geometries of the Fischer groups *Eur. J. Combin.*, **16** (1995), 163–183.

[52] A. A. Ivanov. *Geometry of the Sporadic Groups I: Petersen and Tilde Geometries* (Cambridge: Cambridge University Press, 1999).

[53] A. Ivanov. *The Fourth Janko Group*, Oxford Mathematical Monographs (Oxford: Oxford University Press, 2004).

[54] A. Ivanov and U. Meierfrankenfeld. A computer free construction of J_4. *J. Algebra*, **219** (1999), 113–172.

[55] C. P. J. N. Bray, R. T. Curtis and C. Wiedorn. Symmetric presentations of the Fischer groups I: the classical groups $S_6(2), S_8(2)$ and $3 \cdot O_7(3)$. *J. Algebra*, **265** (2003), 171–199.

[56] C. P. J. N. Bray, R. T. Curtis and C. Wiedorn. Symmetric presentations of the Fischer groups I: the sporadic groups. *Geometriae Dedicata*, **112** (2005), 1–23.

[57] A. Jabbar. Symmetric presentations of subgroups of the Conway groups, and related topics. Ph.D. thesis, University of Birmingham (1992).

[58] Z. Janko. A new finite simple group with abelian Sylow 2-subgroups. *Proc. Natl Acad. Sci. USA*, **53** (1965), 657–658.

[59] Z. Janko. A new finite simple group with abelian Sylow 2-subgroups, and its characterization. *J. Algebra* **3** (1966), 147–186.

[60] Z. Janko. Some new finite simple groups of finite order, I. *Sympos. Math.*, **1** (1968), 25–64.

[61] Z. Janko. A new finite simple group of order 86,775,570,046,077, 562,880, which possesses M_{24} and the full covering group of M_{22} as subgroups. *J. Algebra*, **42** (1976), 564–596.

[62] J. J. Sylvester. Elementary researches in the analysis of combinatorial aggregation. *Phil. Mag.*, **24** (1844), 285–295.

[63] P. Kleidman and R. Wilson. $J_3 < E_6(4)$ and $M_{12} < E_6(5)$. *J. London Math. Soc.*, **42** (1990), 555–561.

[64] F. Klein. Uber die transformation siebenter ordnung der elliptischen functionen. *Gesammelte Math. Abhandlungen*, **III** (1878), 90–135.

[65] J. Leech. Notes on sphere packings. *Can. J. Math.*, **19** (1967), 251–267.

[66] W. Lempken. A 2-local characterisation of Janko's simple group J_4. *J. Algebra*, **55** (1978), 403–445.

[67] S. A. Linton. The maximal subgroups of the sporadic groups T, F_{24} and F'_{24} and other topics. Ph.D. thesis, University of Cambridge (1989).

[68] S. A. Linton. Double coset enumerator. Technical report, University of St. Andrews, **GAP** sharepackage, version 3.4 (1995).

[69] S. S. Magliveras. The maximal subgroups of the Higman–Sims group. Ph.D. thesis, University of Birmingham (1970).

[70] E. Mathieu. Mémoire sur l'étude des fonctions de plusieurs quantités. *J. Math. Pure Appl.*, **6** (1861), 241–243.

[71] E. Mathieu. Sur la fonction cinq fois transitive de 24 quantités. *J. Math. Pures Appl.*, **18** (1873), 25–46.

[72] P. Mazet. Sur les multiplicateurs de Schur des groups de Mathieu. *J. Algebra*, **77**: 2 (1982), 552–576.

[73] M. S. Mohamed. Computational methods in symmetric generation of groups. Ph.D. thesis, University of Birmingham (1997).

[74] S. P. Norton. The construction of J_4. In B. N. Cooperstein and G. Mason, eds., *The Santa Cruz Conference on Finite Groups, Proceedings of Symposia in Pure Mathematics*, vol. 37 (Providence, RI: American Mathematical Society, 1980), pp. 271–278.

[75] R. A. P. R. A. Wilson and J. N. Bray. Atlas *of Finite Group Representations*. http://brauer.maths.qmul.ac.uk/Atlas/.

[76] M. Ronan and S. Smith. Computation of 2-modular sheaves and representations for $L_4(2)$, A_7, $3\cdot S_6$ and M_{24}. *Commun. Algebra*, **17** (1989), 1199–1237.

[77] M. A. Ronan. *Coverings of Certain Finite Geometries* (Cambridge: Cambridge University Press, 1981), pp. 316–331.

[78] E. Schulte and J. M. Wills. A polyhedral realisation of Felix Klein's map $\{3, 7\}_8$ on a Riemann surface of genus 3. *J. London Math. Soc.*, **32** (1985), 539–547.

[79] C. Sims. *An Isomorphism Between Two Groups of Order 44,352,000* (New York: Benjamin, 1969), pp. 101–108.

[80] M. Smith. On the isomorphism of two simple groups of order 44,352,000. *J. Algebra*, **41** (1976), 172–174.

[81] L. H. Soicher. Presentations of some finite groups. Ph.D. thesis, University of Cambridge (1985).

[82] S. Stanley. Monomial representations and symmetric presentations. Ph.D. thesis, University of Birmingham (1998).

[83] J. A. Todd. Representation of the Mathieu group M_{24} as a collineation group. *Ann. di Mat. Pura ed Appl.*, **71**: 4 (1966), 199–238.

[84] J. A. Todd and H. S. M. Coxeter. A practical method for enumerating cosets of finite abstract groups. *Proc. Edinburgh Math. Soc.*, **5** (1936), 26–34.

[85] M.-M. Virotte-Ducharme. Présentations des groupes de Fischer I. *Geom. Dedicata*, **41** (1992), 275–335.

[86] R. Weiss. A geometric construction of Janko's group J_3. *Math. Z.*, **179** (1982), 91–95.

[87] R. Wilson. Maximal subgroups of automorphism groups of simple groups. *J. London Math. Soc.*, **32** (1985), 460–466.

Index